Comparative Evaluations of Innovative Fisheries Management

Comparative Evaluations of Innovative Fisheries Management

Global Experiences and European Prospects

Edited by

Kjellrun Hiis Hauge

Institute of Marine Research, Bergen, Norway

Douglas Clyde Wilson

Innovative Fisheries Management, IFM – an Aalborg University Research Centre, Hirtshals, Denmark

Springer

Editors
Dr. Kjellrun Hiis Hauge
Institute of Marine Research
Nordnesgaten 50
N-5817 Bergen
Norway
kjellrun.hiis.hauge@imr.no

Dr. Douglas Clyde Wilson
Innovative Fisheries Management, IFM –
an Aalborg University Research Centre
North Sea Science Park
Willemoesvej 2
DK-9850 Hirtshals
Denmark
dw@ifm.aau.dk

ISBN 978-90-481-2662-0 e-ISBN 978-90-481-2663-7
DOI 10.1007/978-90-481-2663-7
Springer Dordrecht Heidelberg London New York

Library of Congress Control Number: 2009926870

Printed on acid-free paper

Springer is part of Springer Science+Business Media (www.springer.com)

Acknowledgements

We are very grateful to Elen Hals and Kirsten Klitkou and would like to thank them for their help on proof-reading the book. The CEVIS partnership is also grateful to its advisory committee, consisting of Denis Bailly, Poul Degnbol, Hans Lassen, Bonnie J. McCay, Ragnar Arnason, Niki Sporrong, Nathalie Steins and David Symes, for their unremunerated contributions that did so much to improve the quality of this volume. This study was funded by the Commission of the European Communities Sixth Framework Program, project N° 022686 'Comparative Evaluations of Innovative Solutions in European Fisheries Management' (CEVIS). The book does not necessary reflect the Commission's views on its future policy in this area.

Contents

Contributors

Hendrik Acker Öko-Institut, Novalisstrasse 10, 10115 Berlin, Germany, h.acker@oeko.de

Martin Aranda AZTI Tecnalia, Food and Research Division, Herrera Kaia-Portualdea, Z/G E-20110 Pasaia (Gipuzcoa), Spain, maranda@pas.azti.es

Ragnar Arnason Department of Economics, Oddi v. Sudurgotu, University of Iceland, IS-101 Reykjavik, Iceland, ragnara@hi.is

Francois Bastardie Technical University of Denmark, National Institute of Aquatic Resources, Charlottenlund Slot, DK-2920 Charlottenlund, Denmark, fba@aqua.dtu.dk

Alan Baudron University of Aberdeen, Old Aberdeen, AB24 2TZ, UK, r01arb8@abdn.ac.uk

Richard Bilocca FRI (Iceland), University of Iceland, Reykjavik, Iceland, reydyr@hotmail.com

Jesper Boje Technical University of Denmark, National Institute of Aquatic Resources, Charlottenlund Slot, DK-2920 Charlottenlund, Denmar, jbo@aqua.dtu.dk

Erik Buisman LEI, Alexanderveld 5, 2585 DB Den haag, The Netherlands, erik.buisman@wur.nl

Tammo P. Bult Wageningen IMARES, Haringkade 1, 1970 AB Ijmuiden, The Netherlands, tammo.bult@wur.nl

Anne-Sofie Christensen Institute for Innovative Management (IFM) – An Aalborg University Research Centre, North Sea Centre, Willemoesvej 2, Hirtshtals, Denmark, asc@ifm.dk

Andrea Leme da Silva IFM, Innovative Fisheries Management – an Aalborg University, Research Centre, Willemoesgade 2, 9850 Hirtshals, Denmark, andreale@unicamp.br

Miriam Dross German Advisory Council in the Environment, Reichpietschufer 60, 10785 Berlin, Germany, miriam.dross@uba.de

Søren Eliasen Innovative Fisheries Management, IFM, Aalborg University Copenhagen, Lautrupvang 2B, DK-2750 Ballerup, Denmark, se@ifm.aau.dk

Hans Frost Institute of Food and Resource Economics, Rolighedsvej 25, 1958 Frederiksberg C, DK-Denmark, hf@foi.dk

Dorleta Garcia AZTI – Tecnalia, Txatxarramendi ugartea z/g, 48395 Sukarrieta, Bizkaia, Spain, dgarcia@azti.es

Kjellrun Hiis Hauge Institute of Marine Research, Nordnes, N-5817 Bergen, Norway, kjellrun.hiis.hauge@imr.no

Troels Jacob Hegland Innovative Fisheries Management – An Aalborg University Research Centre, North Sea Centre, Willemoesvej 2, Hirtshals 9850 Denmark, th@ifm.dk

Niels T. Hintzen Wageningen IMARES, Haringkade 1, 1970 AB Ijmuiden, The Netherlands, niels.hintzen@wur.nl

Ayoe Hoff Institute of Food and Resource Economics, Rolighedsvej 25, 1958 Frederiksberg C, DK-Denmark, ah@foi.dk

Emil Kuzebski Sea Fisheries Institute, Gdynia, ul. Kollataja 1, 81-332 Gdynia, Poland, emil@mir.gdynia.p

Bonnie McCay Rutgers University, Department of Human Ecology, SEBS, George H. Cook Campus, New Jersey, USA, mccay@aesop.rutgers.edu

H. Anne McLay FRS Marine Laboratory, 375 Victoria Road, Aberdeen, Scotland, UK, A.McLay@MARLAB.AC.UK

Arantza Murillas AZTI – Tecnalia/Marine Research Division, Txatxarramendi ugartea z/g, 48395 Sukarrieta, Bizkaia, Spain, amurillas@suk.azti.es

J. Rasmus Nielsen Technical University of Denmark, National Institute of Aquatic Resources, Charlottenlund Slot, DK-2920 Charlottenlund, Denmark, rn@aqua.dtu.dk

Geir Oddsson Encargado de Negocios a.i. Embajada de Islandia, Embassy of Iceland, Managua, Nicaragua, geir@iceida.is.

Gudrun Petursdottir Fisheries Research Institute. University of Iceland, Sudurgata, IS 101 Reykjavík, gudrun@hi.is

Jeffrey P. Powell LEI, Alexanderveld 5, 2585 DB Den haag, The Netherlands, jeff.powell@wur.nl

Clara Ulrich National Institute of Aquatic Resources, Technical University of Denmark, Charlottenlund Slot, DK-2920 Denmark, clu@aqua.dtu.dk

Tiit Raid European Commission, DG Joint Research Centre, Institute for the Protection and Security of the Citizen, Maritime Affairs, Via E. Fermi 1, I-21020 Ispra (VA), Italy, tiit.raid@jrc.it.

Carl Rova Luleå University of Technology (LTU), Department of Business Administration and Social Science, Division of Political Sciences, 971 87 Luleå, Sweden, Carl.Rova@ltu.se

Sonia Sanchez AZTI – Tecnalia, Herrera kaia portualdea z/g, 20110 Pasaia, Gipuzkoa, Spain, ssanchez@azti.es

Sten Sverdrup-Jensen Innovative Fisheries Management IFM, Aalborg University Copenhagen, Lautrupvang 2B, DK-2750 Ballerup, Denmark, ssj@ifm.aau.dk

Douglas Clyde Wilson Innovative Fisheries Management – An Aalborg University Research Centre, North Sea Centre, Willemoesvej 2, Hirtshals 9850 Denmark, dw@ifm.dk

Franziska Wolff Öko-Institut e.V., Novalisstr. 10, 10115 Berlin, Germany, f.wolff@oeko.de

Sarunas Zableckis European Commission, DG Joint Research Centre, Institute for the Protection and Security of the Citizen, Maritime Affairs, Via E. Fermi 1, I-21020 Ispra, VA, Italy, szableckis@gmail.com

Abbreviations

ABC	Acceptable Biological Catch
ACE	Annual Catch Entitlement
BR	Biological Robustness
CCAMLR	Commission for the Conservation of Antarctic Marine Living Resources
CCR	Cost Recovery Regime
CEVIS	The Comparative Evaluations of Innovative Solutions in European Fisheries Management
CFP	Common Fisheries Policy
CMB	Community Management Boards
CPUE	Catch Per Unit Effort
CRP	Cost Recovery Programme
CSO	Commercial Stakeholder Organisation
CSY	Competitive Sustainable Yield
CTE	WTO Committee on Trade and Environment
DAS	Deepwater Allocation System
DFO	Department of Fisheries and Oceans
DV	Deemed Value
DVS	Deemed Value System
DWFF	Distant Waters Fishing Fleets
EA	Ecosystem Approach
EAFM	Ecosystem Approach to Fisheries Management
EBFM	Ecosystem-Based Fisheries Management
EC	European Community
ECJ	European Court of Justice
EEC	European Economic Community
EEZ	Exclusive Economic Zone
EFJ	Extended Fisheries Jurisdiction
EU	European Union
F	Fishing mortality
FAO	Food and Agriculture Organisation of the United Nations

FMA	Fisheries Management Areas
FMS	Fisheries Management System
FQA	Fixed Quota Allocations
FRCC	Fisheries Resource Conservation Council
FRML	Fishery Related Mortality Limit
FSRS	Fishermen and Scientists Research Society
GDP	Gross Domestic Product
GMP	Groundfish Management Plan
HCR	Harvest Control Rules
HRC	Human Rights Committee
ICA	Integrated Catch-at-Age Analysis
ICCPR	International Covenant on Civil and Political Rights
ICES	International Council for the Exploration of the Sea
IEF	Innovation Evaluation Framework
IFMP	Integrated Fishery Management Plans
IFQ	Individual Fishing Quota
IFRI	International Forestry Resources and Institutions
ILO	International Labour Organisation
IPOAMC	International Plan of Action for the Management of Fishing Capacity
IQ	Individual Quota
ITQ	Individual Transferable Quota
IVQ	Individual Vessel Quota
MCS	Monitoring Control and Surveillance
MPA	Marine Protected Area
MSE	Management Strategies Evaluations
NAFO	North Atlantic Fisheries Organization
NEAFC	North East Atlantic Fisheries Commission
NGO	Non-Governmental Organization
NIWA	National Institute of Water & Atmospheric Research Limited
NPV	Net Present Value
NWW RAC	North Western Waters Regional Advisory Council
OFL	Overfishing Level
OSY	Optimum Sustainable Yield
PA	Precautionary Approach
PO	Producer Organization
QMS	Quota Management System
RAC	Regional Advisory Council
RAP	Regional Advisory Process
RBM	Rights-Based Management
S&DT	Special and Differential Treatment
SCM	Subsidies and Countervailing Measures
SIS	Sustainability Indicators Systems
SSB	Spawning Stock Biomass

SWW RAC	South Western Waters Regional Advisory Council
TAC	Total Allowable Catch
TACC	Total Allowable Commercial Catch
TAE	Total Allowable Effort
TURF	Territorial Use Rights in Fisheries
UN	United Nations
VMS	Vessel Monitoring System
VPA	Virtual Population Analysis
WTO	World Trade Organization

List of Figures

List of Tables

Chapter 1
Introduction: The CEVIS Idea

Douglas Clyde Wilson and Kjellrun Hiis Hauge

Abstract This book is a product of 'The Comparative Evaluations of Innovative Solutions in European Fisheries' (CEVIS) project. CEVIS was created in response to a call from the European Commission for scientific research on performance evaluations of fisheries management regimes. It quickly became an exploration of how science can aid policy decisions. CEVIS teamed up biologists, economists, and other social scientists to evaluate four fisheries management innovations being considered for Europe: participatory approaches; rights-based regimes; effort control; and decision-rule systems. This introductory chapter outlines the basic ideas, aims and scientific approaches of CEVIS, and offers a brief presentation of the chapters of the book. It provides the reader with an orientation to the book and its origins, hence providing an aid for further reading.

Keywords Biological robustness · CEVIS · Cross-disciplinary · Decision rules · Economic efficiency · Effort control · Fisheries management · Innovation Evaluation Framework · Management costs · Participatory governance · Multi-disciplinary · Rights-based fisheries management · Social robustness · Trans-disciplinary

1.1 Background

The Comparative Evaluations of Innovative Solutions in European Fisheries Management (CEVIS) Project was a three-year exploration of how science can address policy questions at perhaps their most general level. The project was created in response to the following call for proposals for a scientific research project that came out of the European Commission:

D.C. Wilson (✉)
Innovative Fisheries Management – An Aalborg University Research Centre,
North Sea Centre, Willemoesvej 2, Hirtshals 9850 Denmark
e-mail: dw@ifm.dk

K.H. Hauge, D.C. Wilson (eds.), *Comparative Evaluations of Innovative Fisheries Management*, DOI 10.1007/978-90-481-2663-7_1,

> *Performance evaluation of fisheries management regimes: Fisheries management regimes consist of a fisheries management system and an associated system of enforcement. There are a very large number of possible fisheries management systems that differ with respect to the fishery outcomes they generate. This task is to compare and evaluate fisheries management regimes in terms of economic efficiency, robustness with respect of varying conditions and the cost of implementation and operation. Attention should be given to comparing the commonly used 'command and control' management systems with the emerging, decentralized, participatory and rights-based management systems. The attainment of management objectives should be evaluated against the costs of management in the form of necessary research and data collection and enforcement of the management rules.*

The idea is to use science to compare different 'regimes' and their outcomes. These regimes do not have particularly clear definitions or boundaries. Following the call paragraph, a regime is an entity which consists of a combination of two 'systems' – the 'management' system and the 'enforcement' system. It is ambiguous why enforcement should be thought of as having a special status as a separate 'system', as opposed to other components of fisheries management such as the system for providing scientific advice, the system for monitoring fisheries activities, or the system for allocating fishing opportunities. In any case, whatever these things are, some people responsible for fisheries management had decided that they needed to be evaluated and compared in terms of how well they achieved important objectives, ranging from the well defined idea of economic efficiency to the rather murky idea of robustness in respect to a presumably extensive set of unnamed conditions that varied. The group of fisheries biologists, economists and other social scientists who eventually became the partners on the CEVIS project could not resist this call. The opportunity was too great to explore how science could contribute to these critical debates. We believe that the overall conclusions reached by the project (Chapter 12) are strong enough to justify this decision, but the task was a challenging one.

The structuring of management regimes by governments and other stakeholders is a political process. Part of this political process is the selection and definition of the regime's objectives. Once objectives are identified and defined then one can at least make a coherent argument that their achievement can be objectively assessed, indeed there is a whole discipline of evaluation research based on this possibility. First, this requires, however, that the objectives be defined with great, even quantitative, precision, which the proposal call did not do and which was obviously only going to be possible in a very limited sense. Second, at an even deeper level, the proposal call was placing great emphasis on the evaluation of regimes that were emerging and participatory. 'Emerging' implies that the regimes are in a stage of ongoing development while 'participatory' implies that the regimes themselves determine and will quite possibly shift their objectives.

The twin dangers were that we could produce a set of 'results' that were too general and abstract to be useful. Alternatively, in avoiding this danger we could take reductionist approach where we ignored all the phenomena under study except those things that were easily measurable and comparable. This would mean distorting the meaning of those phenomena by allowing our methods to predefine our substance. It would also clearly make it impossible to address anything that was either emerging or participatory.

We felt we could meet the challenge posed by the call by proceeding in a humble and cooperative way. First we needed to acknowledge that different modes and styles of scientific thinking had their contributions to make. To do this we needed to strike the right balance between styles of interdisciplinary cooperation, indeed we felt that learning how to strike this balance would be both our biggest hurdle and perhaps the project's most important contribution. Scientist working together can structure their cooperation a number of ways. First, they can take a 'multi-disciplinary' approach, meaning that they remain entirely within the frame of their own discipline while contributing in their own way to solving a common problem. The advantage here is that the investigation reflects each disciplinary state-of-the-art and develops results about the fisheries management regimes that are based on an orthodox empirical examination of data and are focussed enough to be useful. The disadvantage is that this approach has the least chance of uncovering synergies between disciplines. Second, on the other extreme, scientists can take a 'trans-disciplinary' (Gibbons et al., 1994) approach in which new, common concepts and methods are developed that to various extents reflect the theories and methods of two disciplines. The advantage here is that entirely new questions can be asked and answered. The disadvantage is a high risk that the research questions dissolve into a mush where definitions of concepts are pitched at such a high level of abstraction that the results are nothing but obvious generalities. Third, the middle way is the 'cross-disciplinary' approach in which scientists from more than one discipline work together on a problem, but stay within one discipline's methodology. The role of the scientist from the other discipline becomes offering new insights and raising novel questions.

The basic CEVIS plan was to combine cross- and multi-disciplinary approaches, but to use a trans-disciplinary approach for questions where specific answers could be achieved that way. In the first place, throughout the project, CEVIS made heavy use of a group of scientific advisors made up of four biologists, three social scientists and one economist, all experts in fisheries management. These people attended the plenary meetings, listened to presentations, read draft reports and presented their responses. This was a critical cross-disciplinary activity that made an important contribution to the project. The basic research plan involved a shift from a cross-disciplinary to a multi-disciplinary phase. The idea was to use a cross-disciplinary approach to generate ideas for hypotheses and then use a multi-disciplinary approach to come at the evaluations in several different ways and hence avoid reductionism in the overall project results by combining qualitative and quantitative strategies. The individual disciplines were asked to be focused and realistic about available data in their hypothesis testing in order to avoid trivial generalities. Trans-disciplinary approaches were possible as long as this criterion was met. Whether or not the strategy worked the reader of the book must judge.

The organization of CEVIS began with the selection of the particular innovations we would focus on. We selected four basic types that were receiving the most attention in current discussions of potential changes in European fisheries management at the time: participatory approaches to fisheries governance; rights-based regimes; effort-control regimes; and, decision-rule systems. A reader may wonder, if our criterion was things that were 'receiving the most attention' in Europe, why the precautionary approach and the ecosystem approach to fisheries management are not included. This is because we considered these two things to be critically important criteria for setting management objectives, rather than innovations in management regimes as we are using the term. All management innovations should be judged in terms of both the precautionary approach and the ecosystem approach to fisheries. Other innovations were uncovered and described in the course of the project, particularly as combinations of these innovations, for example instances of participatory governance leading to the development of a rights-based regime as we found in the Canadian case.

The project had two phases. The first phase used a cross-disciplinary approach. During this phase we carried out four in-depth studies of areas outside of Europe where innovative fisheries management regimes have been implemented. These were New Zealand, Canada, Alaska, and Iceland. The visits were made by teams that included at least one social scientist[1] and one natural scientist. Cross-disciplinary teams carried out the research using social science methods based on carrying out and analysing in-depth interviews. They did a literature review of fisheries management in the area and then made visits of approximately two weeks where they interviewed various stakeholders. These areas were chosen because they had implemented at least two of the innovations that CEVIS was interested in investigating. Chapters 2 through 5 of this book are the reports of these studies.

The second phase was carried out in disciplinary working groups and took a basically multi-disciplinary approach. Each working group focused on one of the key objectives identified in the call. In order to get a handle on the objective described as 'robustness with respect of varying conditions' we decided to focus on the 'biological robustness' of the fish stocks and the 'social robustness' of the management institutions. So the disciplinary working groups were four: two run by economists examining the innovations with respect to economic efficiency and costs of management; a group of biologists examining the innovations with respect to biological robustness; and, a group of social scientists examining social robustness. All four groups used data from Europe, including the Faroe Islands. Their assignment was to identify and test specific hypotheses about the relationship between the innovations and their objectives using the methods and data that could be feasibly applied from their discipline. The hypotheses were mainly suggested by the work in the first phase, although we did not seek to exclude suggested hypotheses simply on the basis that they could not be directly linked to the overseas work. Chapters 6 through 9 are the reports of these disciplinary investigations of the innovations.

We used the term 'case' in the development of CEVIS in an analytically loose sense because we wanted to associate the term with the concept of the 'regime'. Developing our analytic strategy, however, required us to be more specific about

[1] In this case the term 'social scientists' include economists, who, of course, are social scientists. However, through most of the book, for clarity's sake, we use the term social scientist to mean non-economic social scientists and mention economists separately.

what we meant by case and to differentiate cases according to management objectives. This specificity should be a function of the scales on which the different general management objectives primarily operate. Therefore the approach we took was predicated on the following levels of analysis:

- Social robustness would be examined at the level of the fishing community;
- Economic efficiency would be examined at the level of the fleet;
- Biological robustness would be examined at the level of the fish stock; and,
- Costs of management would be examined at the level of the polity.

Scale is the most important variable in determining how effectively different types of institutions function (Wilson, 2003). Examining these regimes and their outcomes on multiple scale levels would tell us a much more comprehensive story than we could get either ignoring scale or examining regimes at a single scale level. Clearly no delineation of scales, nor definition of cases, could create crystal clear boundaries. Many of the causal factors related to the management objectives operate on different scales and many related issues arose in the course of CEVIS. For example, legality, which we originally argued was a key aspect of social robustness, also operates on the level of the polity. In the end we separated the legal analysis, a part of which now appears as Chapter 10 rather than as a part of Chapter 8 on social robustness.

1.2 The Idea of an Innovation Evaluation Framework

The underlying question of the CEVIS project has been: What does it mean for science to contribute to a policy discussion? In the original project design we called the product of this reflection the 'innovation evaluation framework' (IEF). As we conceived the project the idea of the IEF was that:

> *An evaluation of any aspect of management be it a policy, a specific measure or an institution, consists of comparing its performance with its objectives. This requires translating both the performance and the objectives into 'indicators' to allow comparison. Therefore an important organizing concept for the multi-disciplinary work in CEVIS is the identification of regime performance indicators related to each of the general management objectives that can be used to evaluate the impact of the innovations to be examined on the performance of the regime. Because this is a multi-disciplinary project, we are using the term 'indicators' broadly here to include measurements and observations of the inputs, key processes and outcomes of management (The CEVIS Project Description of Work).*

Throughout the project activities the different disciplines sought to explain to each other what they were doing in terms of how they were defining their concepts and what indicators they were using as proxies for these concepts. The IEF idea is two-fold. First the IEF answers practical questions about the implementation of the innovations in Europe. In this aspect this entire book is the IEF because it addresses implementation issues for the innovations in detail. The second aspect of the IEF involves a methodological reflection on what kinds of indicators are useful for evaluating policies and what is required to measure them. In order to keep the

IEF as practical as possible we approached this task by asking what indicators we actually had used in evaluating the innovations and what they implied for the ways we were really understanding and defining both the indicators and their objectives. The IEF is presented in the concluding Chapter 12 of this volume in both aspects, first an abstract and discussion of the indicators used CEVIS is presented, and then the main lessons from the overall book are summarized.

The role of science in this kind of broad policy context is not straightforward. A key role of science in contributing to policy discussion is to bring clarity and transparency to factual claims being made. Any policy discussion links facts with particular values and interests, and to the degree that facts can be removed from disputation, negotiations are aided. A critical problem is that facts, values and interests are strongly linked because the selection of relevant facts is a function of values and interests.

These problems are true of both social and natural facts, and social science, natural science and economics each have their own advantages and disadvantages as contributors to policy debates. Social science has developed a large suite of tools for studying meaning based phenomena like governance institutions. Because meaning, and hence subjectivity, is its subject matter, social science can never take an entirely third person, objective perspective in the same way that natural science can. The inherent subjectivity of meaning-based phenomena also severely limits the kinds of predictions that are possible. However, social science can be systematic and reach for coherence of description. Natural science studies material phenomena with tools based on universal criteria that ensure by their very nature coherence and even a powerful form of transparency. Natural science can, in principle, be both objective and predictive. To achieve this, however, it has to construct the policy object as part of the world framed by these tools. This requires an exclusive focus on the material aspects, which is a challenging limitation in seeking to address policy questions where material and social phenomena are mixed. Economics stands somewhere between these two. As a social science it studies human behaviour, but when done well economics limits these studies to behaviour within institutional contexts – such as a business enterprise like fishing – where subjective motivations have been stabilized in the mutually understood model of *Homo economicus* responding to incentives. This allows the economist to adopt a third person, objective perspective in a meaningful way that is beyond the reach of the social scientist observing governance institutions. The economist can describe behaviour objectively, and even predicatively, within these limited institutional contexts.

Fishing policy is based on information from natural science and economics, but trying to assess institutional issues like policy innovations is fundamentally a social science problem. What social scientists have learned over the years is that the inherently subjective nature of institutions makes direct measurement of many important institutional characteristics beyond our abilities, let alone our budgets. There are problems at the level of conceptualization. What is 'participatory management' such that one instance can be compared with another – or with its absence – in order to evaluate outcomes? We know that transparency is as important an institutional characteristic as any, but can one really measure something like 'who gets to look at the books' in a way that the 'who' and the 'books' are the same thing across a large enough sample to analyze one relationship while holding the others constant? There are problems at the level of research design. Complex confounding factors influence outcomes, and many of these factors are so subjective that the predictions or statements that always make up part of hypothesis testing are not possible to make or to test. Sometimes even the most critical substantive factors can be sidelined as 'error'. Therefore social scientists use designs like the multi-site, multi-method case study (Louis, 1982), which we used as the basic structure for CEVIS, as the best substitute.

CEVIS makes an interesting comparison with what is perhaps the best known attempt to systematically study resource management institutions. Long-term research networks are the most promising strategy for rigorous comparison of management institutions. One example is the International Forestry Resources and Institutions (IFRI).[2] This effort has created a common research framework used by a network of researchers that ask the same questions and measure similar biological, social and economic indicators when visiting forest management institutions in 200 sites every 3–5 years. They have developed a large number of rigorous research results about various approaches to forestry management. They are still forced to keep their research focused on specific questions that their network was designed to answer.

If this is the ideal approach, how can we use science to address urgent, current policy debates given limited time and money? A three year Framework Six research project like CEVIS cannot begin to come close to a 200 site long-term panel study, but such projects are actually the most ambitious mechanism the CFP has for such an effort. Indeed, most evaluations of potential management innovations are much more limited desk studies. Even if it were designed and funded, an IFRI-like project could not give policy makers answers within a fast enough time frame to be useful.

To respond to this challenge, CEVIS built a strong scientific team and released these scientists on the problem of evaluating these innovations. We began with a cross-disciplinary approach to generate initial ideas for hypotheses and research questions. Then we used a multi-disciplinary approach, focused on the best available data and with a mandate to contribute new information to illuminate these questions. We created the IEF through reflection on what the team had done. As we hope this book shows, this strategy yielded both substantive results (Chapter 12) and a chance to learn more about how cooperative work among various scientific disciplines can inform policy debates.

1.3 The Selected Innovations

Most fisheries management regimes in developed countries are command and control regimes in which a central agency representing a government makes fisheries management decisions that have the force of law and are enforced by government

---

[2]http://www.sitemaker.umich.edu/ifri/home

agencies. The call paragraph stated that 'attention should be given to comparing the commonly used command and control management systems with the emerging, decentralized, participatory and rights-based management systems' making this approach the baseline for the comparative analyses to be carried out in CEVIS. It was important in approaching this task for us to understand that we were not comparing command and control regimes with *alternative* regimes in the strict sense of that word. The assumption we were making was, in fact, that all of these innovations will take place within an essentially command and control framework for European fisheries management. The reasons for this are threefold:

1) Most fundamentally, in all Western fisheries management regimes the fisheries resource belongs to all citizens, and it is the responsibility of the government to manage those regimes on their behalf. For this reason all proposed innovations are in a final sense commanded and controlled by the government on behalf of the people.
2) Command and control is the most effective basic approach to the management of resources that cover a large geographical scale because it produces relatively predicable outcomes across wide areas. However, an important price is paid for this in both local legitimacy and support and in having to make decisions based on much poorer and less nuanced information than would be possible to achieve working on smaller scales. (Wilson, 2003).
3) Command and control regimes are able to respond and deal with problems where negotiated outcomes are difficult to achieve. In Europe, which faces problems with multiple jurisdictions and competition over resource allocation, there are simply decisions that are best made by central authorities.

While we use the term 'innovations' to indicate that these approaches to management had not been used extensively in Europe at the time we developed the project, these were not new or untested ideas and all of them had been incorporated into modern fisheries management regimes in developed countries. All of them were also being widely discussed within Europe as options for the Common Fisheries Policy (CFP) as the CFP moved towards a more adaptive and ecosystem-based approach to fisheries management. Arguably, most of these innovations had their origin in the 1970s, as tools to expand country influence on the recently incorporated Exclusive Economic Zones, others emerged as contingency measures to stock collapses, others arose as a result of conflicts between fishery sectors and others emerged in search of efficiency.

These innovations were 'ideal types', meaning that in practice they would take many different forms. Indeed, they were given several different operational definitions in the hypotheses testing in CEVIS and these various definitions and related indicators formed the heart of the Innovation Evaluation Framework (IEF) and are a key part of the presentation of the IEF in Chapter 12. Table 1.1 lays out the basics behind the innovations.

Table 1.1 Characteristics of the Four Regime-Level Innovations

General type	Participatory governance	Rights-Based approaches	Effort control	Decision rule systems
Main approaches	Management of particular fisheries through industry groups.	Individual quotas including ITQs.	Direct regulatory control of fishing effort.	Harvest control rules reduce the reliance on politics in implementing management measures.
	Larger scale management through stakeholder representation.	Community quotas including locally controlled individual quota systems.	Marine zoning and area management including marine protected areas (MPAs).	Non-predictive adaptive systems seek to avoid the need to make specific predictions about stock dynamics.
The status of the innovation in EU fisheries management when CEVIS was conceived	The first approach has a long history in a small number of geographically limited cases. The second approach was just beginning with the Regional Advisory Councils (RACs).	The first approach was found in Europe in pelagic fleets while the second kind was found in some producer organizations.	These two approaches had had relatively little application in Europe compared with other parts of the world, but this was changing.	The first approach had recently become important, especially in respect to stock recovery plans.

1.4 Participatory Governance

A crucial potential source of legitimacy is the various forms of participation by fishers and other stakeholders in making management decisions. When the focus is on participation by fishers a commonly used term was 'co-management' (Wilson, Nielsen, & Degnbol, 2003). Co-management mobilizes several assets to aid effective management. One is facilitated access to information (Pinkerton, 1989) including aid in the enforcement of fisheries regulations. Others are increased legitimacy through increased transparency in decision-making (Jentoft, 1989), greater accountability for officials (Magrath, 1989), and increased sensitivity to local perspectives (Pomeroy & Carlos, 1997). The weight of the evidence from global experiences with co-management generally and strongly supports the hypothesis that co-management makes management more effective (Wilson et al., 2003).

The relationship between participation and management, however, is complicated. First, there was the critical question of who legitimately participates (Wilson & McCay, 1998). Moreover, participation alone does not increase satisfaction with policies. In a survey of stakeholders, Hunt and Haider (2001) found no relationship between participation and satisfaction with forestry policy. While participation may increase a sense of inclusion and ownership, often depending on how the participation was done, people often participate in a process because they were opposed to a policy in the first place. The case study literature points to many instances where public participation in science-based policy has been unhelpful. In some situations apparent explanations for risks were deceptive, leading to public reactions based on an unrealistic appraisal of the situation (Collins & Evans, 2002). Culture gaps between experts and lay people lead to communicative breakdowns that exacerbate mistrust (Kaminstein, 1996).

Participatory governance has been institutionalised in a number of ways. This is a difficult challenge because of the tensions between needing the participation of stakeholders and the legal principle, almost universal in the West, that management is carried out on behalf of the entire public and not for the benefit of user groups. In general (Table 1.1) it is useful to distinguish between small-scale co-management efforts and those carried out on a regional or larger scale (Wilson et al., 2003). On a small scale, a bay for example, nearly all stakeholders are able to participate in face-to-face or almost face-to-face discussions. Stakeholders have a chance to air their differences and a good deal of legitimacy is made possible by the participation of local and county-level governments. This model is in some use in Europe, there are several co-management efforts, for example, doing an ecosystem approach on the North Sea related to the EU Natura 2000 initiative. On a larger scale, where questions of representation come into play, participatory governance has proven more difficult. The United States has perhaps the most developed system with the Regional Fisheries Management Councils, which began in the 1970s. This experiment has gone through difficult growing pains. The initial nearly complete exclusion of conservation groups, for one thing, led to US fisheries management being considerably hampered by a large number of law suits. Members are appointed by state governors and this has severely undercut the advantages of co-management as most people in the fisheries still felt unrepresented in management. This experience is very relevant for us as the CFP was beginning to experiment with regional level, though purely advisory, co-management with the Regional Advisory Councils (RACs).

There has been a tendency in the literature to assume that participatory governance is always a good thing. Experience has shown that, in spite of a generally positive record, it is not always a good thing and there are a number of factors in both design and implementation that affect outcomes (Wilson et al., 2003). The application of this innovation needs to be carefully examined in respect to any management regime for which it was being considered.

1.5 Rights-Based Approaches

Rights-based approaches to fisheries management mimic terrestrial property rights by allocating a right to the fisheries resource, *in situ*. There are many forms but a basic difference among them is whether the allocation is to individuals as private property, or to a group as a form of common property.

Basic economic theory demonstrates that the individual form, especially individual transferable quotas or ITQs (see Chapter 11), increases economic efficiency. The ITQ system allocates shares of the total allowable catch (TAC) to fishermen who are subsequently allowed to buy, sell or lease quota shares among themselves. Because ITQs create some degree of ownership over a quota share, and hence the control of fishing practices, the race for fish is ended and fishermen have an incentive to minimize costs and maximize revenues. Consequently, efficiency is promoted through the pursuit of economic self-interest. Allocation, formerly an expensive component of fishery management, becomes the function of the quota share market. Less efficient producers tend to sell their quota share and leave the fishery, reducing the level of fishing capacity in the fleet. Thus, ITQs are perceived primarily as a measure of avoiding over-investment and generating resource rent. ITQ programmes have been operative for some years now in Iceland, New Zealand, Australia, Canada and the US.

ITQs have several drawbacks. They make management regimes less flexible, tending to lock the system into single-species quota-based management because security of tenure is an important source of their benefits and because property rights of any type are difficult to take away once granted. They do not directly address some nature conservation objectives such as maintenance of biodiversity. Some also contest the merits of ITQs for fisheries in temperate waters because of equity and distributional effects. Copes (1997) remarks that the 'theoretical case for superiority was highly dependent on gross simplifications imbedded in the implicit or explicit assumptions, which remove the ITQ mode for the real world of fisheries'. He is troubled by the social inequities that ITQs tend to create, for instance between generations of fishers. In the case of Iceland, Helgason and Pálsson (1998) showed that quota rights became geographically concentrated, and they argued that this removed an important part of the economic base from a number of coastal communities.

For others, however, ITQs remain the solution. Davis (1996) points out in a summary of studies in a special journal issue on ITQs that various scholars have associated ITQs with important management goals such as resource conservation, economic efficiency, fisheries sustainability, and fisheries co-management. ITQs remain a contested issue in fisheries management in most countries where the system has been introduced. ITQ systems do address important issues, such as how to smoothly reduce the excess harvesting capacity that puts pressure on both fish stocks and fishing profits. They are largely insensitive to the social and cultural impacts on communities. They can also make the barriers of access for young newcomers very high.

The alternative approach to rights-based management, group rights, is in use in Alaska, Canada and in Europe in the form of quota allocations to Producer Organizations in the United Kingdom. In both the Canadian and UK cases, a number of the community groups managing these quota allocations do so by creating their own internal ITQ system. This is a particularly interesting innovation in that it combines participatory governance with a rights-based regime and, arguably, creates an ITQ system that achieves most of the benefits of ITQs while mitigating the negative effects.

1.6 Effort Control

CEVIS examined effort control in two basic forms: as allocated fishing effort among fishers, e.g. the allocation of days-at-sea and as marine protected areas (MPAs) which reduce or eliminate fishing effort in particular areas, either specific times or all the time. Both of these kinds of effort control had been gaining attention in Europe as they were seen by some people, especially in the industry, as working very well in neighbouring areas outside of the CFP, especially in the Faroe Islands. The Faroe Islands' management regime played an important part in evaluating this innovation. They have not been seriously considered in Europe until relatively recently because effort is difficult to calculate clearly enough to be used as the basis of allocation among member states. More recently, however, ways of calculating fishing effort based on kilowatt-days have begun to be used by the Commission, particularly in recovery plans where limits on effort were seen to be unavoidable. Effort controls are, in fact, a requirement of recovery plans. Kilowatt-days were being allocated to member states and can be distributed to and, in principle, transferred among vessels by the member state governments. These effort regulations are introduced in addition to TAC regulations.

As areas where fishing and other human activities are restricted or prohibited, MPAs range from highly protected nature reserves to large multi-use areas with modest limitations on specific types of human activities. As a fisheries management tool MPAs have gained increasing popularity over the last couple of decades and some consider their establishment as a necessary condition for successful fisheries management. MPAs are expected to reduce fishing on spawning stocks and recruits, to increase fish abundance within the protected area and to promote spillover of the increased fish abundance into neighbouring areas where it may lead to improved catches. By reducing fishing effort MPAs can contribute to ecosystem conservation and may enhance or preserve local biodiversity. Their introduction is therefore often supported by conservation organizations (Halpern & Warner, 2003). Once they are established MPAs typically require less biological information than other management tools and they may therefore be a more cost-effective way to conserve fish stocks than either TACs or effort control.

Despite these advantages, MPAs have been met with criticism both within and outside the discipline of ecology. One criticism was that their protection was limited to relatively stationary species and that they do little to protect migratory species. MPAs may trigger redistribution and concentration of fishing effort in adjacent areas, potentially leading to overexploitation. Previous experiences with MPAs show that few have fulfilled expectations. In an assessment of MPAs around the world Kelleher, Bleakley, and Wells et al. (1995) thus found that less than 31% of the MPA's surveyed could be classified as achieving their management objectives. The lack of success has been suggested to be caused by inappropriate MPA size and design, by a lack of economic and social science input in their establishment, by insufficient stakeholder participation and involvement, and by inadequate institutional capacity for monitoring and enforcement (Halpern & Warner, 2003).

1.7 Decision Rule Systems

CEVIS set out to examine two forms of decision rule systems (Table 1.1), harvest control rule systems, which aim at reducing the reliance on political processes in decision-making on management measures, and non-predictive adaptive systems, which aim at reducing reliance on predictions about stock dynamics. Decision rule systems set out to be self-binding mechanisms, which were *inter alia* applied to overcome the urge of politicians to harvest short-term benefits at the expense of long-term benefits. Decision rule systems are supposed to transfer decision-making power from politicians to a system of more or less 'automatic' responses to certain developments or situations. In the EU, the introduction of multi-annual recovery plans, which was an important part of the provisions of the new basic regulation of the CFP implemented from 1 January 2003, was the beginning of wider use of harvest control rule systems. During the course of the CEVIS project the use of this approach continued to expand within European fisheries, at least in principle, and the practices has increasingly become the goal. Partly as a result of this increased interest, and also as a result of the increased reliance within biology and economics on simulation modelling, most of the evaluation of this approach that we did in CEVIS took the form that assumed that such rules were in play or that the ICES precautionary framework was applied.

Non-predictive adaptive approaches constitute qualitatively different decision rule systems. Instead of aiming at predicting the results of certain management measures and having rules according to these predictions, these systems focus on monitoring the system (in a broad sense) and adapting to developments and changes, which are discovered by means of generally agreed indicators of the state of different elements of the system. At the time CEVIS was organized, this approach was being implemented in Europe for the first time in the current recovery plan for Southern hake (*Merluccius merluccius*). A well known example of a rather different non-predictive adaptive system was the 'Traffic Light method', which has been applied in the advisory process for the Northwest Atlantic shrimp stock and on trial basis for some groundfish stocks in the Scotia-Fundy region, Canada. The basic element of this method was a broad range of indicators, which represent estimates of certain attributes of the fish stocks and the fishery. These indicators, which must be carefully described, validated and generally accepted by the concerned interests, can be categorised as stock assessment indicators, indicators of ecosystem effects of fishing, indicators of economic and social outcomes and, finally, indicators of regulatory compliance. The CEVIS research on Canada (Chapter 3) found that the Traffic Light approach had turned out to be very difficult to implement and had lost favour for a while, but was beginning to be taken up again in a modified form.

1.8 Outline of the Book

Chapters 2 to 5 take a closer look at how various innovations have been implemented in four countries outside EU: New Zealand, Canada (Nova Scotia), US (Alaska) and Iceland. All of these countries have reputations for highly innovative fisheries management, and examining the lessons they might have for the Common Fishery Policy seemed an excellent way to launch CEVIS. Extensive literature reviews, capped by study tours to these countries were made to gather information regarding the four evaluation criteria defined in the CEVIS project: biological and social robustness, economic efficiency and management costs. Local managers, scientists and stakeholders, mainly from the fishing industry and conservation nongovernmental organizations (NGOs), were interviewed to examine what they see as the best practices for an effective implementation of the specific innovation studied.

The team writing Chapter 2 on New Zealand consisted of Martin Aranda and Anne-Sofie Christensen. The innovations they studied were the quota management system (QMS) and participatory governance. The QMS and Iceland's individual transferable quota (ITQ) system were the earliest and the most comprehensive and fully transferable rights-based systems. The QMS was later combined with participatory governance. The team argues that qualities intrinsic to the right such as high transferability, security and durability have been determining factors in outcomes. The New Zealand experience suggests that introduction of high quality property rights is the path to be followed for increasing economic efficiency. However, property rights applied as comprehensively as in the New Zealand system do create social costs and managers must be careful in dealing with related objectives.

Clara Ulrich and Douglas Clyde Wilson visited Canada and co-authored Chapter 3. The aim of their visit was to investigate both rights-based and participatory management in Nova Scotia. The focus was on the ITQ system and the Community Management Boards, in which communities, industry and government work together to develop and enforce regulations. The Chapter focuses in particular on two important areas of innovation. One is a community management board that has developed its own internal ITQ system that has worked well in balancing economic and social objectives. The other is the new ways that Department of Fisheries and Oceans is developing scientific advice through experimenting with new kinds of indicators and new kinds of relationships with stakeholders. These innovations have improved biological robustness by increasing the feeling of ownership and responsibility for the resource and improving the commitment to scientific advice.

The State of Alaska, U.S., has introduced a number of innovations in fisheries management. Franziska Wolff and Kjellrun Hiis Hauge cover two of these in depth in Chapter 4: the Tier System and the pollock cooperatives. The Tier System is a decision rule system that is applied to all the groundfish stocks in federal waters off Alaska. The pollock cooperatives combine rights-based management with self-governance. They argue that overall the innovations have made a positive difference. The Tier System has helped lead to abundant fish stocks through a mainly precautionary approach to management while the coop system has created the means for making the pollock (*Theragra chalcogramma*) fishery highly profitable.

Chapter 5 is by Anne-Sofie Christensen, Troels Jacob Hegland and Geir Oddson, who collaborated on the CEVIS study of Iceland. The aim of the visit to Iceland was to evaluate the individual transferable quota shares system with its many ancillary management innovations such as harvest control rules, closed areas, and community quotas. They found, *inter alia*, that the flexibility that Iceland has built into its ITQ system is essential so that the system can match the fluctuations of unpredictable fisheries. The case illustrates the importance of a strong enforcement and monitoring framework. They also argue that the harvest control rules have been set up in such a way that the TAC setting system is quite robust to both economic and biological changes.

These chapters that describe the systems in the four innovative countries visited in the first phase of CEVIS were cross-disciplinary studies on one or two innovations. In contrast, the next four chapters are disciplinary or trans-disciplinary studies focussing on a success criterion related to their discipline. Each evaluates several innovations testing hypotheses for which they had relevant data and an opportunity to generate useful results. The hypotheses were mainly generated from the visit experiences and discussed at one of the CEVIS project meetings. The hypotheses are tested using information on fisheries in Europe, in this case including the Faroe Islands.

The members of the CEVIS biological team that contributed to the chapter on biological robustness include Francois Bastardie, Alain Baudron, Richard Bilocca, Jesper Boje, Tammo Bult, Dorleta Garcia, Niels Hintzen, J. Rasmus Nielsen, Gudrun Petursdottir, Clara Ulrich, and Sonia Sanchez. Their investigations, related in Chapter 6, describe how a range of innovative management alternatives may influence biological robustness in various fisheries in the Baltic Sea, the Western Shelf, the Faroe Islands and the North Sea. They used simulation models to evaluate their hypotheses and they developed a number of clever ways to understand and evaluate biological robustness and how it could be related to innovations such as participation and effort control. The main approach they took was to relate these innovations to the information available to fisheries management and its implications for biological robustness. One interesting result was that new information obtained through participatory approaches could best increase biological robustness by reducing the bias of fishery information rather than increasing its precision. Another was that a two-step management system using a decision rule that allows TAC adjustment based on recent information on the state of the stock improves biological robustness.

In Chapter 7, Erik Buisman, Hans Frost, Ayoe Hoff, Arantza Murillas, and Jeff Powell, who make up one of CEVIS' two economics teams, evaluated hypotheses related to economic efficiency looking at information about fleets fishing in the same areas as the biologists examined in Chapter 6. The two chapters differ in that some of the innovations were defined differently, but they used similar bio-economic simulation models. The focus of this chapter was on the net present value of the various fisheries and innovations, and the results were based on simulations. Among their findings is that the introduction of participatory governance, understood as a way to improve information, improves economic performance. They found mixed results about effort control where the impact on economic performance depended on various factors in how restrictions were structured. Marine protected areas reduce economic performance in the short and medium term even while having a positive influence on recovery of fish stocks.

In Chapter 8, the CEVIS social science team, Anne-Sofie Christensen, Martin Aranda, Bonnie McCay, H. Anne McLay, Carl Rova, Andrea Leme da Silva and Franziska Wolff, present a conceptual framework for evaluating social robustness and apply it to the analysis of four European innovation case studies. They define social robustness as a combination of acceptance by stakeholders and a capacity for institutional learning. The innovations represent a range of systems that incorporate both rights-based management, including transferable effort allocations, and participatory governance, which they examine in four cases taken from the Baltic Sea, the Faroe Islands, the North Sea, and the Western Shelf. They found that stakeholder acceptance in the Baltic case, where management was approached in a traditional top-down fashion and where there is little evidence of complex institutional learning, is lower than in the others. This has led to problems with acceptance and compliance. In respect to institutional learning they found the interesting result that rights-based management facilitated high institutional learning among the rights holders, but this learning was typically geared towards strengthening their rights.

Chapter 9 is the last disciplinary chapter focussed on hypothesis testing. In the chapter our second team of CEVIS economists, Sarunas Zableckis, Tiit Raid, Ragnar Arnason, Arantza Murillas, Søren Eliasen, Sten Sverdrup-Jensen and Emil Kuzebski discusses the costs of management. They examine these costs in terms of administration, research and enforcement costs. The aim of the study was to assess whether cost levels changed as a result of implementing innovative management regimes. Based on both literature and expert interviews, analyses where carried out for selected European fisheries in the Baltic Sea, the Faroe Islands, the North Sea and the Western Shelf. The task proved very challenging because they had to use different kinds of data, which were of varying, and often low, accessibility. They were not able to come to conclusions with respect to effort or participation. However they were able to conclude that rule-based systems in the form of harvest control rules will likely not reduce research or monitoring costs. They also suggest that transferable quotas may reduce control and enforcement and overall administrative costs.

The last two chapters before the conclusion play special roles in strengthening our analysis of rights-based management.

In Chapter 10, Miriam Dross and Hendrik Acker examine the legal aspects of implementing ITQs. Legal conformity was initially seen as part of social robustness, but we chose to treat this specific issue in a chapter of its own. The chapter draws on global experiences with ITQs and presents laws and regulations that need to be considered when implemented this innovation in EU fisheries. Legal norms in general and in the European Community in particular are compatible with the introduction of ITQs. Initial allocation, however, must be carefully considered and has been challenged in the courts. The considerations involved include rules about non-discrimination, the free choice of occupation and protection against deprivation of property. For these and other reasons it is critical to make sure that these systems are implemented with sufficient flexibility to allow the system to evolve.

Chapter 11 is a special addition to the book as it is the only chapter that is not a result of the CEVIS research plan. In the chapter Ragnar Arnason not only strengthens our analysis of rights-based management, he takes a very different philosophical tact than the overall CEVIS project did. Rather than seeking a middle ground to avoid the twin dangers discussed above, his argument suggests that the most important aspects of fisheries management can indeed be reduced without distortion. In his approach, economic efficiency is the key goal that makes other goals possible. This allows him to propose a method through which relatively simple theoretical comparisons can be made that arrive at straightforward conclusions about the implications of various characteristics of property rights. This is a clear alternative to the CEVIS approach, and its inclusion will allow the reader to make more informed judgements about the advantages and disadvantages of the approach taken by the project.

In Chapter 12, Douglas Clyde Wilson, Kjellrun Hiis Hauge and Martin Aranda draw together conclusions based on the findings from the other chapters. The main conclusions are summarized and the overall findings in relation to the innovation evaluation framework are discussed. This includes a retrospective discussion of the choices that were made about how key concepts were defined and represented within the hypothesis testing by the various perspectives, skills and backgrounds that made up the multi-disciplinary CEVIS project.

CEVIS was an experiment in how multi-disciplinary scientific work can contribute to a policy debate. We chose one of many possible approaches to the questions we were posed. We invite the reader to take a critical look at how we proceeded and at the conclusions we reached. Fisheries management and environmental management in general are above all political processes. At the end of the day, it is the mechanisms for ensuring transparency, systematic reflection and institutional learning that will determine how well we manage a sustainable relationship to the sea. We hope that CEVIS has contributed to these mechanisms.

References

Collins, H. M., & Evans, R. (2002). The third wave of science studies: studies of expertise and experience. *Social Studies of Science, 32*(2), 235–296.

Copes, P. (1997). Social impacts of fisheries management regimes based on individual quotas. In G. Pálsson & G. Petursdottir (Eds.), *Social implications of quota systems in fisheries* (p. 593). Copenhagen: Nordic Council of Ministers, TemaNord.

Davis, A. (1996). Barbed wire and bandwagons: A comment on ITQ fisheries management. *Reviews in Fish Biology and Fisheries, 6*, 97–107.

Gibbons, M., Limoges, C, Nowonty, H., Schwartzman, S., Scott, P., & Trow, M. (1994). *The new production of knowledge.* London: Sage Publications.

Halpern, B. S., & Warner, R. R. (2003). Matching marine reserve design to reserve objectives. *Proceedings of the Royal Society Series B, 270*, 1871–1878.

Helgason, A., & Pálsson, G. (1998). Cash for quotas: disputes over the legitimacy of an economic model of fishing in iceland. In J. G. Carrier & D. Miller (Eds.), *Virtualism: a new political economy* (pp. 117–135). Oxford: Berg.

Hunt, L., & Haider, W. (2001). Fair and effective decision making in forest management planning. *Society and Natural Resources, 14*, 873–887.

Jentoft, S. (1989). Fisheries co-management. *Marine Policy, 13*, 137–154.

Kaminstein, D. S. (1996). Persuasion in a toxic community: rhetorical aspects of public meetings. *Human Organization, 55*(4), 458–464.

Kelleher, G., Bleakley, C., & Wells, S. (1995). *A global representative system of marine protected areas*. Four volumes. Washington: The World Bank.
Louis, K. S. (1982). Multisite/Multimethod studies: An introduction. American Behavioral Scientist, *26*(1), 6–22.
Pinkerton, E. (1989). Introduction: Attaining better fisheries management through co-management – prospects, problems and propositions. In E. Pinkerton (Ed.), *Co-operative management of local fisheries* (pp. 3–36). Vancouver: University of British Columbia Press.
Pomeroy, R. S., & Carlos, M. B. (1997). Community-based coastal resource management in the Philippines: A review and evaluation of programs and projects 1984–1994. *Marine Policy, 21*(5), 445–464.
Magrath, W. (1989). *The challenge of the commons: The allocation of nonexclusive resources*. World Bank: Environment Department Working Paper #14.
Wilson, D. C. (2003). Conflict and scale: A defence of community approaches in fisheries management. In D. C. Wilson, J. R. Nielsen, & P. Degnbol (Eds.), *The fisheries co-management experience: Accomplishments, challenges and prospects* (pp. 193–211). Dordrecht, The Netherlands: Kluwer Academic Publishers.
Wilson, D. C., & McCay, B. J. (1998). How the participants talk about participation in mid-Atlantic fisheries management. *Ocean and Coastal Management, 41*, 41–61.
Wilson, D. C., Nielsen, J. R., & Degnbol, P. (Eds.). (2003). *The fisheries co-management experience: Accomplishments, challenges and prospects*. Dordrecht, The Netherlands: Kluwer Academic Publishers.

Chapter 2
The New Zealand's Quota Management System (QMS) and its Complementary Mechanisms

Martin Aranda and Anne-Sofie Christensen

Abstract The New Zealand's Quota Management System (QMS) is one of the first individual transferable quota systems (ITQs) and the most referred example of implementation of right-based management in fisheries. In New Zealand various groundbreaking measures on fisheries management have been introduced. New Zealand does not share resources with neighbouring countries. All fisheries are under the full jurisdiction of the government and thus no external factors have affected the QMS system since its introduction. In addition, the government's aim of achieving economic efficiency has determined that government intervention is low. The QMS has evolved being strongly market-based although the government changed the design of the QMS in its early stages due to stock collapses and Maori claims. The QMS has allowed the introduction of mechanisms implemented to reduce management costs, that are now entirely borne by the industry and tools aiming at providing flexibility to the system such as the deemed value instrument and the annual catch entitlement (ACE). Participation is another of the major improvements of the QMS. Indeed the management process is consulted to a wide variety of stakeholders who actively participate in input giving even in scientific matters. Although the system aims at reducing government intervention, drastic decisions of fishing closures are still being taken by the government. The aim of this chapter is to evaluate New Zealand's QMS system in terms of biological robustness, cost-effectiveness of management, economic efficiency, and social robustness. The chapter is based on two sources of information: desk studies and a field study trip.

Keywords QMS · Individual transferable quotas · Enforcement · Cost-recovery · Participation

---

M. Aranda (✉)
AZTI Tecnalia, Food and Research Division, Herrera Kaia-Portualdea, Z/G E-20110 Pasaia (Gipuzcoa), Spain
e-mail: maranda@pas.azti.es

K.H. Hauge, D.C. Wilson (eds.), *Comparative Evaluations of Innovative Fisheries Management*, DOI 10.1007/978-90-481-2663-7_2,

2.1 Introduction

The rise of the Extended Fisheries Jurisdiction (EFJ) during the 1970s encouraged states to devise mechanisms to occupy and exploit maritime jurisdictions extended up to 200 nautical miles. Before this change, many countries in the world had maritime domains up to 12 nautical miles. Outside these boundaries, fishing was carried out by industrialised fishing nations able to operate distant water fishing fleets. After extending their maritime jurisdictions, many nations encouraged the development of their national fishing industries.

New Zealand was not an exception and established the 200 nautical miles Exclusive Economic Zone (EEZ) off the country's littoral in 1978 (Harte, 2000). The emergence of a wide Exclusive Economic Zone (EEZ) encouraged the government to promote growth of the local fishing industry, especially the offshore sector. The government of New Zealand soon recognised that these protective measures generated overcapacity and a rush for fish with the subsequent overexploitation.

The seeds of the New Zealand's ITQ model, locally known as Quota Management System (QMS), can be tracked back to ideas that emerged elsewhere in the world such as theoretical models developed by Canadian economists Moloney and Pearce in 1979 (Dewees, 2006). In spite of this, the QMS is shaped by the particular characteristics of New Zealand and its fisheries, which did not have a long tradition before the QMS inception. New Zealand consists of two main islands and does not share any resources with neighbouring countries. Thus, the country can make sovereign decisions with regard to fisheries management without much external influence.

Since the introduction of the EFJ, New Zealand's fisheries have changed substantially. This is the result of a long process of learning and adjusting steered by a government's belief that the market has to guide the evolution of this industry. The government's aim of reducing management costs and a permanent seek for adding value to the industry has changed the appearance of New Zealand's fisheries. The current fishery system is based on property rights and has many complementary mechanisms, such as private research, a cost recovery programme, active stakeholder participation and auxiliary instruments such as the paper trail tool and the discount rate instrument to reduce sea mammal bycatch. All these mechanisms and instruments will be described and evaluated throughout this chapter.

2.2 Research Methods

Twenty years have gone by since the inception of QMS, and both foreign and domestic researchers have described the New Zealand experience with ITQs. We have gathered insight through a literature review carried out as the first step to approach the QMS case. The literature review allowed us to become acquainted with the case and to identify key sources of information that led us to key people and institutions. The second stage was a study trip, which placed us in closer touch with the case and

Table 2.1 Professional Affiliations and Academic Background of the Interview Participants

	Representative of						
Profession	Government	Industry	Research	Green	Academic	Customary	Total
Biologist	3	1	2	2	1	1	10
Economist	5	1	1		1		8
Anthropologist			1				1
Fisherman		2					2
Journalist	1			1			2
Total	9	4	4	3	2	1	23

its actors, thus enabling us to seek for the sources of success or failure during the QMS implementation.

The interviews took place in three cities (Wellington, Auckland, and Nelson), where main management and harvesting activities are based. Interviews took place in November 2006. Prior arrangements were made to interview four key representatives from the Ministry of Fisheries, the Seafood Council, the conservationists and the academic realm. These representatives directed us to a broader group of stakeholders.

Twenty-three people were interviewed (see Table 2.1) and two kinds of questions were asked: (1) Open questions on how the system had evolved during the last 20 years from their perspective and, (2) Specific questions focusing on those aspects in which the informant contributes his/her best according to his/her background and the new information he/she was providing.

2.3 The New Zealand Quota Management System (QMS)

2.3.1 Background

When in 1978 the government extended the maritime jurisdiction to 200 nautical miles, a range of fish stocks came under national control. Before the emergence of the New Zealand's maritime jurisdiction, foreign fleets exploited offshore fisheries facing little control on fishing activities. In the early 1980s, the country had a low yielding fishery since overexploitation led inshore fisheries into crisis, and licensed foreign fleets largely dominated the deep-sea fishery within the EEZ (Harte, 2000). The government issued financial aid and tax reductions to encourage the development of the offshore fleet. Stakeholders used economic support to develop larger and more efficient offshore capacity, which finally was diverted to the already depleted inshore fisheries (Strakker, Kerr, & Hendy, 2002). Consequently, in 1984 the government announced a moratorium in financial assistance to fisheries (Gibbs, 2008).

In the Fisheries Act 1983, the government introduced a quota-based mechanism to manage the seven deepwater fisheries, also known as the Deepwater Allocation

System (DAS). This can be considered the precursor of the current ITQ system (called QMS for quota management system). Fixed catch quotas for the deepwater fisheries were allocated for ten years and were not transferable (Lock & Leslie, 2007). One of the goals of the DAS introduction was to encourage and secure the development of the deepwater fishery (Clark, 1993). In 1985, quotas allocated for deepwater species were granted in perpetuity. After a lengthy appeal process for inshore fisheries that lasted for 12 months, both deep and inshore fisheries were brought into the QMS from October 1986, following the Fisheries Amendment Act 1986 (Lock & Leslie, 2007).

In a more general economic context, the New Zealand's government introduced many changes in the early 1980s since economic crisis called for immediate and drastic actions. The Minister of Finance, Roger Douglas, propelled the introduction of liberal measures in many key economic activities aka *rogernomics* (Dewees, 2006). The general liberalisation plan included telecommunications, postal services, health services, education, etc. The general economic plan aimed at making the economy more competitive and open by lowering tariff barriers and dismantling subsidies. One senior manager illustrated the radical decision of privatising a public asset with this statement: *the inception of the QMS was a brave decision back then, but seen in context there were many brave decisions at that time in history.*

According to Connor, the QMS implementation was preceded by a substantial consultation process before finalising the allocation process in legislation (Connor, 2001a). This process included everyone's interests but the Maori's. A key management officer informed us that the consultation process consisted of a poll carried out among all boat owners in all fisheries, which ended up in a majority supporting the implementation of ITQs in New Zealand fisheries. The consultation process aimed at raising support and commitment from the fishermen. To achieve this aim, the government produced documents outlining the proposal and held meetings around the country.

2.3.2 The Introduction of the QMS

On October 1st 1986, the QMS was extended from the deepwater fisheries to all inshore and offshore fisheries. This system performed a fixed fish tonnage allocation to be held in perpetuity (Symes & Crean, 1995). These rights were allocated for free to the existing participants, they were transferable, and imposed a 20% limit in ownership for inshore stocks and a 35% limit for deep-water stocks.

The QMS assured the right to use the resource, while the fishing permit remained as the right of access. Initial fixed amounts of fish were allocated according to historical catch. Although rights were allocated for free, requirements for initial allocation were rather demanding. Rights were allocated to holders of fishing permits in May 1985. To receive permits under the new QMS, fishermen were required to demonstrate that they received 80% of their income or NZ$10,000 from fisheries in the fishing year 1982/1983. 2,260 permit holders (46%) could not meet this requirement and were considered part-timers. Thus they were excluded from the rights allocation (Strakker et al., 2002).

Fishermen were left with the sole decision of keeping their rights or transferring them. Rights were considered an asset from the very beginning. Social scientists criticised the informative process as poor, and many boat owners sold their quota because they found the process of keeping control of their catches and other formalities, extremely complex: *Some fishermen didn't even bother getting quota. Others sold fairly quickly to companies, understanding they would be able to lease them back.* Fishermen therefore decided to sell their rights to big companies. Leasing back hardly happened and many were expelled from the system.

In 1990, the original specification of the QMS in fixed tons was changed into percentages of the Total Allowable Commercial Catch (TACC) (Connor, 2001b). Regarding characteristics of the property right, it seems that a combination of high property rights qualities such as permanent durability and high transferability, have most likely been the seeds of the steady growth of New Zealand' fishing industry. Bess and Harte (2000) report that during the first ten years of the QMS the positives outcomes were a rise in industry profitability, high levels of investment and improved stock abundance due to developments in assessment and recovery strategies.

In the beginning, the setting of the TACCs was faced with limited knowledge about stock abundance and distribution. Other difficulties were encountered in finding the criteria on how to allocate rights among stakeholders. An inexperienced Monitoring Control and Surveillance system (MCS) had to face the challenge of controlling and keeping track of activities such as quota busting and black marketing. In spite of this, New Zealand enjoys some advantages such as having a limited number of harbours where vessels can offload and hence, reasonable control can be carried out. This fact emerged as an advantage for the implementation of the system.

The enforcement apparatus effectively backed up the implementation process. Enforcement and punishment actions were strong. For instance, penalties for quota busting were hard, including immediate confiscation of boat and gear. Other difficulties such as monitoring of fishing activities in fishing areas also needed time to be addressed. New Zealand responded to these challenges by implementing innovations to enhance the MCS by installing the first satellite fishing tracking system in the world, the Vessels Monitoring System (VMS), in 1994.

2.3.3 The Core of the QMS

2.3.3.1 Characteristics of Property Rights

Property rights comprise six characteristics: transferability, durability, quality of the title, exclusivity, divisibility and flexibility (Scott, 1988). Informants pointed out that all these characteristics are inherent to the QMS. Economists mainly aiming at economic objectives such as development of offshore fisheries, reduction of the government intervention and rise of exports, designed the system. Most informants considered the system to be successful as it has added a meaningful source of income to New Zealand and have allowed the development of a modern fishing industry. The property rights of the QMS have brought about the restructuring of the industry that is currently in the hands of mostly vertically integrated companies. These factors have brought about the growth of fish exports.

As pointed out before, the initial allocation of quota was carried out in fixed tonnages of fish. When the government decided to cut Total Allowable Commercial Catch (TACC) for orange roughy because of stock declines, the property rights holders found the means to oppose decisions and even to challenge the government. The government reacted by expressing rights as percentages of the TACC. This decision improved resource protection since individual quotas vary according to resource status. In addition, this new allocation mechanism protected the government from being challenged for further reductions on quotas due to stock declines (Strakker et al., 2002).

Decisions regarding TACC cuts are surrounded by controversy; stakeholders have on several occasions opposed substantial reductions of TACCs and even taken the government to court. In September 2007, the Minister of Fisheries, Jim Anderton, decided to reduce the TACCs in the orange roughy fishery from 914 tonnes to 870 tonnes in waters of the Bay of Plenty. This was challenged by the fishing company Antons Fishery in the High Court (Independent Financial Report). A similar announcement in late September 2006 led to a judicial process started by the same fishing company. On that occasion, the Minister did not defend the case in court. Instead, he introduced the Fisheries Amendment Bill that would give the Minister full power in resource sustainability. At the time of writing, this proposal has not yet been approved by the parliament. In a different case, the decision to cut down quotas for hoky by 10%, to 90,000 tonnes has been welcomed by the two giants of the New Zealand industry, Sealord and Sanford. They requested the Minister to lower the TACC to 80,000 tonnes. Yet smaller operators requested the Minister to keep the TACC for hoky at 100,000 tonnes since a substantial reduction would harm small operators. A process of negotiation has taken place and the government has managed to counterbalance both factions' interests and sustainability goals.

Some other rules launched by the government establish the limitation on quota ownership. A governmental officer pointed out that *theoretically five companies could own the entire fishery.* The government position is consistent to this market-based approach, and the above officer stated that if market determines that only the five fittest companies own the fishery, it would enhance a more accurate monitoring and consequently, it would reduce costs. The market-based solution has also allowed reducing costs that are huge in other countries such as the collection of economic data on fleets characteristics and operations. Overcapacity is not considered a problem by managers, and neither subsidies nor are decommissioning schemes are carried out to counteract fleet inefficiency.

In 2001, New Zealand implemented the Annual Catch Entitlement (ACE), a mechanism that stands as a remedy to soften changes in quota ownership. ACE allows stakeholders to buy and lease the share for the current year (which varies according to stock status) in such a way that quota owners do not need to transfer their right forever, but they are able to transfer or lease the share that corresponds to the current year. Thus ACE allows small operators to lease quotas from quota owners. ACE is said to have smoothed the process of ownership change.

The philosophy of low government intervention has also established that the cost for research and management must be recovered from the rights holders. Some costs for recreational and customary activities are still paid by the government, but all other costs are covered by the industry. The QMS is widely accepted by right holders; however, it faces some scepticism from conservationist and academics on conservation issues and social implications respectively. The fact that right holders support the QMS is considered advantageous for its application. A comment by one industry representative illustrates this fact well *Many people are satisfied with the system because they got something for nothing. It means that now they are owners of a quota that they can trade. In the past they just had an allowance to fish.*

2.3.3.2 The Enforcement System

ITQ systems have to be backed up by effective judicial systems able to punish infractions on the established rules (Arnason, 1992). In New Zealand, such judicial apparatus together with an efficient MCS system are the backbone of the QMS.

The Ministry of Fisheries' infrastructure includes patrols, a boat tracking system supported by satellite and an experienced staff, in cooperation with the military forces. Major offences include falsifying of records, misreporting, dumping, illegal fishing and declaration of catches from other areas than those where boats are allowed to fish. In the latter case, the Ministry uses forensic science and DNA analysis to determine whether or not fish have been caught in a given area. Enforcement staff compares catch compositions from vessels with observers on board with those from vessels without observers in order to identify misreporting and possible dumping.

There may be an important amount of misreporting in offshore fisheries. An environmentalist respondent pointed out that *fish caught by vessels with observers on board on average are smaller than fish caught by boats without observers.* In spite of this, it seems that enforcement officers consider stakeholders essential in identifying non-compliant activities and active in denouncing them. A manager pointed out *We depend on quota holders to prevent dumping activities because it is their assets that are being eroded.*

Informants considered punishment to be draconian. The sanctions include confiscation of fishing vessels and gear, withdrawal of licenses and quotas, penalties and sometimes even imprisonment. Several of the informants pointed out that discarding of species with low economic value is likely to take place in spite of the deemed value system, which is the management tool designed to prevent discarding. The MCS system controls the paper trail system and fulfilment of the technical measures such as mesh size, size limits, area restrictions and limits imposed on effort in the squid fishery. Technical measures, however, are not a major issue in the management of New Zealand fisheries, which according to a management officer is in line with the philosophy of market-based regime. Costs of enforcement are recovered from the industry, whilst enforcement costs for recreational and customary fishermen are paid by the government.

2.4 Mechanisms Complementary to the QMS

The QMS system in New Zealand is best known for the distribution of quotas through ITQs, but a number of management mechanisms complement the ITQs in the regulation of the fisheries. The QMS is not solely market-based; some parts of the system are highly regulated by the government. One of the most remarkable features of the QMS is that parts of the system (e.g. research, administration, fisheries observers, etc.) are paid by the industry through *the cost recovery programme*. Another noticeable mechanism is the system of *deemed value*, a fee that allows fishermen to overfish the TAC in order to prevent discarding. The prices of both schemes are calculated and set annually with the TAC. These two mechanisms and the *paper trail* and *discount rates* are further discussed below. Active participation and consultation are other features that accompany the system and that are in line with both the government's philosophy of openness and the stakeholders' sense of ownership.

2.4.1 Participation

The QMS is steered by the Ministry of Fisheries, but an active process of consultancy with stakeholders is taking place. Consultancy can be tracked back to the days prior to the QMS inception when the Ministry's officers were sent to harbours to discuss with fishermen the possibility of introducing an ITQ system in New Zealand (Connor, 2001a).

Early in the process decisions started to change the face of New Zealand fisheries, while participation started to shape up when companies gathered in commercial stakeholder organisations (CSOs) under the umbrella of the New Zealand Seafood Council (SeaFIC). Industry representatives pointed out that the organisation of the industry is complex and that there is poor collaboration between CSOs. Industry participates actively in discussion papers such as the initial position review of the TACCs, conversion factors, and final advice papers.

Stakeholders, including industry, conservationists, and indigenous people, participate together with the Ministry, scientists and other government departments in the research planning process as part of the planning groups and coordinating committee. Stakeholders and scientists working for stakeholders revisit and contest the outcome from the stock assessment working groups in the plenary held annually in May. The main outcome of the plenary is the Plenary Report, which is the basis for management recommendations. Stakeholder participation is said to bring about stakeholder understanding of research needs and improve assessment with meaningful input, but stakeholders' participation is also said to be complex and time consuming.

It is interesting to see that the industry's participation includes supplying input to the setting of some technical measures such as excluding devices for sea lions. Participation also involves conservationist groups. They have an increasing role, but the lack of funding was pointed out as one of the main reasons for the conservationists to lack active participation. A management officer recognised the need for conservations involvement and that maybe the government should help the greens to participate, otherwise their representation will drift away from the intention.

Stakeholders actively participate in the proposal of management objectives. This aspect internalises responsibilities and legitimates management decisions. Stakeholders' involvement in the process of management is such that they have proposed banning bottom trawling and dredging from 31% of the EEZ and 6% of the territorial sea. The Ministry of Fisheries needs a range of inputs from stakeholders to assist them in making good fisheries management decisions. Each year the Ministry and stakeholders undertake a research planning process that results in the Proposed Fisheries Research Plan. The Minister of Fisheries as part of the Ministry's work plan or Statement of Intent approves the revised document. A substantial portion of the costs of many of these research projects is recovered from the commercial fishing industry.

2.4.2 The Cost Recovery Programme (CRP)

The cost recovery programme (CRP) that was introduced in 1994 aims at recovering the costs of management, including enforcement and research for all commercially exploited stocks. The principles supporting CRP request individuals to pay for the exploitation of resources from which they are benefiting (Stokes, Gibbs, & Holland, 2006). The government pays for the costs of public interest that involves customary and recreational fisheries. Costs for multi-sector fisheries, which are fisheries such as snapper that are shared by recreational and commercial fisheries, are borne by the industry and the government.

Stokes et al. describes the various objectives aimed at by the actors involved in the CRP (Stokes et al., 2006). For the government, objectives are; efficiency, accountability and end of the dominance of the research service provider. For the industry objectives are reduction of costs and interest in the services provided by the research contractors. For research providers objectives are independence in collaborating with either industry or the government and competitiveness determined by the range of potential providers of research services.

Peacey says that CRP allows the Ministry to recover about 30% of the annual budget (30 million NZ$) (Peacey, 2007). One of the main advantages of the CRP is to ensure focus on cost-effective research methods. Yet the system is administratively complex, and scientific merits are clouded by cost considerations because the more economically attractive offer does not necessarily mean the better science. Harte sees among the various advantages of the CRP, the improvement of accountability and transparency in the delivery of management services, involvement of industry in the determinations of management services and increasing efficiency in the delivery of services (Harte, 2006).

The way the CRP works through consultation is a huge driver in the QMS since CRP is a comprehensive system of commercial fishermen paying the expenses of fisheries management, research and enforcement. Generally speaking, scientific advice regarding TACC reduction may not be welcomed by an industry that may be inclined to reduce research. In turn, conservationists suggest that the industry's financing of research somehow directs research to the most profitable species.

2.4.3 The Deemed Value Instrument

Discarding is often the biggest problem of an individual quota system, but banning discarding without allowing certain flexibility into the system may not work in mixed fisheries. New Zealand wanted to create an incentive for fishers to avoid targeting outside the TACC. For this purpose, the deemed value system (DVS) was created twenty years ago. The idea behind DVS was to create an instrument that would encourage fishermen not to target above the TACC, instead of dumping catches. This system is applied when fishermen exceed their quota (and cannot/will not buy or rent more quotas). Thus they have to pay the deemed value (DV) to the government.

The DV is set annually and is balanced so that the fishermen will neither gain nor loose economically from exceeding the quotas; hence there are no economic incentives to discard and no incentives to keep fishing after the quota is caught. The DVS is flexible: If a fisherman goes fishing without ACE or quota, or overfishes, he has to pay DV of the catches on the 15th of the following month. Until then he can buy ACE or quota to fit his catch and hence not pay the DV. If he pays DV, but buys ACE or quota within the end of the year, he can have the DV refunded.

The setting of the DV is based on economic calculations of prices. To set the DV so that the system obtains the required effects is almost impossible as the DV is interactive with the prices of fish and of quota/ACE. When the DV is set too high, the DVS undermines the quota/ACE prices by encouraging people not to buy quota or to discard. When the DV is set too low, the fishermen have incentives to overfish the TAC within the legal boundaries.

This system was one of the most criticised features of the QMS. An economist stated that he did not consider DV to enhance sustainability because it allowed the TACC to be overfished; he considered the DV to be another tax. A biologist said that DV had resulted in an economic invitation to exceed the TACC. There were many polarised opinions regarding the DV as being an effective instrument of management, and conservationists pointed out that due to crews paying DV, they tended to discard when the DV is too high.

A management officer informed us that the DV instrument had been changed many times, and was thus slowly improving. At the time of the field trip, a working group composed of industry and the government representatives had sent out a discussion paper on this issue. The question is, however, whether this tool is suitable for mixed fisheries. In a mixed fishery, it is more likely that low value species will be dumped, and that fishermen will pay DV for the high value species, which will at least allow them to recover the paid DV.

2.4.4 The Paper Trail System

Shortly after the inception of the QMS, managers realised that strong enforcement measures were required for the QMS to be effective. In that context, a computerised system for monitoring fish trading was created. The aim of this system, known locally as the paper trail, was to stop the black marketing of fish by registering and tracking all fish brought into and sold in New Zealand. The paper trail system has been a pillar of the QMS since then.

The paper trail system is currently being widely criticised. Management officers considered that the paper trail is dated because of problems with compliance due to emerging ways of regulation circumvention. Some fishermen criticised the paper trail system for not working well: *the paper work is too heavy and complicated and it is followed by substantial fines if it is not done right.* A fishermen representative stated that the Ministry was reluctant to give instructions on how to fill out the papers, as the Ministry did not want to risk getting sued if wrong advice was given. Hence, the Ministry recommends the fishermen to seek legal advice if they have doubts on how to fill out the papers.

Other people pointed out that the paper trail system had caused damage to local communities: The paper trail system is based on inspection; hence it requires using the larger landing sites. As New Zealand geographically covers a large area, most landing sites were closed down in order to enable the enforcement of the system. Not only the landing sites are controlled –the receivers of the fish are also controlled. A large number of fish receivers (e.g. local fish markets, restaurants, etc.) were refused license, as the amounts of fish were too small. One of the results is that it is impossible to buy freshly caught fish in many remote local areas of New Zealand.

2.4.5 The Discount Rate Instrument for Reducing Sea Mammal ByCatch

The discount rate can be seen as a system of incentives for fishermen to adopt technical measures to avoid bycatch into the fisheries. The system of discount rate applies to squid trawlers in the southern waters of New Zealand. The problem in the squid fisheries is that sea lions are often caught as they feed on squid. Often the sea lions are killed in the process, and the rate of survival for the sea lions that manage to escape from the fishing gear is low.

In order to reduce the number of sea lions killed, the industry was proactive in developing new technology of excluding devices by employing people from overseas to help develop technical measures. *Fishermen are active in developing gear for avoiding seabirds and sea mammals* said a key management officer enthusiastically. Fishermen managed to come up with an excluding panel that releases the sea lions by a slide leading up to the top of the trawl.

The squid fisheries are managed by a dual system: the normal TACC system and a maximum allowance of killed sea lions per vessels. The system of maximum allowance of killed sea lions per vessel works through calculated averages: For example, the Minister can decide that the fishery related mortality limit (FRML) is 200 sea lions a year. This setting of the FRML can be based on both political and biological objectives. Often the greens in New Zealand have strong protective attitudes towards sea mammals and work intensively to reduce the FRML. From the FRML the Minister can calculate backwards: He knows that the bycatch rate for sea lions equals 6 sea lions per 100 tows. The squid fisheries have to end either before 3.333 tows are made or when the TACC is caught. The result is that the squid fisheries are stopped before the TACC for squid is caught. The system of discount rate fits into this system. If the fishermen voluntarily install the excluding panel in the trawl, they get 20% extra tows as about 20% of the sea lions survive an encounter with the excluding panel.

2.5 The Outcomes of the QMS Implementation

2.5.1 Fishing Industry Development

Since the introduction of the QMS, a substantial increase has taken place in both quantity of harvest and its economic value for many species, for example, hake (see Fig. 2.1). During the first years, the QMS allowed for the rise of employment mainly in the processing sector due to the fact that rights allow a long term planning horizon that stimulates investment in technological improvements, hence diversifying and adding value in a competitive processing sector (Annala, 1996; Batstone & Sharp, 1999). Many companies prefer to process abroad, which reverts the trends in employment levels in the inshore sector.

Generally speaking, security of right tenure and other characteristics such as exclusivity, perpetuity and transferability of rights in New Zealand's QMS have encouraged the planning of operations, technological improvements and research and development. These positive spillover effects have spread onto other sectors outside the QMS such as aquaculture. The case of the Greenshell™ one of the 'star' products of New Zealand aquaculture is a good example of the latter (Bess & Harte, 2000). Furthermore, the investment in innovation has allowed for the development of some highly vertically integrated fishing companies that compete worldwide (Bess, 2006).

Most of the fishing industry growth was experienced in the offshore sector between 1986 and 1989. At that time, local fleet lacked offshore capability and charter vessels carried out the fishing. Between 1990 and 1992, a sharp increase in exports was registered and local companies invested heavily in deep-water

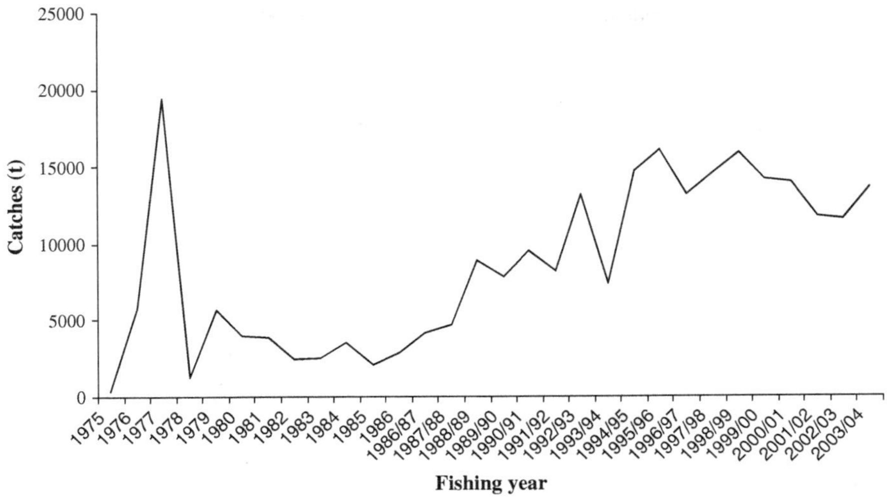

Fig. 2.1 Evolution of the Hake Fishery During the Last 30 Years. Notice that after the QMS inception the fishery has experienced a substantial growth in terms of catches. Data source: Ministry of Fisheries

capabilities and sea farming (Bess, 2005). It seems that the introduction of the system has been fleet developmental, probably because it did not start out facing an overcapacity problem (Batstone & Sharp, 1999). The inshore fleet, which comprises boats < 12 and 12–24 m, has experienced significant restructuring, including vessels replacement, change in ownership, new gear configuration, and changed targeting. Capacity of the core inshore fleet in the range 12–24 m has been kept constant from the mid 1970s, while the < 12 m sector has experienced a significant reduction of circa 70%. This fact seems to be an effect of a shift towards larger vessels. Connor reports an increase in the overall fleet size of about 43% during the period 1987–1998, and this is mainly due to the growth of the offshore fleet (especially > 33 m), which was built to replace charter vessels and to increase specialisation (Connor, 2001a).

The 24–33 m segment has gone from a few boats in the mid 1970s to a significant sector of the fleet. These boats are being devoted to offshore species. Since the fleet changes have been more developmental, it seems that the predominant change in the industry has been quota concentration without meaningful capacity reduction. Connor argues that gains in efficiency were located outside the harvesting sector; for instance returns to scale in the processing and export sector, synergies between the inshore and offshore operations and new and larger companies (Connor, 2001a).

Stewart et al. studied quota concentration in New Zealand and elaborated a profile of exiters. They found that most exiters were boat owners without involvement in processing. These stakeholders had years of involvement in the industry and made a rather quick decision of leaving (Stewart, Walshe, & Moodie, 2006). According to the study above, exiters left the industry voluntarily and for a variety of reasons other than loss of competitiveness. It seems that quota concentration has been addressed in the design of the QMS through mechanisms that impose quota limitations with the aim of avoiding excessive quota concentration in few hands (Strakker et al., 2002).

Many informants suggested that some sort of capacity expansion had taken place in some sectors of the fleet and consequently spillover effects on international waters (e.g. Commission for the Conservation of Antarctic Marine Living Resources – CCAMLR) or foreign EEZs (e.g. Chile or Namibia) have been generated. The capacity reduction in the inshore sector is evident, but capacity has been expanded in the offshore sector because of incentives for entrepreneurs to exploit deepwater resources that were fairly abundant during the first years of the QMS. The management officers interviewed pointed out that in a comprehensive ITQ system like the one in New Zealand, overcapitalisation cannot be a concern for the management but an issue for the firms having to take decisions in order to succeed, which include decisions on investing heavily in fishing capacity. Consequently, in New Zealand there are no subsidies for fuel, decommissioning schemes, vessel construction or renovations programmes. New Zealand as well as Australia has not taken any action in implementing the International Plan of Action for the Management of Fishing Capacity. Australia and New Zealand are two of the countries in which property rights have been widely adopted. Yet the plan has been widely subscribed by most FAO member states (Pascoe, 2007).

When the QMS was introduced, the < 12 m inshore fleet shrunk and consequently labour suffered accordingly. The economic system absorbed the impact by alternative labour opportunities in, a then expanding, fish processing sector. It seems as if the property right brought security and allowed for long planning horizons and, consequently, investment in processing facilities was possible. Nowadays, in many cases resource availability has dropped and firms are seeking to reduce costs by processing fish in China and other countries in the Pacific Region, where labour costs are lower, or by chartering Ukrainian fishing boats that have lower operating costs. With prestige conquered in the global fish market, some firms such as Sealord process their products abroad, thus there is a tendency to do away with fishing and processing capacity. Good natural conditions for aquaculture and prestige in the world market for New Zealand's seafood are allowing for a rapid expansion of aquaculture. Informants pointed out that, however, there was an increasing synergy between the fishing and the aquaculture due to the fact that aquaculture occupies a large part of the inshore area and that this may generate conflict.

2.5.2 Indigenous People

Proper fisheries management and restructuring of the fleet were the obvious challenges for the quota management system in New Zealand. QMS inception aimed at economic efficiency, and social objectives were not taken into account from the start. The QMS was not designed to respond to the Maori people's rights claims for exploiting natural resources that were stipulated in the Treaty of Waitangi. Later on, and after legal challenges in the Treaty of Waitangi Tribunal, the government found the means to combine traditional Maori claims with the modern capitalistic management system.

New Zealand has a different history compared to other colonised countries as it was not conquered and forced under the Crown: The Maoris signed the Treaty with a representative of the English Crown in 1840. The Treaty of Waitangi states that the Maori have rights to their natural and cultural resources – including fishing resources.

In 1957, the United Nations (through the International Labour Organization) adopted Convention No. 107 of 1957 to be applied to indigenous and tribal populations in independent countries aiming at protecting these people. Since then, indigenous peoples have increased the political impact of their countries claiming back rights and resources lost in and after the colonisation. Whether the adoption of these conventions on supra-national level was a result of increased focus on indigenous peoples' lost rights or the other way round is hard to say. The Maori had made many claims in vain, but things started moving under a left wing government in the early 1970.

In 1975, the Waitangi Tribunal was established to make recommendations to the government on how the Waitangi Treaty should be applied in current political matters. In spite of this, the part-time fishermen – many of which were Maori – were not eligible for the initial allocation of quotas in 1983. The initial allocation concerned 29 species corresponding to more than 80% of commercial fisheries.

Numerous Maori organisations protested against the allocation of fishing rights and applied for injunction in the High Court, which was granted in 1987. The Waitangi Tribunal assured the Maori that the Treaty of Waitangi guaranteed the Maori full rights to their traditional fisheries (Waitangi Tribunal, 1988). The government and the Maori reached an agreement on the fishing rights – the government arranged for buy-back schemes to be finalised by the end of 1992. The Parliament passed the temporary Maori Fisheries Act in 1989. From a non-commercial customary Maori perspective, this settlement of a share of the ITQ was not satisfying. Hence, the government established the fisheries task force to advise on appropriate legislative change and reform. The task force saw a need for the Maori to be involved in the management of the fisheries. The task force suggested two components for addressing the issues of traditional fisheries: a harvesting right and exclusive rights. Hence, the customary fisheries were ensured to the Maoris by offering them exclusive rights to certain inshore areas.

Others see the situation differently. Boast sees it from a legal perspective (Boast, 2000). He argues that the system is rooted in political pragmatism rather than in the legal constitution, and goes further into saying that its complexity, especially as to customary fisheries, has prevented a clarification of the Maori fishing rights. Dewees argues that the Maori people have had a hard time adjusting their fisheries from the traditional fisheries, as fisheries now require a new set of skills in order for fishermen to manoeuvre in the bureaucratic system (Dewees, 1997). Representatives from the Ministry of Fisheries, Hooper and Lynch, argue that the process sketched above is an expression of the recognition of and provision for the rights of the Maori and the coastal communities (Hooper & Lynch, 2000).

2.5.3 Fishing Communities and Recreational Fisheries

With regard to fishing communities, social scientists pointed out that many communities have disappeared because QMS propelled a movement in which the small boat owners rejected their rights to avoid great bureaucratic processes in the hope of leasing their rights back later. Leasing back has not occurred to the extent expected. The social scientists interviewed suggest that information was not sufficient and that the small-scale operators were not advised to be cautious in the use of their rights. Management officers recognise that the QMS has had a negative social impact on fishing communities. A key management officer pointed out that *the QMS from a macro economic perspective is good; if you want to maintain communities, it is bad.* Managers did not consider this a failure of the QMS since *social objectives were not in the original agenda.*

What is clear is that introducing such a comprehensive ITQ system transforms the face of fisheries. The actors did not foresee QMS outcomes in the beginning. The most efficient companies that had better management capabilities to plan their operations absorbed many other small operators. Other informants pointed out that the complicated paper trail system required fewer points of offloading to make it easier to handle. Consequently, many small offloading points were closed. Local fishermen were not allowed to sell fish locally, as restaurants were not considered to be fish receivers. Stewart et al. (2006) suggests that the impact on employment was not high since leaving fishermen were absorbed by the fishery system where they found alternative labour opportunities. In addition, former boat owners took the decision to exit fisheries on their own.

Both the government and the public in general consider recreational fisheries as an important source of satisfaction. It provides fresh fish for home consumption and is an important source of income for fishing communities with neighbouring fish spots. It creates jobs in retailing, entertainment and services. The economic value is estimated at $973 million for the major recreational species (Lock & Leslie, 2007).

The management of recreational fisheries is a polemic issue in New Zealand. Recreational fishing is considered a threat to resource sustainability by many groups such as commercial fishermen and conservationists as it is not strongly regulated. This activity is a deeply rooted tradition, considered by New Zealanders as a birthright. Recreational fisheries are practised by New Zealanders from a wide range of ethnic and social backgrounds (Hawkey, 1994 quoted by Lock & Leslie, 2007). This is well understood by the government, and the non-economic and economic values of recreational fishing are being taken into account when estimating the recreational share of the TAC.

It seems that a conflict for space and resources between increasing recreational fishing, marine farming, conservationists, commercial and customary fishing will increase in the near future. Lock and Leslie (2007) see it necessary to create a management mechanism to facilitate interaction among these factions.

2.5.4 Resource Status and Assessments

Many argue that the QMS is not a success in terms of biological sustainability. However, the abundance of some low mobility species such as rock lobster, scallops and abalone populations has increased. Such increases are likely due to the QMS, the participatory approach and the co-management of these fisheries. A fisherman, however, stated that *there is too much recreational fishing for the quota system to work; e.g. in the rock lobster fishery about 25–50% of catches are not reported.*

Assessment of inshore resources is relatively accurate since their biology is well known and data used in the assessment is not undermined to the extent of offshore species, as illegal activities are easily detected and denounced by inspectors, fishermen and the general public. In the interviews, recreational fisheries were acknowledged by all fractions as a factor that threatens stocks' health due to scarce regulation. A conservationist commented: *The quota system is being undermined by the other kinds of fisheries which are not included in the QMS.* Recreational fishing, which cannot be measured, is considered an impediment to evaluating resource status in the inshore fisheries. The extent of recreational fishing is evaluated through voluntary surveys, but there is a belief that many recreational fishermen are not aware of the daily limits (Lock & Leslie, 2007). In addition, customary fishing can also be considered an impediment to accurate assessment of the status of inshore resources and their management. Customary fishing is not subject to strict size restrictions, bag limits and other management measures (Bess & Rallapudi, 2006).

It is difficult to assess the impact of the QMS implementation on slow growing deepwater populations such as the orange roughy. TACC-setting demands a certain amount of data and knowledge, but this is especially difficult as regards exploitation of deepwater resources where assessment is expensive and difficult (e.g. orange roughy reflects sound poorly in acoustic surveys due to lack of swimming bladder). The initial orange roughy TACC was based on limited data and educated guesses based on a review of the – at that time rather scarce – grey literature on the dynamics of other orange roughy stocks in the world. The growth rate was overestimated, and fish behaviour was misunderstood. Consequently, the TACC was overestimated and the fleet overfished the resource. Overfishing was exacerbated by the fact that this species gathers in compact aggregations to feed, and this behaviour makes the species highly vulnerable. Furthermore, scientists suggested that this species may not have a steady recruitment. As the orange roughy population was seriously threatened (see Fig. 2.2), the government decided to cut down the initial TACC for orange roughy, which generated law suits between the industry and the government in 1990–1991.

Out of the 592 stocks, 220 stocks are managed by means of a TACC estimation using catch records, 75 through a TACC based on catch per unit effort (CPUE) analysis, and 75 (about 8 species) through a TACC estimation on full stock assessment including acoustic and trawl surveys. Species included in the latter are among the most profitable species; snapper, hoki, orange roughy, rock lobster and oyster among others. Stock assessments are carried out by 13 stock assessment working groups. The Ministry of Fisheries runs the process in which the National Institute of Water & Atmospheric Research Limited (NIWA) participates. The industry also participates

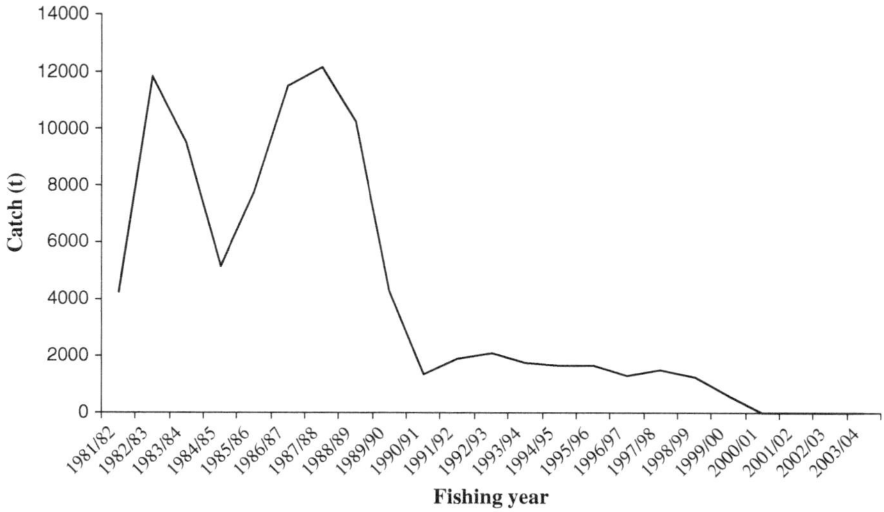

Fig. 2.2 The Rise and Fall of the Orange Roughy Fishery. The fishery collapsed in 2001 and was finally closed. Data source: Ministry of Fisheries

in hiring leading international researchers. Assessment services for resources that are commercially exploited are purchased by the Ministry and then recovered from stakeholders.

The plenary for discussing stock assessments and proposed TACCs is carried out on 31st of May so that the fishing season is ready for October. This is usually a lengthy process because of the various consultations to be carried out among the groups concerned. According to a management officer, there is a need to increase the budget for research. New Zealand has a water area eleven times its land surface. Hence, much research is needed to perform the proper stock assessment. The stock assessment process gives room for research purchased directly by the industry, for example, tagging for rock lobster, fine scale harvest data for abalone, acoustic surveys of orange roughy, catch sampling, habitat mapping and development of excluding devices for sea lions (Peacey, 2007).

A representative of the official sector argued that: *one of the main outcomes of the QMS has been conservation ethics.* An example of this is the industry's suggestion to ban bottom trawling and dredging operations in 31% of the EEZ. It is said to be the largest marine protective action ever proposed within a nation's EEZ (Bess & Rallapudi, 2006). Even though it sounds like a good example of increasing environmental ethics, it is worth pointing out that the areas the industry are proposing to include in the ban are deep-water areas in which fishing is unfeasible. However, since fishing technology is progressing so quickly, the interviewed representative pointed out that industry has probably lost future opportunities. This issue has also generated polarised opinions between conservationists, some of them approving the industry's proposal since the industry suggests banning trawling in areas in which trawling has never been carried out. Other factions of the greens considered that this is just a first step in a negotiation process in which the industry will request access to seamounts and other closed areas.

2.6 Evaluation of the New Zealand QMS

The purpose of CEVIS is to evaluate innovations with regard to four criteria: Cost of management, economic efficiency, biological robustness and social robustness. The findings are summarised in the table below where (–) means 'decrease', (+) means 'increase', (+/–) means 'not clear relation' and blank means 'no apparent relation':

The most important innovation in New Zealand's fisheries management has been the introduction of *Property Rights* with high transferability, durability, security and exclusivity. These property rights features (see 1 in Table 2.2) have caused a rise in economic efficiency allowing the most efficient actors to remain in the fishery and to increase competitiveness through value adding of products. Biological robustness has increased for some species, but not in the case of deep-water species where the lack of a solid knowledge basis plus heavy exploitation, was the reason for the

Table 2.2 Overview of the Four Evaluation Criteria

	Economic efficiency	Biological robustness	Social robustness	Costs of management
(1) Property rights				
High transferability, durability, security and exclusivity	+	+/–	–	
(I-A) Main related processes				
Fleet capacity changes	+		–	+
Need for effective monitoring		+	–	+
The industry's search for value adding	+	+/–		
The industry's seek for participation		+	+	+
The government's introduction of deemed value measure		+/–	+	+
The government's introduction of ACE measure	–	+	+	+
The government's implementation of the paper trail measure		+	–	+
(2) Participation				
The industry's involvement in research		+	+	
The government's consultation to stakeholders		+	+	+
The industry's technical solutions to by catch problems		+	+	
(3) Devolution of responsibilities				
Cost-recovery system		+/–	+	–

decline of the orange roughy fishery in the Challenger area. Social robustness has diminished because the ITQ system has excluded less efficient fishers and negatively impacted on small fishing communities.

The property rights characteristics of the QMS have generated some other processes that have changed the face of the New Zealand fisheries (see 1-A in Table 2.2). *Fleet capacity changes* show different patterns in the different sectors of the fleet. In the inshore sector, property rights inception has determined a reduction of capacity, but in the offshore sector, it has resulted in excess capacity. This may be the result of the search for species with higher abundance (e.g. orange roughy) to supply international markets. It is likely that the sense of ownership generated by the property right approach has encouraged a *search for improving of monitoring* and *seeks for participation*, which is also generated by the cost-recovery system. The *deemed value system* is criticised for being an invitation to overfish, but it has also been a good measure to avoid discarding.

ACE has smoothed the process of ownership change and it has also caused a positive impact in terms of social robustness because many small operators can lease out quotas from rights owners. In terms of economic efficiency, the ACE impact could be considered negative, as the introduction of *ACE* has slowed down the pace of concentration of rights by the most efficient operators. The management of ACE, however, seems to be hard to implement due to the great amount of information there is to deal with, a fact that substantially increases costs.

In turn, the *paper trail system* is considered positive in terms of biological robustness since actors are discouraged to misreport. However, social scientists pointed out that many points of offloading were closed, as they were considered too small. This affected small-scale commercialisation of fresh fish and the small communities. Paper trail is considered expensive and complicated.

Participation is the second main management mechanism in New Zealand. Many positive outcomes are found related to participation (see 2 in Table 2.2). *Industry involvement in research* through participation in working groups, hiring of international experts and support in data collection are considered to increase biological robustness. *Consultation* of management issues increases social robustness since actors, including conservationists and customary groups, feel part of the process, which makes them feel part of the system and enhances stewardship. However, consultation also makes the process complicated and costly. *Technical alternatives to problems of bycatch,* for example, involve the fishermen in determining meaningful input and strengthen bonds with managers.

Finally, *Devolution of Responsibilities* is the mechanism that has allowed substantial reduction of management costs, even if some aspects of management are still covered by the government (see 3 in Table 2.2). *Cost recovery* has meant wider involvement of the industry in all processes of management. Hence, it has caused a positive impact in terms of social robustness. The system is criticised for its focus on the species of higher economic value. Consequently, the *cost recovery* system has caused a negative impact in terms of biological robustness for some species.

2.7 Conclusions

Although it is difficult to draw conclusions about which factors have been the main forces in shaping a rather successful New Zealand's QMS, it is noticeable that qualities intrinsic to the right such as high transferability, security and durability have been determining factors. These characteristics have enhanced the quality of the titles that have developed a sense of ownership. The sense of ownership has, in turn, generated stakeholders' involvement in management and enhanced competitiveness. This participatory aspect has been among the driving forces in developing the New Zealand fishing industry. It has given rise to a growing concern about how to improve the management system by implementing measures such as the cost-recovery system.

Withdrawal of government's subsidies have enhanced industry inventiveness, which has been expressed in the development of products through research and development, improvement of sea and land capabilities, and the expansion of export markets. Although many of the persons interviewed point at the impact on equity as one of the negative outcomes of a system that gives efficient actors the opportunity to prosper at the expense of the inefficient actors, it seems that the New Zealand system has offered the exiters alternative economic opportunities. The issue of indigenous people and their degree of involvement in the system also means that New Zealand has taken decisions to respect customary rights. Therefore the integration of the Maoris into the system is also acknowledged to be successful.

New Zealand's experience tells us that managers should take the introduction of property right systems into careful consideration by clearly defining what the objectives are, from the beginning. Trade offs are to be carefully taken into account. If a country or region aims at economic efficiency as its overall objective, the New Zealand experience tells us that introduction of high quality property rights is the path to be followed. However, if social concerns are the objective, property right cannot be applied comprehensively as in the QMS and the quality of the property right will diminish, thus reducing economic efficiency. Hence, it is in the hands of the managers to analyse the trade offs and to attempt to counterbalance the management objectives and tailor made management actions according to each given fisheries context.

Acknowledgments The authors wish to express their gratitude to all individuals and institutions in New Zealand for devoting their time and expertise in helping us to reach our goals. Any error remains the sole responsibility of the authors.

References

Annala, J. H. (1996). New Zealand's ITQ system: have the first eight years been a success or a failure? *Reviews in Fish Biology and Fisheries, 6*(1), 43–62.

Arnason, R. (1992). Theoretical and practical fisheries management. In E. Loayza (Ed.), *Managing fishery resources* (pp. 3–10). Proceedings of a Symposium Co-sponsored by the World Bank and Peruvian Ministry of Fisheries, held in Lima-Peru, 1992. Washington: World Bank Discussion Papers 217.

Batstone, C. J., & Sharp, B. M. H. (1999). New Zealand's quota management system: The first ten years. *Marine Policy, 23*(2), 177–190.
Bess, R. (2005). Expanding New Zealand's quota management system. *Marine Policy*, *29*(4), 339–347.
Bess, R. (2006). New Zealand seafood firm competitiveness in export markets: The role of the quota management system and aquaculture legislation. *Marine Policy, 30*(4), 367–378.
Bess, R., & Rallapudi, R. (2006). Spatial conflicts in New Zealand fisheries: The rights of fishers and protection of the marine environment. *Marine Policy, 31*, 719–729.
Bess, R., & Harte, M. (2000). The role of property rights in the development of New Zealand's seafood industry. *Marine Policy, 24*, 331–339.
Boast, R. P. (2000, March). Maori Fisheries 1986–1998: A reflection. *Indigenous peoples and the law: An online institute of law affecting indigenous peoples*. Retrieved from http://www.kennett.co.nz/law/indigenous/1999/48.html
Clark, I. (1993). Individual transferable quotas: the New Zealand experience. *Marine Policy, 17(5)*, 340–342.
Connor, R. (2001a). Initial allocation of individual transferable quota in New Zealand fisheries. In R. Shotton (Ed.), *Case studies on the allocation of transferable quota rights in fisheries* (pp. 222–250). Rome: FAO Fisheries Technical Paper, no. 411.
Connor, R. (2001b). Trends in fishing capacity and aggregation of fishing rights in New Zealand under individual transferable quotas. In R. Shotton (Ed.), *Case studies on the effects of transferable fishing rights on fleet capacity and concentration of ownership* (pp. 151–185). Rome: FAO Fisheries Technical Paper, no. 412.
Dewees, C. M. (1997). Fishing for profits: New Zealand fishing industry changes for "Pakeha" and Maori with individual transferable quotas. In G. Palsson & G. Peturdottir (Ed.), *Social Implications of quota in fisheries* (pp. 91–106). Proceedings of a seminar held in the Vesterman Islands in May 1996. Copenhagen: TemaNord.
Dewees, C. M. (2006). *Historical observations of New Zealand*. Retrieved September 12, 2007, from University of California Davis, New Zealand fisheries study tour Web site: http://nzfishtour.ucdavis.edu/tour2006.cfm?displaysection=5.
Gibbs, M. (2008). The historical development of fisheries in New Zealand with respect to sustainable development principles. *The Electronic Journal of Sustainable Development, 1*(2), 23–33, Retrieved September 10, 2008, from http://www.ejsd.org/public/journal_article/7.
Harte, M. (2000). Industry Perspectives: Taking the Initiative for the Management of New Zealand's Commercial Fisheries. In R. Shotton (Ed.), *Use of property rights in fisheries management* (pp. 270–273). Proceedings of the FishRights99 Conference, Fremantle, Western Australia, 11–19 November 1999. Rome: FAO Fisheries Technical Paper 404/1.
Harte, M. (2006). Funding commercial fisheries management: Lessons from New Zealand. *Marine Policy, 31*, 379–389.
Hawkey, D. (1994). *Property rights, ITQs, and the slice of the fish pie: An appraisal of fishery culture and conflict in the Northland region*. Auckland: University of Auckland. Working Papers in Economics, no. 17, April 1994.
Hooper, M., & Lynch, T. (2000). Recognition of and provision for indigenous and coastal community rights using property rights instruments. In R. Shotton (Ed.), *Use of property rights in fisheries management* (pp. 199–205). Proceedings of the FishRights99 Conference, Fremantle, Western Australia, 11–19 November 1999 Rome: FAO Fisheries Technical Paper 402/2.
Independent Financial Report: Fisheries firm heads back to court as quotas cut. (n.d.). Retrieved September 26, 2007, from http://www.stuff.co.nz/4214523a23399.html.
Lock, K., & Leslie, S. (2007). *New Zealand's quota management system: A history of the first 20 years*. Wellington: Motu Working Paper 07–02.
Moloney, D. G., & Pearce, P. H. (1979). Quantitative rights as an instrument for regulating commercial fisheries. *Journal of the Fisheries Research Board of Canada*, *36*, 859–866.
Pascoe, S. (2007). Capacity Analysis and fisheries policy: Theory versus practice. *Marine Resource Economics*, *22*, 83–87.

Peacey, J. (2007). *New Zealand fisheries: How research underpins existing management and priorities for the future*. Presentation of the Ministry of Fisheries of New Zealand at the MariFish Symposium: Future demands for fisheries research; Roskilde, Denmark; June 7 and 8, 2007. Retrieved November 15, 2007, from http://www.marifish.net/News/2007-08-07.htm

Scott, A. (1988). Conceptual origins of rights based fishing. In P. Neher, R. Arnason, & N. Mollet (Eds.), *Rights Based Fishing* (pp. 11–38). NATO ASI Series. Series E: Applied Sciences-Vol. 169. Dordrecht: Kluwer Academic Publisher.

Stewart, J., Walshe, K., & Moodie, B. (2006). The demise of the small fisher? A profile of exiters from the New Zealand fishery. *Marine Policy, 30*(4), 328–340.

Strakker, G., Kerr, S., & Hendy, J. (2002). *A regulatory history of New Zealand's quota management system*. Wellington: Motu economic and public policy research trust.

Stokes, K., Gibbs, N., & Holland, D. (2006). New Zealand's cost-recovery regime for fisheries research services: an industry perspective. *Bulletin of Marine Science, 78*(3), 467–485.

Symes, D., & Crean, K. (1995). Privatisation of the commons: The introduction of individual transferable quotas in developed fisheries. *Geoforum, 26*(2), 175–184.

Treaty of Waitangi. (n.d.) Retrieved July 28, 2007, from http://www.treatyofwaitangi.govt.nz/

Waitangi Tribunal (1988). Report of the Waitangi Tribunal on the Muriwhenua fishing claim. Retrieved July 28, 2007, from http://www.waitangitribunal.govt.nz/scripts/reports/reports/22/796C25F5-7F61-4F4F-8C50-EED0218276F3.pdf

Chapter 3
Rights-Based Management and Participatory Governance in Southwest Nova Scotia

Clara Ulrich and Douglas Clyde Wilson

Abstract In the late 1980s the ground fish fishery in Atlantic Canada suffered a massive collapse. This collapse and some institutional factors, including a massive cut in the budget of the Department of Fisheries and Oceans, led to a number of management innovations. The chapter focuses on the substantial expansion of both rights-based management and participatory governance and the ways these two changes interacted with one another. The most common form of rights-based management in Nova Scotia is ITQs. However, the smaller boats fishing with fixed gears are using community quotas instead. One community from this group, the one with the largest fishery, has developed an internal ITQ system to allocate its community quota and this approach has proven successful at mitigating some of the social costs of ITQs while retaining most of the economic benefits. Participatory governance in Nova Scotia also extends to some extent to the way scientific advice is developed and used. Overall, this process has improved social robustness, by reducing the feeling of industry of being ignored. It has also improved biological robustness, by increasing the feeling of ownership and responsibility for the resource and improving the commitment to scientific advice.

Keywords Rights-based fisheries management · Participation · Participatory fisheries science · Community management · Nova Scotia

3.1 Introduction

3.1.1 Background to the Case Study Innovations

3.1.1.1 Changes from Round Fish to Invertebrate Fisheries

The state of fisheries in Atlantic Canada is still very much a result of the collapse of the Northern Cod (*Gadus morhua*) stock and the depletion of most other groundfish

C. Ulrich (✉)
National Institute of Aquatic Resources, Technical University of Denmark, Charlottenlund Slot, DK-2920 Denmark
e-mail: clu@aqua.dtu.dk

K.H. Hauge, D.C. Wilson (eds.), *Comparative Evaluations of Innovative Fisheries Management*, DOI 10.1007/978-90-481-2663-7_3,

stocks in the late eighties and early nineties. Following the collapse and a subsequent moratorium imposed on commercial fishing for cod in 1992, industry restructuring and social dislocation in coastal communities across the Atlantic coast led to approximately 40,000 persons out of work (Harris, 1995, cited in Potts, 2003).

Invertebrate fisheries, on the other hand, have become increasingly important. Frank, Petrie, Choi, and Leggett (2005:1621) argue that the Scotia Shelf ecosystem has experienced a *trophic cascade* driven by what they describe as the *virtual elimination* of the structural influence of commercial fish species on the ecosystem. One result was a marked increase in the abundance of small pelagic fish and benthic macroinvertebrates, especially northern shrimp (*Pandalus borealis*) and snow crab (*Chionoecetes opilio*). According to the Nova Scotian government the landed value of invertebrate fisheries in 2004 was $596 million or 80% of the overall landed value from all species.

3.1.1.2 Institutional Changes in Canadian and Nova Scotian Fisheries Management

In addition to and to some degree in response to the ecological changes, major changes took place in the early 1990s in the way that Nova Scotian fisheries are managed. These changes are interrelated and driven by a complex mixture of management ideology, the changes in the fishery, and bureaucratic imperatives.

One major change was an accelerated shift to quota management through individual quotas based on historical participation in the fishery. Fishing rights or privileges within a quota system (individual quotas (IQs) and individual transferable quotas (ITQs)) have existed in Atlantic Canada since the 1980s. Our discussion here focuses on two fleets, both of which are based on an individual quota system but structured very differently. Since 1991, the inshore (<65′) mobile gear groundfish fishery, which mainly uses otter trawls, has been managed under an ITQ system. The inshore (<45′) fixed gear fishery, which uses long lines, gillnets and hand lines, is managed through community quotas based on the aggregations of individual quotas.

A second major institutional change in Nova Scotia fisheries management was in response to severe cutbacks (around 30%) in the budget of the Department of Fisheries and Oceans (DFO). Management costs and some management responsibilities were broadly transferred from government to industry including monitoring, surveillance and day-to-day management of the quota system.

A third major change, which followed directly from the introduction of the individual quota approach, was the development of an effective monitoring system. The heart of the system is a privatised 'hail-in hail-out' monitoring system for fish landings, operated by independent companies and using electronic logbooks. This is a requirement for the fishing licence. This kind of system began with the <65′ ITQ system but is now ubiquitous across Atlantic Canada. The role of DFO is policing the system to ensure full compliance but not to be involved directly.

A fourth change is adoption of a new legal framework for fisheries management that occurred for the whole of Canada. In response to the cod crisis, the 1996 Oceans Act commits Canada to integrated, ecosystem-based precautionary management. The Oceans Act is *an extraordinary piece of legislation* (Haward, Dobell, Charles,

Foster, & Potts, 2005:17), that expanded the role of DFO to integrate all ocean use activities and users rather than simply fisheries only, at the federal, provincial, territorial and local levels. A key part of this has been the development of Integrated Fishery Management Plans (IFMP) as operational tools for achieving consistency in management processes since 1995 (Auditor General, 1999, cited in Potts, 2003). Later, the Species at Risk Act came into force in 2003 and has increased the focus on bycatch species. It has evolved into a very restrictive law-driven constraint on fisheries management.

Finally, the last major institutional change in the background of this case study is a general move towards a more participatory approach to fisheries management, albeit a participatory approach firmly under the control of DFO. Both of the fleets we examine closely below have their own industry advisory committee. The community management boards, examined at length below, are perhaps the most ambitious example of a participatory approach to management.

3.2 The Rights-Based System

3.2.1 A Brief History of the Innovation

The first implementation of individual quotas in Canada began on an experimental basis in 1982 with 'enterprise allocations' for the greater than 65′ offshore mobile gear fleet. This continued on a trial basis until 1989. Quota allocation for the remaining mobile gears and the >45′ fixed gear fleets began in February 1990 with a Working Group of representatives from the catching sector and the ground fish industry associations, provincial governments, and DFO. The Working Group met with fishing communities in the summer of 1990 to explain the programme and hear the views of licence holders. The programme began on 1 January 1991 (Liew, 2001).

Further modifications were carried out by an IQ Management Committee, which was created in late 1991, and later became the ITQ Management Committee. They quickly made some major changes such as making the IQ system permanent and allowing permanent transfer, thus creating a true ITQ system. Working with DFO they designed the self-financed dockside monitoring system (Apostle, McCay, & Mikalsen, 2002).

Vessel owners were given the option of joining the ITQ system, fishing under a competitive quota reserved for fixed gears, or joining a 'generalist' category that would also fish under a competitive quota. Of the 455 eligible vessels 325 chose to remain in the ITQ system. Their number then dropped quickly, and estimates at the turn of the century were around 100 (Apostle et al., 2002).

3.2.2 Structure of the Individual Access Rights

3.2.2.1 Core Fishers

Since 1976, the overall fishing for all species has been limited through a licensing system. To acquire a licence you have to be a full time fisher but the definition of full time fisher varies by region. The ownership of fishing quota, fishing licences and the basic access rights are technically separate issues in the Nova Scotia inshore (<65′) fleet because being a 'core fisher' and holding a licence are not the same thing. The status of the core fisher was created in 1996 and included 700 individuals identified in the mid nineties as being, as one manager put it in an interview, a *bona fide professional fisher*. The criteria are that the fisher must: *(a) be the head of an enterprise; (b) hold key licences (or, for some Scotia-Fundy fishers, a vessel-based licence); (c) have an attachment to the fishery; and (d) be dependent on the fishery.* Our DFO respondent told us that what they were *really deciding was who was really dependent on the industry and who was dabbling at it.*

3.2.2.2 Fleets and Quota Allocation

In Nova Scotia groundfish are allocated to individual fleets as shown in Table 3.1. There is an attempt to make these 'sharing arrangements' as stable as possible. The Groundfish Management Plan (GMP) shown in the table covered two years; the subsequent one covered five. Shifts or swaps of quota between fleets have taken place but they are considered extraordinary actions.

3.2.2.3 Historical Participation and its Problems

It was quite striking that the dominant subject in the early part of nearly all of our interviews was the problems in the early 1990s with the introduction of the IQ system and especially the distribution of the initial IQs. While economic theory might suggest that the best way to allocate IQs, at least from the point of view of society's overall economic welfare, would be to auction them, the distribution is almost always based on the 'historical participation' of individual fishers in the fishery in question.

The argument quickly became what 'history' one was going to base the allocation on. IQ systems are almost always introduced in fisheries that have been under other kinds of management systems for a long time and these other management systems have partly determined who was going to have the largest and smallest

Table 3.1 Nova Scotia Groundfish Fleets and Their Allocations – 2000

Gear	Fleet	Management system	Active licences	Cod allocation in percentage	Haddock allocation in percentage	Pollock allocation in percentage
Fixed	<45″	Community	883*	55	25	28
	45″–65″	ITQ	20	5	4	1
	>65″	ITQ	11	1	1	0
Mobile	<65″	ITQ	131	32	56	23
	65″+	EA (ITQ)	35	7	13	49

*Includes 47 active licences in New Brunswick
Scotia-Fundy GMP 2000–2002 (DFO, 2004).

fishing 'histories'. In the Nova Scotia case the prevailing record keeping system in particular turned out to be critical. Before 1986 DFO had kept very sparse records of catches in the inshore fleet. So most suggestions about when 'history' should begin started at that point. After that there was any number of ways that history could be defined. Groups formed around the years that would give them the best allocations. A Shelburne fisher explained how his group wanted a *straight forward 1986–1993 and nothing else.* But another group was formed to lobby for 1989–1993 years.

Different gear types had kept different kinds of records. In the 1980s there had actually not been very much control, especially in relation to the smaller boats using fixed gears (hand liners in particular). One respondent explained that many people had been more interested in getting unemployment benefits than in recording fish landings. They would ask their friends to put their fish in the friend's name so they could get the benefits. *They were cheating the system and cutting their own throats at the same time*. Whole areas were disadvantaged for technical reasons. A man from the port of Digby explained that in his area fish for salt processing was not counted, nor was the fish that they had been selling to the mobile gear fleet.

In the end a number of accommodations were made, and formulas were developed for estimating under-recorded catches. It was a painful experience that still seems to play the role of foundational myth for the current Nova Scotia fisheries management system. Apostle et al. (2002) offer a quote from one fisher describing what these meetings were like that seems an apt summary: Fishers *were looking at the generated numbers and realising they were going to end up with 60 tonnes of fish, and realising they were finished. It was a really tense, tough, emotional time and we did that for a year.*

3.2.3 Impacts of the Rights-Based System

3.2.3.1 Enabling the Transition to a More Sustainable Fishery

A central point that one fisheries manager strongly emphasised was that people tend to conflate impacts of the individual quota system, the hail-in hail-out monitoring system, the cutbacks in the overall magnitude of the quota driven by the ecological situation, and the transferability of the ITQs. All these things are lumped together and called 'the ITQ system'. His argument was that the huge drops in numbers of active boats, processing plants and the geographical concentration of fishing activity (see Section 3.2.3.2) were all going to happen from cuts in the overall quota with effective enforcement whether or not ITQs were in place. People involved in the fixed gear fishery were seeing large numbers of fish plants being closed down and blaming the ITQ system for this, when in reality a number of those plants had been kept alive by black landings and were no longer viable because of the new enforcement system. Similar changes were seen under community management boards where no ITQs were in play. What the ITQ did was to determine the process by which fishing and processing capacity was reduced, not the reduction itself.

Another of our respondents, a commercial fishing representative, supports this view. He believes that the main reason why there was a good deal of reluctant support at the time was that people believed it was the only way to avoid chaos and mass bankruptcies. He suggests that as a mechanism to reduced capacity, the ITQ is good because it allows fair-trading and real value to transfer. A manager suggested to us that the central question that ITQs pose for fishers is what they really want to do with their business. They can decide to have the groundfish be a supplement to what they are doing with lobster (*Homarus americanus*) or harpooning, or to fish for groundfish full time.

The process is not over. The smaller and less efficient operations continue to be marginalised.

The cost of fishing has gone up. Global competition is intense. More and more of the costs of management have been placed on the industry. DFO is requiring increased monitoring and observer coverage. Other government agencies are putting pressure on fishers for 'professionalisation', meaning more training and required certifications, greater investments in safety precautions, workmen's compensation, and insurance. Meanwhile the groundfish resource is still very small by historical standards.

3.2.3.2 Geographical and Organisational Concentration

The design of the Nova Scotia ITQ system was heavily influenced by the fact that the communities involved were very dependent on fishing (McCay, Creed, Finlayson, Apostle, & Mikalsen, 1996). This led to the requirements that ITQ holders be bona fide fishers and a rule that no one could own more than two percent of the total quota (Apostle et al., 2002). However, these authors conclude that concentration of ownership has increased since 1990 in spite of the provisions to avoid this. Within a short time after the implementation of the ITQ system, Creed, Apostle, and McCay (1994) found vertical integration within the community they studied. Only two or three out of 30 mobile-gear vessels there were not tied to one of the fish plants. In their ethnographic investigation of the impacts of ITQs on the Scotia-Fundy mobile gear ground fish sector Creed et al. (1994) found that people in communities with significant quota became gatekeepers to the fishery. This changed relationships in ways no one liked, even the gatekeepers themselves (Apostle et al., 2002). These authors gave also evidence of clear geographical concentration with a big drop in the cod landed in eastern and central Nova Scotia.

Two other policies that are in place to limit organisational and geographical concentration in Nova Scotia fisheries are the Owner-Operator Policy and the Fleet Separation Policy. Both policies are aimed at separating processing and harvesting (DFO, 2004). The first requires licence holders using vessels less than 65 feet to fish their licences personally. The second restricts corporations from holding any new fishing licences for inshore vessels. The fleet separation policy was in place before the ITQ system was introduced. Industry views regarding both policies are highly polarised (DFO, 2004). The problem from the perspective of the inshore fishers as well as many in the general public was 'trust agreements', under which a licence holder enters into an agreement with a third party to control the use of a licence. Opposition to the trust agreements, which are alleged to allow processors to control harvesting, is very strong. However there are also proposals to make the owner-operator and fleet separation policies more flexible without limiting the use of trust agreements.

3.2.3.3 Retirement and Recruitment

The entry and exit of fishers into the fishery is an important area of concern among our respondents in respect to the ITQ system, as well as the Community Management Boards discussed below. Several respondents emphasised that ITQs facilitated the retirement of fishers by providing them with an asset they could sell when leaving the business. One respondent who was deeply involved in the Community Management Boards considered a desire to leave the fishery to be perhaps the main determinant of people's attitudes towards ITQs. People who are planning to keep on fishing are generally opposed to the ITQs system because their increased costs through taking on debt to buy quota would be greater than their benefits. But those who wish to leave fishing say yes to ITQs because it provides a mechanism for doing this. Another respondent said that he thought it was more common to sell a licence in order to buy a licence in another fishery than to sell a licence in order to retire.

Respondents pointed out, however, that the market for small licences is currently weak. Transfers in the inshore fishery have traditionally tended to take place between a father and his son or other relative. *But if you look at the papers there are licences for sale everywhere.* A young man can qualify as a 'professional fisher,[1]' but then he must buy a licence and it may not be possible to use the licence to secure a loan from a bank. The ITQs have not usually been recognised as assets for the purpose of loan collateral but very recently court cases have suggested that the licence, i.e. the access right itself, is an asset in the legal sense. ITQs have been argued to be a block to a young person getting into the fishery because of the cost of quota, but in Nova Scotia licences for non-ITQ species (e.g. lobster or crab) are just as expensive. With such high costs of entry the only real choice young fishers have is to go to the processing plants for a loan. This then ties them to that plant and is one source of the 'trust agreements' discussed above. Finally, many young people in Nova Scotia are choosing to go out west drawn by the oil boom in Alberta. This has implications for both finding future boat owners and finding adequate crew now.

3.2.3.4 Crew

ITQs have changed some of the share systems used through which crew members are paid. Owners of larger firms have placed the cost of ITQ on 'the top of the lay', i.e. the cost of the quota is considered a cost of fishing and deducted from the crew share, not only from the share of the ITQ owners (McCay et al., 1996). A respondent from the industry explained that while some investors in quota are still

[1] The various designations can be confusing. Professional fisher is a qualification that depends on a training certificate while core fisher is a separate designation that was used to limit and finally eliminate part time fishing. The core fisher status can be purchased along with a fishing licence.

very concerned with communities and the quality of life others are focussed only on maximising profit. Consequently return to crew members is less today than it was 20 years ago: *Once people started buying the quotas they had another debt and the less reputable ones would shovel that cost on to the crew members.*

Other factors are at work as well. Changes in skill requirements are one. One respondent who works in the industry told us that even in the 1970s a generous portion went to the crew because they took a risk: they had to be skilled and they had to manage the trip. But now fishing has become safer and electronic equipment and reduced quotas are reducing the level of skill required. Another fisher explained *there are four of us in our boat. I used to carry seven. This is because of the lower number of fish we have to catch. I used to fish 7–8 days hard, but now I can't so I only take four. The way they are paid is being changed because you have no fish to catch and you have to buy fish before you come then it has to be paid for.*

3.2.3.5 Markets, Quality and Price

One claim that was made during the introduction of the ITQ system is that it would improve fish quality because fishers could fish more slowly and time their fishing in relation to the market. Some evidence exists of increases in quality. The prices obtained by the inshore mobile sector of cod and haddock (*Melanogrammus aeglefinus*), but not pollock (*Pollachius virens*), have converged with those of the fixed gear fleet that traditionally got better prices because of higher quality (Apostle et al., 2002).

Among our respondents, however, even those who are very supportive of the ITQ system expressed some disappointment that the improvement in quality and price has not been as great as they would have liked. One explained that the market is not well organised in Nova Scotia. Fish buyers in New York and New England are the main drivers and this has kept Canadian prices low. He argues that Nova Scotia is hurt by the lack of vertical integration created by the fleet separation and owner-operator policy. This weakens the ability of Canadian firms to resist the influence of the American market and set their own prices. The reward for quality is not really worth the investment.

3.2.3.6 Fishing Behaviour and Conservation

Evidence for a link between ITQs and stewardship is not readily evident and what is there gives mixed signals (Apostle et al., 2002). People are becoming more concerned with enforcement, as protection of their investments. The <65′ ITQ fleet did decide to voluntarily adopt a square mesh net. Creed et al. (1994) also heard reports of increased compliance, even claims that illegal landings had almost disappeared. Some observer data suggests that discarding, dumping and high grading have increased in the ITQ fleet (Apostle et al., 2002). However, an analysis of violation statistics found that the ITQ system seems to have had a strong downward impact on both the number of violations and the severity of the offences (Apostle et al., 2002).

ITQs have raised questions about their direct implications for conservation. An ITQ is probably perfect in a single-species context, argued one respondent from the fishing industry, but bycatch is the Achilles heel of the ITQ system. In a multi-species context such as Nova Scotia, bycatch makes an ITQ system a *nightmare* from a business perspective. A Nova Scotia fisher can be dealing with up to six quota species as well as other species with bycatch restrictions. A fisher has a basket of holdings of quota and catches more of one species and less of another. The economic theory would assume the market would operate and you would buy or sell this quota. But quite quickly it becomes apparent that it is easier to discard the fish you have caught than it is to buy quota to cover it. While there are certainly people on shore who had the quota needed to cover the incidental catch, they will be asking three and four time its market value because they know that the fisher will not be able to catch the target species without some quota for the bycatch. This same observation was confirmed by other respondents from the industry.

The most interesting aspects of rights-based management in Nova Scotia have emerged in its interplay with the reforms toward greater participation that have taken place, especially in the form of the Community Management Boards. It is to this subject that we now turn.

3.3 The Community Management Boards

3.3.1 A Brief History of the Innovation

The Community Management Boards (CMB) were formed for the management of the small vessel (<45′) fixed gear fleet. The Boards were formed in the wake of organised protests focussed on resistance to the introduction of ITQs. Charles, Bull, Kearney, and Milley (2005) suggest that this happened because the fixed gear fishers did not like what they saw happening in the inshore mobile fishery after ITQs were introduced. Community-level organising at the county level had begun by 1995 and that was a bottom-up development, not a DFO innovation (Kearney et al., 1998). The Sambro community requested an experiment with a ‘community quota’ allocation, which was approved in 1995. They ensured the plan would be enforced in a democratic way by designing a Fishing Conservation Harvest Plan adopted by fishers through a formal contractual agreement. The contract shifted much of the management responsibility from DFO to the Association (Loucks, 1998).

The fishers understood that they had to demonstrate full compliance if the co-management approach was going to work. This was the first community quota in Atlantic Canada and the first time a group of fishers in the Scotia Fundy region signed a contract committing themselves to a specific harvesting regime. It required that they hire, for one percent of the catch, one of the independent monitoring companies that were involved in the ‘hail-in hail-out’ system originally set up to monitor the ITQ fleets. DFO would also do random monitoring, and if violations were detected the contract would be cancelled.

One of our respondents was a manager who was involved in these activities from the DFO side. He explained that DFO had become very frustrated trying to develop a single management plan for everyone. They were continually running into problems such as different fishing seasons, differences in tides, etc. DFO started to address this by basing management on gear types. They created a gill net group, a long line group and a hand line group. They had the fishers choose which group they would belong to and used available data to divide the quota. The system worked more or less well for different groups depending on the fishing history information and other factors, but overall it was not a very satisfactory system.

It was in this context that the Sambro experiment was requested by the community and agreed to readily by DFO. During the first year the other communities ran through their quota while Sambro kept right on going through the year. The shift of responsibility to the group, Loucks (1998) argues, resulted in high community cohesion. The Sambro community purposely under fished their quota by five percent.

In the fall of 1995 the other community groups held a meeting and invited people from DFO. Two hundred people came and said they wanted to try community management. It was not easy to arrange. The most contentious area was Shelburne County, which is by far the most important area in Nova Scotia for fixed gear; nearly half of the fixed gear fishing takes place there. Sinclair, O'Boyle, Burke, and Peacock (1999) describe the complexity of the Shelburne County fishery, with more than 800 fixed-gear <45′ licences. They suggest that the fishers were basically forced to organise. Shelburne County could not come to an agreement about whether to use IQs, ITQs, or competitive quotas, and in the end DFO had the divide them into two management boards, Shelburne A and Shelburne B. This arrangement continues to this day. Finally, DFO formalised 'community quota regions' throughout Nova Scotia in 1996.

DFO is very satisfied with the division of responsibilities. *The boards do a whole bunch of things we did before, we have downloaded responsibilities. We had very little support before in trying to manage fisheries, now they can do their own thing,* explained a manager who had been involved in the process. The boards have the responsibility for defining entitlements on how to harvest the assigned allocation (Peacock & Hansen, 2000). The communities have taken a number of approaches, which range from a competitive fishery (by gear type) within an overall community-quota on a per species basis, to an industry developed and delivered ITQ initiative.

For most of the CMBs the shifting of fishers between boards is not an issue because it is based on county of residence. The two Shelburne boards are the only ones within the same county. Once a community quota is created, if people want to move between boards the board must approve the decision and decide upon the conditions.

The management boards are all operated differently, which is part of the idea of local control. Charles et al. (2005:8) identify that the following characteristics are shared by all or most management boards:

1. The boards were established and are run by fish harvesting organisations, and strive for inclusive decision-making processes;

2. The boards sub-allocate the community quota among different gear types and devise rules for all licence-holders in the form of a community management plan;
3. The management plans are enforced through contractual arrangement between the board, the licence-holders, and the catch monitoring companies;
4. Management plans are consistent with basic conservation requirements and rules set out by DFO;
5. Management boards have infractions committees to judge alleged violations of management plans and impose penalties;
6. Seasonal adjustments are made including the sale or trade of unused quota between different management boards;
7. Individual licence holders can still choose to fish under a generic management plan devised by DFO for the whole Scotia-Fundy region instead of under a community management plan devised for local conditions.

The Community Management Boards are organised around previously existing fishermen's associations (often reflecting the three gear groups of long lines, hand lines and gill nets) and are influenced by other place-based networks. Shelburne A for example is made up of three previously existing groups and Shelburne B is made up of five. The 'Fixed Gear Committee' represents all of the boards in meetings with DFO. Each board has three representatives on this committee; one for each gear group, and a CMB is required to present a unified position to the meeting when an issue is decided.

3.3.2 The CMBs and the Costs of Management

No quantified information exists on the implications of the Community Management Boards for the costs of management. We believe it is a reasonable hypothesis that many costs are more cheaply born by community groups than by government agencies or even an industry group such as the ITQ Management Group for the <65′ mobile gear fleet. In comparing the costs of management between the CMBs and the ITQ system a DFO manager said that they have some significant costs dealing with quota transfers within the mobile fleet. This activity is not necessary for DFO to do on the CMB side because they manage the fisheries internally according to their harvesting plan. Indeed, such harvesting plans, which are themselves a contract with DFO, can be considered an important innovation as Europe is considering how to apply 'results-based management' such as this in fisheries. Only when quota is shifted between CMBs must DFO must keep track of the exchange. Even the CMB that uses individual quotas (Shelburne B) does so internally and this creates no costs for DFO. DFO does have responsibility for making sure that the harvesting plans are honoured. Most of the information needed for this is developed by the privatised hail-in hail-out monitoring system, which also involves little or no costs to DFO. When asked if he is sure that DFO's costs for the CMB system are less than that for the ITQ system his response was *I'm sure it is because it is a lot less paper work.*

3.3.3 The CMBs, Sanctions and Compliance

Once a fisher has finished his quota for one species he must stop fishing for groundfish. The same is true for the group quotas in the CMB. This is a major incentive for group organisation. As one fisher explained '*some fellow would just go to Georges [Bank] and catch all the cod they could catch and shut everything down for everyone. We do not want radicals shutting everything down*'. Thus, the groups have an incentive to develop strict enforcement mechanisms. With respect to penalties the fishers are tough. In CMBs the penalties, which are normally reductions in quota or time at sea, are harsher than those the government would impose, there is no appeal, and the enforcement is quite effective (Peacock & Hansen, 2000).

When a fishing enterprise goes over its assigned catch or otherwise breaks a CMB rule it is taken before an infractions committee. One respondent explained how this worked in his CMB. The membership of the infractions committee is different every time it meets and is formally anonymous. The infractions committee reviews a case file that is also anonymous. Sanctions begin with warnings for small infractions but can become serious. The largest sanction has been five years with no contract, which means that the person is forced into Group X. Group X is made up of people who are not affiliated with a CMB. Group X has almost no management services, its quota is fished competitively, and when the Group X quota is exhausted the entire group's fishery is shut down for the year.

3.3.4 The Case of Shelburne B

We focused a good deal of our short stay in Nova Scotia on the Shelburne B CMB because it is an interesting example of combining a community approach with a rights-based system. This CMB has chosen to use an internal transferable quota system to solve their allocation problems. Community involvement helps avoid the quota busting, high grading, and misreporting that are the common problems associated with quota-based fisheries management.

The system is based on an operational triangle between the CMB, the hail–in hail–out monitoring company, and the fishing vessels. The monitoring company reports the landings to the CMB who keep track of the uptake of each individual's quota share. DFO sets the quotas for the CMB as a whole, approves the CMB's Fishing Conservation Harvest Plan, audits the monitoring companies, and keeps track of the small amount of quota that may shift between CMBs. While the Shelburne B CMB meets once a year as a whole, operationally, it is a group of five full time managers, hired by the five fishers' associations that make up the CMB. This group consists of three fisher's wives and two ex-fishers. One of these people told us that she does about 300 fish swaps among her group in a season, assuming the five associations are of comparable size this indicates a total of 1500 swaps in Shelburne B as a whole. She also indicated that a large part of her job is monitoring fisheries management issues and representing her group, for example at meetings called by DFO.

The internal sales of quota are not usually permanent, but in Shelburne B fishers can still sell out their licences and retire. These sales are made directly to other fishers within Shelburne B and registered with the Shelburne B board. They are not registered with DFO as they are considered an internal CMB matter, and, as such a fisher cannot choose to sell quota to fishers outside the CMB, not even to Shelburne A, the other CMB in Shelburne County. This is in contrast to the official ITQ system in which DFO bears the costs of recording when quota changes hand. A DFO respondent said *We could not care less what they do as individuals.* If a fisher decides to stop fishing he can get a reasonable price from the community. These prices, however, are not as high as he would get if it were an official ITQ system. One fisher explained *If I want to retire I can sell my licence and then they [the buyer] would become a part of the group [the CMB].*

The Shelburne situation, including the formation of two CMBs, Shelburne A and B, has seen a lot of conflict. One fisher described the two ideological camps that led to the formation of A and B. On the one side are those who want to chase fish wherever they are. They see the others as lazy, and on the other side are those who want to wait for the fish to come closer so they don't steam as far. They say the others are greedy and wasteful. These two attitudes are expressed in disagreements about management and were, in the opinion of several respondents, an underlying reason for the division of Shelburne into Shelburne A, the 'lazy' CMB, and Shelburne B, the 'greedy and wasteful' CMB. While this conflict was very intense ten years ago the groups have settled and seem now to coexist. *We are not enemies,* one respondent told us. At the beginning of each year a Shelburne fisher has the right to choose which group to belong to and if there is a decision to shift CMBs that person's quota moves with him. Over the years there has been some movement between the groups. Most of the movement is toward the B side. Some changed simply because they thought the group was working better, but more of the shift was through people selling licences. The B group, being the more business oriented, was simply the more likely to be the purchasers.

The Shelburne B set-up, according to two respondents active in its management, gives any member of the community some access, even if it is only a small share because of little fishing history. Another respondent described the benefits of the system this way: *you go the time of the year you want, if you want to go, you go, and you set the fish aside [to fish later] if you want to go swordfishing. When you want to go you can go, that is how the IQ will make things work for an individual if you are small like we are.*

The community basis of the system seems strongly rooted. It is supported locally and changing it would also require a major shift in management policy and increased costs for DFO. Our respondents explained that the reason their group has been mainly against adopting a full-scale ITQ system that allowed trades outside the community is because it would harm the smaller communities in the county. They believe that the quotas would be bought up by larger-scale fishers, such as the <65′ mobile gear fleet, and would never come back to their area. Another respondent from Shelburne B believes that equity is what fishers really want out of a management system, and it was the inequality under the quota system, a perception rooted in the struggles over 'history' as the basis of allocation, that caused so much resistance. In community management you get local enforcement. If he were to go down a dock with, for example, undersized lobster, he would be confronted by fishers with various ways of expressing their anger over an infringement that they see as affecting them personally, *but if you cheat on a [non-community] quota you are just cheating the government.*

A substantial group within Shelburne B would like to move to a full ITQ system. The process that DFO has set up for making such a decision is a demanding one and it does not look like this group has the support to prevail in the near future. The true ITQ groups are also a bit resentful of the Shelburne B system as they have the benefits of ITQs without their costs. These costs are the allocation fee and the more extensive dockside monitoring that the DFO requires of the ITQ fleets.

3.4 Participatory Approaches to Science and Management Decision

3.4.1 History of Innovation

3.4.1.1 Increased Industry Participation in Science

The collapse of groundfish stocks and, among other causes, the perceived role of science in it through uncertain and potentially overly optimistic stock assessment (Walters & Maguire, 1996; Shelton & Lilly, 2000) have sharpened the mistrust of the industry over the traditional ways of providing scientific advice to management, and they have demanded the chance to participate in the scientific process. Furthermore, the development of new fisheries almost from zero, with emerging data-poor target species, has weakened the established model-based scientific system, because of the needs of new methods for scientific advice. This has created incentives for industry participation and better use of industry's knowledge. Finally, drastic cuts in DFO budget have also reduced the possibilities of scientific surveys and analyses. As a consequence, a major trend in Atlantic Canada over the last fifteen years has been towards an increasing participation of industry in management advice, and an educational process as the industry becomes more involved in stock assessment and research.

3.4.2 Participation in Stock Assessment Processes

3.4.2.1 Groundfish Stocks

In spite of some flaws in the traditional Virtual Population Analysis (VPA)-based assessment methods revealed by the groundfish stock collapse (Hutchings & Myers, 1994; Walters & Maguire, 1996; Shelton & Lilly, 2000; Shelton, 2005), these methods are still used for stock assessment and scientific advice to management for most fish stocks. Alternative indicator-based approaches were tried (see Section 3.4.5) but have not replaced the existing system. All groundfish stocks are handled in the same way. It has meant that industry participation has been subordinate and conditioned to the existing scientific system. Potential criticism from the industry to the scientific hypotheses first requires in-depth understanding of the scientific methodology. As such, the participation of industry is formal.

An initiative jointly created by the industry and DFO scientists was the so-called 'sentinel fishery', a survey mostly designed for maintaining information flow during fishery closures, primarily in Atlantic Canada and Gulf of St Lawrence. Sentinel fisheries have succeeded not only in providing crucial information for stock assessments, as a supplement to research vessel surveys, but also in becoming well-established and accepted among fishers, playing an instrumental role in creating a more co-operative atmosphere between scientists and fishers (Charles, 1998).

Some other initiatives were launched in the nineties by the industry alone in order to provide alternative surveys that would supplement the scientific surveys used in assessment. A main one is the so-called 'ITQ survey' performed by the trawler fleet >45′ entitled to ITQ in area 4X. DFO used to have a regular trawl survey, but which could not sample along the shore in shallow waters because of the size of the research vessel. The industry proposed to cover that area and started a systematic survey with scientifically validated protocols in 1996. The costs are fully born by the industry, through some unallocated quotas that are used for science instead of being redistributed to each quota owner. The survey has been added to the scientific survey and is used in stock assessment. *This is a success story, with willingness and commitment from both parts. The industry makes good and objective job, and the science branch has been willing to modify their methods*, said a fisherman engaged in that survey. Similarly, an industry halibut (*Hippoglossus hippoglossus*) survey has been in force in ten years. It was initially proposed by scientists but was designed in collaboration between the scientists and the industry. The results of that survey have been quite consistent with scientific findings, giving fishermen confidence in assessment results.

However, including industry-based surveys in the assessment is not always straightforward, if the results differ significantly from scientific findings. A longline fleet also launched a survey, as their perception of stock abundance was the opposite of that observed with trawl-based surveys (*we see cod and few haddock, they see haddock and fewer cod*) said a longline fisherman. The industry survey lasted six years but was never included in the assessment.

3.4.2.2 Invertebrate Stocks

The situation is quite different for invertebrate stocks, as usual assessment tools cannot be used. Every invertebrate species is so unique that there has to be a new technique. This has lead to a system whereby individual DFO scientists are dedicated almost full time to one particular stock over several years. Assessment methods vary from stock to stock, depending on data available and scientist's background, but also on industry demands and funds. A number of scientific studies are paid by the industry as a source of knowledge for their own goals.

The specificity of invertebrate species is also found in the industry exploiting them. Most invertebrate stocks (lobster excepted) have smaller spatial distribution and mobility than groundfish stocks do and are targeted by a limited number of specialised fishermen forming a rather homogeneous and cohesive group. As a result, the full-time involvement of a DFO scientist on a stock exploited by a limited number of stakeholders often lead to close collaboration and high commitment between the scientist and the industry. These long-lasting relationships strengthen the trust and credibility of science and ease the data collection process.

However, this system leads to two types of issues. First, the scientists may get too accustomed to their routine work and do not have the chance to compare with methods used on other similar species. This leads to some inconsistencies between stocks, which may not be so problematic (*We are sort of disjointed and inconsistent according to some, but it is not inconsistency. It is specificity to a situation*, claimed a respondent), but it still raises issues regarding science quality and equity. DFO works now towards improved communication between scientists assessing invertebrate stocks, with support from statisticians and modelling experts. The second issue relates to scientific independence and integrity. The high level of interaction with industry creates the risk that industry puts pressure to obtain the scientific evidence they want. There have been some instances where industry trusted the scientist and followed his or her recommendation of decreasing catches. These were cases where a real relationship of trust existed: *I said they should cut back, and (...) they said OK. They said you were with us when we went up and this was, importantly, based on my history with this group of people*, said a DFO scientist. But this is not true for all cases, and there are suspicions of scientific manipulation, as well as claims over secrecy, lack of transparency and absence of peer-review of scientific results, also acknowledged by the professionals. Some industries pay parts of the salary of the scientist involved and may even be involved in their selection.

3.4.2.3 Participation in the Regional Advisory Process (RAP)

As most stocks in the Scotia-Fundy area are under sole Canadian jurisdiction, their assessment is under the responsibility of DFO and not the North Atlantic Fisheries Organisation (NAFO) and is conducted within Regional Advisory Process (RAP) meetings. The RAP was established in 1993 as an open forum for peer review of scientific findings on the status of fisheries and marine mammal resources involving industry, stakeholders, and outside scientific experts in the review process. These meetings have been opened to industry and NGO participants as a way to improve collaboration between stakeholders and science after the groundfish collapse. And indeed, this has facilitated the dialogue with the industry and improvement and acceptance of scientific results. The industry feel involved, and feel that they have to be (*If you don't ask question they will say whatever they want*, said a fixed gear representative). When they do not agree with scientific findings, they try to come with evidence to support their hypotheses. The industry is involved in helping to write the evaluation report, and industry's comments and concerns are recorded (*We have to go in with them. They are fair and willing to listen, it does not mean they will change the report, but I don't feel slighted,* said another fixed gear representative). Some communities are also organising local science meetings with DFO scientists prior to the RAP, in order to collect information from fishermen who do not attend the RAP. These open meetings also help the industry to understand the difficulty and complexity of stock assessment, and that uncertainties are inevitable. But it is clear that some mistrust is still there, although less radical than before. In particular are concerns among the fixed gear industry, that because most of the science is based on surveys using mobile gears that do not accurately catch species such as cusk (*Brosme brosme*), pollock, hake (*Merluccius bilinearis*) and halibut, that these assessments may be inaccurate.

The major issue of such open science meetings is the risk of distortion because of political issues. The industry may put pressure on the meetings to get the results they want. The quality of interaction with industry depends on the level of the stock. *At one part they were invited to the assessment meeting, they started bringing lawyers and it became a very political discussion,* said a DFO respondent. *They are going to change and go back to an invitation only meeting, reenforcing that participation is about bringing scientific inputs. It got to the point where people did not want to chair the meeting as they were afraid of being sued.*

3.4.3 Participation in Other Scientific Work – The FSRS

A notable initiative launched in Nova Scotia in the aftermath of the groundfish collapse was the creation of the Fishermen and Scientists Research Society (FSRS), a voluntary organisation for collaborative research and co-education of fishermen and scientists and the first of its kind in the world. The initiative was initially supported financially by the government, which also provides continuous office facilities. But it is now an independent non-profit society, financial support for which includes industry funds and governmental research grants. FSRS promotes collaborative science relevant to the long-term sustainability of the fishery, fishers having a key role in identifying research priorities. The Society stays away from controversial management issues, being prohibited by law from engaging in lobbying and other management activities. In 2007 it counted 367 active members, mostly fishermen and scientists.

The FSRS has played a key role in the educational process of the industry and in the restoration of the credibility of science. Fishermen trust data they collect themselves. *How can you argue about something you collected*, explained the FSRS manager. FSRS worked towards increased understanding of the scientific rationale for data collection protocols and increased participation in RAP meetings. It also taught the scientists to give timely feedback on their project results. The main success was in promoting communication, discussion and dissemination, which helped *humanise* each group in the eyes of the others.

In spite of these positive initiatives, some of the industry respondents, although part of the educated elite, were not very supportive of the FSRS. In particular, its status as a non-profit organisation requires a constant chase for grants and funds for maintaining its existence, which distorts its image of non-profitability. Secondly, most of the initiatives are still proposed and piloted by scientists. The FSRS is still perceived by some as a governmental body, which did not necessarily support industry's own initiatives such as the 'ITQ survey'.

However, it is clear that in spite of criticisms, the FSRS has existed over fifteen years, surviving the massive DFO cuts in research programmes. This longevity is the main proof of success, as the Society would not have survived without support from the industry.

3.4.4 Industry Involvement in Management Decisions

3.4.4.1 Harvest Control Rules and Management Plans

Traditionally, management decisions about single groundfish stocks Total Allowable Catches (TACs) were taken partly based on clear Harvest Control Rules (HCR) such as $F_{0.1}$. Shelton (2007) showed that the management strategies have however changed over time, including changes in reference points and time-scales. This is due both to an increasingly complex and restrictive legal framework for fisheries management (see Section 3.1.1.2), and to increasing participation of industry in management decisions and scientific understanding.

Management decisions for groundfish are taken as part of the Groundfish Management Plan established for the period 2002–2007. Annual fishing plans are developed in consultation with the fishing industry and are reviewed annually. TACs have been fairly stable over the recent years, reflecting general commitment towards stability and long-term sustainability *as the crucial starting point for improving relationships with industry, stakeholders and other resource users*, stated fisheries Minister Hearn (http://www.dfo-mpo.gc.ca/media/newsrel/ 2006/hq-ac07_e.htm).

Shelton (2007) acknowledged that this weakens the use of the scientific knowledge, as decisions are now taken ad-hoc. Indeed, there is a clear reluctance on the part of both the management bodies and the industry to use clear and pre-agreed harvest control rules, as the final management decision comes about through consensus and negotiation. The final decision process is not always fully clear and transparent. *Management has fisheries roundtable discussions, but I don't see that there is an open process for taking science advice and moving to decisions, which is why these discussions bleed into our science meetings,* said a scientific respondent. Lack of consensus across various industry groups undermines the possibilities for real co-management, and the decision power still resides with the Minister of Fisheries and Oceans. *Co-management was a concept a few years ago, but not now,* deplored a groundfish industry representative.

To improve the transparency of the decision process, The Science Branch is currently trying to introduce simulation-based Management Strategies Evaluations (MSE) that aim at identifying management strategies robust to various sources of uncertainties. First trials were conducted in 2007 on Arctic surfclams and ocean quahogs (Boudreau & O'Boyle, 2007). This is still too new to get real feedback on such a process and the immediate feelings about this approach are mixed. *Many questions industry has are with the whole picture, and if you hear some of their questions then maybe you hear this idea about strategies and decision rules,* said a scientific respondent. But *regional management folks were a bit negative, I did not know if this was distaste for formalised management or just that they don't like something new, they do like to be flexible in how they use advice.*

3.4.4.2 The FRCC

A particular initiative of increased participation of the industry in the decision process was the creation in 1993, right after the groundfish collapse, of the Fisheries Resource Conservation Council (FRCC), to form a partnership between scientific and academic expertise, and all sectors of the fishing industry. This occurred during the same period as the initiatives around RAP open meetings. The Council consists of 12 members, with an appropriate balance between 'science' and 'industry'. Members are chosen on merit and not as representatives of organisations. Together, they make public recommendations to the Minister of Fisheries and Oceans on conservation measures for the Atlantic fishery. Up until 2005, the Council's primary focus was to provide annual advice on total allowable catches and other conservation measures related to Atlantic groundfish stocks. It worked as a '*depoliticised advisory process, providing written public recommendations to the minister*, which then should be able to justify publicly why if it doesn't listen to FRCC', explained an industry member.

Recently, the Council took a very different direction, focusing on long-term conservation strategies including other key species such as snow crab and lobster, and looking at sustainability issues from an ecological, economic, social and institutional perspective.

In spite of its laudable mandate, some critics were raised about FRCC, mostly because of significant conflict of interest problems. '*They kept reappointing these people and you kept seeing obviously manipulated quota allocation*', said a groundfish industry representative. Needs for consensus can create dangerous 'hostage' situations if a party brings conflicts of interests in. But it has nevertheless given a real frame for co-management with a legitimate mandate to the industry, and remains a major institution in the region.

3.4.4.3 Conclusion

Some progress has been towards industry participation into final management decisions based upon agreed scientific advice. Charles (1998) illustrated how the opening of the scientific process to industry helped reduce uncertainty in cod stock status in area 4X. However, fifteen years of co-management have also shown some limits, as decisions will always result from a combination of legal framework and management objectives on one hand, and politically charged negotiations on the other hand. A transparent and legitimate decision cannot always be reached by consensus, especially when industry groups are numerous and heterogeneous.

3.4.5 Indicators and the EBFM

The initial choice of innovation with regard to science was the use of alternative tools for providing scientific advice and moving away from the traditional model-based and forecast-based methods. Canada is moving towards integrated management with clear objectives accounting for ecosystem and socio-economic sustainability, which are established through agreement (shared stewardship) with a number of stakeholders. This has naturally led to a growing need to identify reliable and measurable sustainability indicators systems, SIS, using 'pressure-state-response'-type frameworks (OECD, 2001), and Canada has experienced a decade of development and exploration of these systems.

3.4.5.1 The Precautionary Approach Framework

Shelton and Rivard (2003) described the history of development of the precautionary approach (PA) since the cod collapsed. Over the 10 years following the collapse, Canada has been engaged in a process of developing a precautionary framework that is consistent with the 1999 United Nation Fisheries Agreement (UNFA). Development of this framework has been given high priority since the concerns raised in 2002–2003 that post moratorium TACs had been unsustainable and were jeopardising stock recovery. The term 'precautionary approach' is to be used to refer to situations that can result in harm that is serious or difficult to reverse (impaired productivity), but not to situations of reduced yield and economic inefficiency. In 2007, the PA framework was routinely implemented in a way similar to ICES procedures, on a single-species basis with traffic-light based coloured zones as indicators for management advice.

3.4.5.2 The Traffic Light Approach

The Traffic Light Approach (TLA) was developed in the Maritimes as a method to incorporate PA and decision rules in fisheries management, following initiatives from Caddy (1998). DFO Maritimes initiated an investigation of the TLA in 1999 (Halliday, Fanning, & Mohn, 2001). It was to be used as part of stock assessment, broadening the approach to include non-traditional information. The key appeal of the TLA is a means of visualisation of indicator data as a series of traffic lights categorising indicators in relation to target and limit reference points. The TLA was initially designed for implementing the PA in data poor situations, but was thus adapted to data rich situations. The main interests of the method are the ability to include all new sources of information, and a way to propose a visually pleasing and transparent process for communication and understanding among users. *You say 'this is all the information we have fellows, now you know as much as I do' and we can start talking about all the inconsistencies*, reported a DFO scientist.

In 2007, the TLA was part of routine stock assessment only for the small eastern Scotian Shelf shrimp stock (DFO, 2005), with a summary indicator being a simple average of equally weighted indicators (Fig. 3.1):

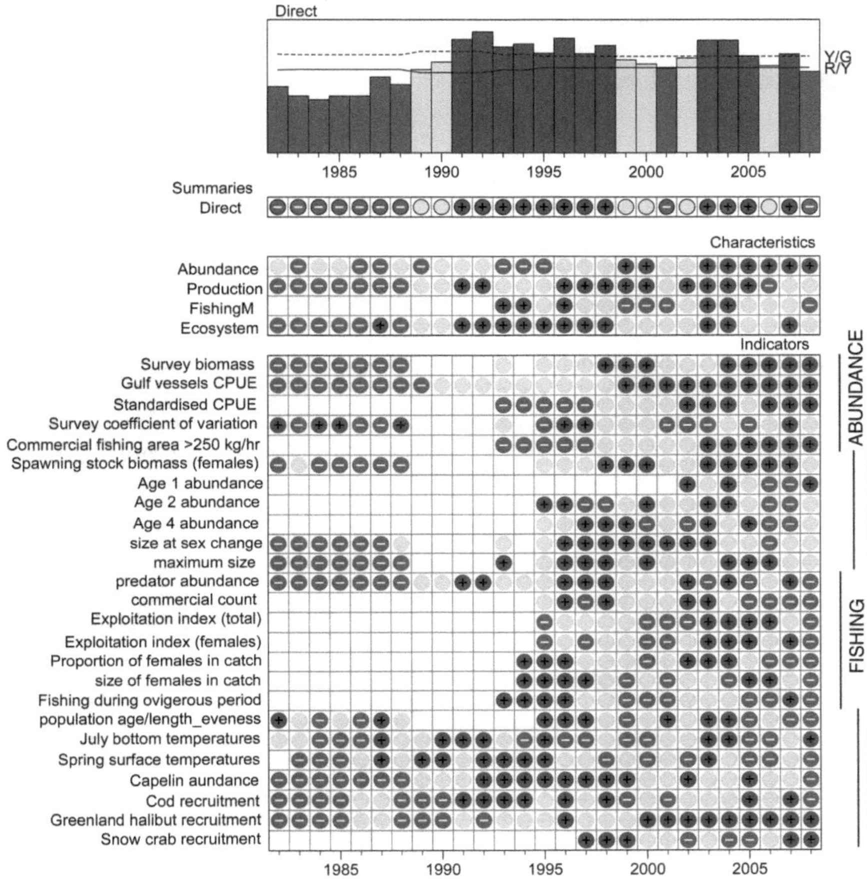

Fig. 3.1 Example of Traffic Light Analysis (DFO, 2005)

The success of the method for a small stock of shrimp was partly explained by the strong relationship of trust between the industry and the scientist. The industry trusted the management strategies proposed by the scientist, without arguing of scientific uncertainty for requiring higher quotas.

The traffic light method was applied as a trial basis for some Scotia-Fundy groundfish stocks. As such complete trust in scientific advice as in the shrimp case does not always exist, it was felt necessary to formalise the method and the harvest control rules that could be applied from it, in a desire to propose objective and transparent indicator-based management decisions. Main criticisms, also from the industry side, dealt with the oversimplification of the results, the loss of information and the need for more formal and causal mechanisms, as well as the issue of combining disparate lights into summary lights (integration): *First is happy with red, the other is happy with green, if you make yellow as a compromise nobody catches anything*, said a fixed gear representative. Halliday et al. (2001) conducted a thorough analysis of the technical aspects of the method. The trials made to make the method more quantitative lost their simplicity without solving the issue of integration, and its use did not proceed beyond the pilot stage (Koeller, 2004). As Koeller (2007) noticed, this final product was essentially a compromise between two irreconcilable philosophies, and collapsed under its own complexity. In 2007, however, during our study tour, the method was gaining a revived interest, and was to be tested on two invertebrate stocks of primary importance, the Gulf snow crab and the Northern shrimp.

However, the simplistic approach of the TLA, as shown in the shrimp case, suggests that summary statistics may track 'stock health' more comprehensively and usefully than individual indicators, and might be more precautionary than traditional methods (Koeller, 2004). In particular, it accounts for some other parameters than traditionally used in assessment, which could be indicative of stock status (Hutchings & Myers, 1994), and it avoids relying on comprehensive models.

3.4.5.3 Ecosystem Approach to Fisheries Management (EAFM)

The renewed interest in indicators is related to the ecosystem approach, to which Canada is committed by law. Indeed, Choi, Frank, Petrie, and Legget (2005) tried to use a similar traffic light approach as a descriptive tool for the Scotian shelf, choosing indicators in collaboration with scientific experts from the various relevant fields but without trying to combine indicators for potential management action. This seemed to work well to track dramatic changes, as the Scotian shelf has experienced over recent years, but not so much for weak changes.

Many years of discussion about implementation of the ecosystem-approach in fisheries management have lead to some progress. Influential scientists are moving towards a pragmatic and urgent approach based on current knowledge, rather than on developing comprehensive models trying to include all ecosystem processes. DFO Science is developing an ecosystem science framework (DFO, 2007, http://www.dfo-mpo.gc.ca/science/Publications/Ecosystem/index-eng.htm), that integrates advice and support within its five main programmes (Fisheries, Aquaculture, Oceans, Habitat, Species-At-Risk). The framework includes a number of key components reflecting the highest priority management and policy challenges, as well as the multi-functional nature of an ecosystem science approach.

However, operationalisation of such a framework is not straightforward, as traditional stock assessment cannot easily provide all the required information. DFO is changing its strategy from scientists focused on single management issues to a team approach that brings together a wide range of skills, but suffers from limited human and financial resources. In most cases, assessment and management meetings are still attended almost uniquely by traditional fisheries management science groups and industry representatives, with under representation from environmentalists and ecosystem scientists.

3.5 Conclusions

3.5.1 Rights-Based Management in Nova Scotia

The benefit of the ITQ system is that it provided a mechanism for removing capacity from the fishery in a way that reduced the inevitable disruption in fishers' lives and livelihoods by providing a transparent system for the reallocation of value. The ITQs smoothed the process by which fishing and processing capacity was reduced, but they were not the main engine of the reduction itself. The main engines were much smaller quotas and the introduction of effective enforcement. Most of the unfairness that was experienced stemmed from the initial allocation of individual quota based on the reconstructed historical participation rather than the system developed for trading those individual quotas.

The ITQ system in Nova Scotia has had the same negative impacts that have emerged in other similar areas. It has intensified the organisational and geographical concentration of the industry, which would likely have accompanied capacity reduction, however. It has shifted more of the burden of reducing excess capacity to crew members than is perhaps fair. Attempts to reduce these negative impacts through the design of the system and closely related policies have not been very effective and remain controversial. The impacts of the system on conservation are both unclear and mixed, but from a legal and institutional perspective it has reduced potentials for adaptive management by locking ecological realities that evolve either naturally or as a result of greater scientific understanding – for example the definitions of particular fish stocks – into hard institutional boxes.

3.5.2 Participatory Management in Nova Scotia

3.5.2.1 The Community Management Boards

The Community Management Boards have developed an international reputation as an experiment in fisheries co-management. All of the respondents we interviewed were very supportive, some even quite proud, of the CMB system. Even pro-ITQ industry respondents considered the CMBs to be as good a deal as they can expect to get at this time. The CMBs seem to have worked particularly well from the perspective of DFO. They have greatly reduced taxpayer costs while giving them effective local institutions for working with the fishing industry.

We found the Shelburne B experiment to be particularly interesting. On the one hand the feared loss of a local fisheries base through industry concentration precipitated by tradable individual quotas has not happened. On the other hand, the CMBs that have not allowed the transfer of IQ among members have had problems dealing in a fair way with exit from the fishery. They have also no doubt paid a considerable cost in economic efficiency in comparison to a formal ITQ system, as is evidenced by the lower price that Shelburne quota gets in comparison with the

<65′ mobile gear quota. A DFO respondent emphasised that the CMBs takes on management costs that are borne by the Canadian taxpayer in the mobile gear ITQ, where the considerable fees paid by the mobile fishers do not cover all the costs of administering the quota system.

The CMBs are only one of the institutional platforms for fishers' participation in fisheries management in Nova Scotia. They have developed into an important resource that contributes to the success of other initiatives, such as the individual quotas and the monitoring system.

3.5.2.2 Participation in Science

The scientific process has dramatically changed over the last fifteen years. A real effort has been made toward transparency and openness for effective governance. The Science Branch has been willing to improve dialogue and communication with the industry, and to integrate it in the management process. On the other side, the industry has been willing to participate at own costs, and has gone through a real educational process to be able to be proficient in collaboration with scientists. Regular meetings between scientists and industry have created certain situations of long-lasting and personal relationships with high levels of commitment and trust, especially in invertebrate fisheries, but this cannot be generalised to all fisheries.

Participation of industry in the stock assessment process has, however, not always been straightforward. A degree of mistrust is still present between the two worlds, especially when scientific results are based on comprehensive models and with input numbers based on extrapolation of sampling data. This has not solved the uncertainties in stock assessment results, especially when industry and scientists perceptions of stock trends go in opposite direction. In comparison with the previous system, which was completely closed to industry participation, the open process must avoid going the other way, with a too large role accorded to the industry and thus undermining the role of science. Some uncomfortable situations were observed, with industry putting pressure on science meetings' outputs for political reasons, especially when the level of scientific uncertainty is high. However, it is clear that this system, although not perfect and not free from political issues, is strongly felt as preferable to the old system prevailing before the collapse, and no respondent would consider reversing back to it.

Overall, this process has improved social robustness, by reducing the feeling of industry of being ignored. It has also improved biological robustness, by increasing the feeling of ownership and responsibility for the resource and improving the commitment to scientific advice.

References

Apostle, R., McCay, B. J., & Mikalsen, K. (2002). *Enclosing the commons: Individual transferable quotas in the Nova Scotia fishery*. St. John's: ISER Books, Faculty of Arts Publications, Institute of Social and Economic Research.

Auditor General. (1999). Auditor general report: Chapter 4. *Fisheries and oceans. managing Atlantic shellfish in a sustainable manner*. Ottawa: Office of the Auditor General: 35pp.

Boudreau, P., & O'Boyle, R. (2007). *Proceedings of the Maritime Provinces Regional Advisory Process on Assessment and Management Strategy Framework for Banquereau Arctic Surfclam and Ocean Quahogs on Sable Bank and in St. Mary's Bay*. 17–18 January 2007 and 4–5 April 2007. CSAS, Canadian Science Advisory Secretariat, Proceedings Series 2007/008.

Caddy, J. F. (1998). Deciding on precautionary management measures for a stock and appropriate limit reference points (LRPs) as a basis for a multi-LRP Harvest Law. *NAFO SCR Doc.*, No 8, SN 2983, 13p.

Charles, A. T. (1998). Living with uncertainty in fisheries: Analytical methods, management priorities and the Canadian groundfishery experience. *Fisheries Research*, *37*, 37–50.

Charles, A. T., Bull, A., Kearney, J., & Milley, C. (2005). Community-based fisheries in the Canadian Maritimes. In T. McClanahan & J. C. Castilla (Eds.), *Fisheries management: progress towards sustainability*. Oxford: Blackwell.

Choi, J. S., Frank, K. T., Petrie, B. D., & Legget, W. C. (2005). Integrated assessment of a large marine ecosystem: A case study of the devolution of the eastern Scotian Shelf, Canada. *Oceanography and Marine Biology: An Annual Review*, *43*, 47–67.

Creed, C., Apostle, R., & McCay, B. J. (1994, September 22–27). *ITQs from a community perspective: The case of the Canadian Scotia-Fundy Groundfish Fishery*. Paper presented at the 1994 Statutory Meeting of ICES, St. Johns, NL.

DFO. (2004). *Preserving the Independence of the Inshore Fleet in Canada's Atlantic Fisheries – A Discussion Document*. Atlantic Fisheries Policy Review Public Consultations, Retrieved from http://www.dfo-mpo.gc.ca/afpr-rppa/home_e.htm

DFO. (2005). Northern Shrimp on the Eastern Scotian Shelf (SFA 13-15). DFO *Canadian Science Advisory Secretariat, Science Advisory Report* 2005/060.

DFO. (2007). *A new Ecosystem Science Framework in support of Integrated Management*. Department of Fisheries and Oceans, http://www.dfo-mpo.gc.ca/science/Publications/Ecosystem/index-eng.htm

Frank, K. T., Petrie, B., Choi, J. S., & Leggett, W. C. (2005). Trophic cascades in a formerly cod-dominated ecosystem. *Science*, *308*, 1621–1623.

Halliday, R. G., Fanning, L. P., & Mohn, R. K. (2001). *Use of the traffic light method in fishery management planning*. CSAS, research document 2001/108.

Harris, L. (1995). The east coast fisheries. In B. Mitchell (Ed.), *Resource and environmental management in Canada* (pp. 130–150). Toronto: Oxford University Press.

Haward, M., Dobell, R., Charles, A. T., Foster, E., & Potts, T. (2005). Fisheries and Oceans Governance in Australia and Canada: From Sectoral Management to Integration? *Dalhousie Law Journal, 26*(1), 5–45.

Hutchings, J. A., & Myers, R. A. (1994). What can be learned from the collapse of a renewable resource? Atlantic cod, Gadus morhua, of Newfoundland and Labrador. *Canadian Journal of Fisheries and Aquatic Sciences 51*, 2126–2146.

Kearney, J., Bull, A., Recchia, M., Desroches, M., Langille, L., & Cunningham, G. (1998, May 10–14). Resistance to Privatisation: Community-Based Fisheries Management in an Industrialised Nation. *Proceedings of the International Workshop on Community-Based Natural Resource Management*, The World Bank, Washington, DC.

Koeller, P. (2004). *The brief history of traffic lights in the Northwest Atlantic*. Paper presented at the SCOR Symposium on Ecosystem Management, 31 March–3 April, Paris.

Koeller, P. (2007). *Ecosystem-Based Psychology, or, How I Learned to Stop Worrying and Love the Data*. Paper presented at the ICES Annual Science Conference, Helsinki, Finland CM 2007/R:06.

Liew, D. S. K. (2001). Initial allocation of quota rights in the scotia-fundy inshore mobile-gear groundfish fishery. In R. Shotton (Ed.), *Case studies on the allocation of transferable quota rights in fisheries* (pp. 75–85). FAO Fisheries Technical Paper No. 411, Rome, FAO 373pp.

Loucks, L. (1998). Sambro community quota fisheries management: A case of innovative community based decision-making. In L. Loucks, T. Charles, & M. Butler (Eds.), *Managing our fisheries, managing ourselves* (pp. 54–58). Halifax: Gorsebrook Research Institute for Atlantic Canada Studies.

McCay, B. J., Creed, C. F., Finlayson, A. C., Apostle, R., & Mikalsen, K. (1996). Individual transferable quotas (ITQs) in Canadian and US fisheries. *Ocean and Coastal Management, 26*(1–3), 85–116.

OECD. (2001). *Towards sustainable development: Environmental indicators*. Paris: OECD, 26 pp.

Peacock, F. G., & Hansen, J. (2000). Community management in groundfish: A new approach to property rights. In R. Shotton (Ed.), *Use of property rights in fisheries management*. FAO Fisheries Technical Paper 404/2.

Potts, T. W. (2003). *Sustainability indicators in marine capture fisheries*. PhD thesis, University of Tasmania, 393pp.

Shelton, P. A. (2005). Did over-reliance on commercial catch rate data precipitate the collapse of northern cod? *ICES Journal of Marine Science, 62*, 1139–1149.

Shelton, P. A. (2007). The weakening role of science in the management of groundfish off the east coast of Canada. *ICES Journal of Marine Science, 64*, 723–729.

Shelton, P. A., & Lilly, G. R. (2000). Interpreting the collapse of the northern cod stock from survey and catch data. *Canadian Journal of Fisheries and Aquatic Sciences, 57*, 2230–2239.

Shelton, P. A., & Rivard, D. (2003). *Developing a precautionary approach to fisheries management in Canada – the decade following the cod collapses*. NAFO SCR Doc. 03/1, Serial No. N4808.

Sinclair, M., O'Boyle, R. N., Burke, D. L., & Peacock, F. G. (1999). Groundfish management in transition within the Scotia-Fundy area of Canada. *ICES Journal of Marine Science, 56*, 1014–1023.

Walters, C., & Maguire, J.-J. (1996). Lessons for stock assessment from the northern cod collapse. *Reviews in Fish Biology and Fisheries, 6*, 125–137.

Chapter 4
Abundant Fish Stocks and Profitable Fisheries off Alaska – A Study on Harvest Control Rules and Pollock Cooperatives

Franziska Wolff and Kjellrun Hiis Hauge

Abstract In this chapter, we discuss the success of two innovative fisheries management regimes in Alaska, United States of America: (1) the Tier System, a harvest control rule that defines the upper limit for total allowable catch for all groundfish stock in the federal waters off Alaska, and (2) the system of industry cooperatives in the Bering Sea walleye pollock (*Theragra chalcogramma*) fisheries that combines rights-based management with self-governance. The empirical basis for the assessment comprises expert and stakeholder interviews as well as secondary literature. The Tier System and the quota-setting process contribute to precautionary harvesting, and are well accepted among most stakeholder groups. The cooperative model scores high with regard to economic performance. It also has some positive impacts with regard to management costs but is socially rather contentious; in particular among environmental NGOs and industry stakeholders that do not participate in a cooperative. The cooperatives only indirectly impact on pollock stocks and the respective ecosystems.

Keywords Harvest control rules · Impact assessment · Participatory management · Pollock cooperatives · Rights-based management · Tier System

4.1 Introduction

Worldwide, we hear of fisheries in crisis, of declines in stocks, degraded marine ecosystems and contingent impacts on fishing industries and communities. This seems to be different in many of the fishing grounds off Alaska, especially in the Bering Sea. The Alaskan pollock fishery, for instance, was the first large whitefish fishery worldwide to have become certified as sustainable by the Marine Stewardship Council.

F. Wolff (✉)
Öko-Institut e.V., Novalisstr. 10, 10115 Berlin, Germany
e-mail: f.wolff@oeko.de

K.H. Hauge, D.C. Wilson (eds.), *Comparative Evaluations of Innovative Fisheries Management*, DOI 10.1007/978-90-481-2663-7_4,

In this chapter we describe and discuss two innovative management regimes in Alaska fishery management, which are exemplary of successful fisheries management at least in specific respects. The first of those innovations, the Tier System, defines the maximum acceptable biological catch (ABC) and the overfishing level (OFL), where the former sets the upper limit for the total allowable catch (TAC). The Tier System consists of six tiers of harvest control rules, each with a different level of data quality requirements. It applies to all groundfish stocks in the federal waters off Alaska. The second innovation is a system of industry cooperatives that jointly harvests walleye pollock (*Theragra chalcogramma*) in the Bering Sea and Aleutian Islands area. The cooperatives combine rights-based management – exclusive harvesting privileges are allocated to industry sectors, which may fish cooperatively – with self-governance within the cooperatives. Micro-management issues formerly tasked to fisheries managers are now managed by the fishery participants themselves. We assess the success of these two innovations in terms of biological robustness, economic performance of the fleets, management costs and stakeholder acceptance (as a core element of 'social robustness', as defined in Chapter 8). The empirical basis for the assessment comprises a total of 22 expert and stakeholder interviews (see Table 4.1) as well as secondary literature. The semi-structured interviews were carried out in early 2007 in Juneau, Anchorage and Seattle with key representatives of the fishing – more specifically the pollock – industry, fisheries management, environmental organisations and academia.

Table 4.1 Expert and Stakeholder Interviews

	Fishing industry (with focus on pollock industry)	Fisheries management (NMFS, Council)	Environmental organisations	Academia
No. of interviews[1]	8	10	3	1

[1]Note that an interview would often comprise more than one interviewee.

4.2 Background

The fishing grounds off Alaska are well known for their abundance, especially with the Eastern Bering Sea covering a huge continental shelf, which makes it one of the most productive marine ecosystems in the world. Commercial fishing started in the 1860s and for more than hundred years was foreign dominated. The initial focus was on salmon, Pacific cod (*Gadus macrocephalus*), sablefish (*Anoplopoma fimbria*) and halibut (*Hippoglossus stenolepis*), and later on Pacific herring (*Clupea pallasi*). After World War II, the presence of (especially Japanese and Soviet) long-distance fleets intensified, which targeted crab and groundfish species. This included walleye pollock which today is the principal fishery. Until the mid-1960s, these fisheries were virtually unregulated. The enactment of the Fishery Conservation and Management Act in 1976 (renamed Magnuson-Stevens Act in

1996) was a major turning point (NOAA, 2002; NPFMC, 2006). The Act unilaterally extended the nation's fisheries management jurisdiction to 200 nautical miles, later called the Exclusive Economic Zone (EEZ). It set up a participatory management system that allows advice to be developed at regional level through fishery management councils. In these, industry, public administration, scientists and non-governmental organisations jointly provide fisheries recommendations to the Secretary of Commerce through the director of the National Marine Fisheries Service (NMFS, also called NOAA[1] Fisheries Service). Management responsibility for the commercial fisheries off Alaska is shared by a mix of state and federal institutions. Apart from the NMFS, these include the North Pacific Fisheries Management Council, hereafter referred to as 'the Council', the Alaska Department of Fish and Game and its Fisheries Commission.

Fishery management plans (FMPs), developed by the Council, form the basis of the management of the federal fisheries off Alaska. A central – and unique – management feature is upper limits on total catches of all groundfish species ('caps', see Section 4.3). In addition to a number of technical measures, time and area restrictions, harvesting of target stocks is restricted by stringent bycatch provisions. These may include closure of whole fisheries, even if the TAC for the target species is not yet taken. Bycatch management is based on a near real-time in-season process and an extensive observer programme pertaining to both vessels[2] and shoreside processors in the groundfish fisheries. Over the past few decades, access to many fisheries off Alaska has been limited, both through a licence limitation programme and a series of rights-based management schemes, called 'limited access privilege' or 'rationalisation' programmes (Section 4.4).

4.3 The Tier System and the TAC-Setting Process

The Tier System is a set of harvest control rules (HCR) and forms the basis for setting the upper limit of the TAC for all groundfish stocks and some bycatch species (see Table 4.2) managed by the Council. It is therefore central in the fisheries management plans for the Bering Sea Aleutian Islands (BSAI) and Gulf of Alaska (GOA) areas. Further, the Tier System is the Council's interpretation and operationalisation of the fisheries management strategy as laid down in the Magnuson-Stevens Act.

The TAC is set annually for each groundfish stock in the BSAI and GOA areas. TAC-setting is based on the following pillars:

[1] National Oceanic and Atmospheric Administration.

[2] Regulation requires at least one observer onboard during all fishing operations for vessels longer than 125 feet, and is less strict for smaller vessels.

Table 4.2 Various Figures Related to the Tier System for the Groundfish Stocks in the BSAI Area

	2007			2008	
Species or complex	ABC	TAC	Catch	Max F_{ABC}	Tier
Pollock EBS	1,394,000	1,394,000	1,350,000	0.22	1b
Pollock AI	44,500	19,000	2,488	0.2	3a
Pollock Bogoslof	5,220	10	0	0.017	5
Pacific cod	176,000	171,000	172,000	0.22	3b
Sablefish	5,790	5,790	2,170	0.084	3b
Yellowfin sole	136,000	136,000	119,332	0.19	1a
Greenland turbot	2,440	2,440	1,946	0.51	3a
Arrowtooth flounder	158,000	20,000	11,700	0.24	3a
Northern rock sole	198,000	55,000	37,013	0.19	1a
Flathead sole	79,200	30,000	19,500	0.28	3a
Alaska plaice	190,000	25,000	19,411	0.59	3a
Other flatfish	21,400	10,000	25,176	0.13/0.06/0.2	5
Pacific Ocean perch	21,900	19,900	17,800	0.059	3a
Northern rockfish	8,190	8,190	3,940	0.045	3a
Shortraker	424	424	318	0.023	5
Rougheye	202	202	163	0.019	5
Other rockfish	999	999	635	0.023/0.068	5
Atka mackerel	74,000	63,000	56,620	0.33	3a
Squid	1,970	1,970	1,190	n/a	6
Sharks	68,800	37,400	26,500	n/a	6
Skates				0.075	3a/5
Sculpins				0.14	5
Octopus				n/a	6
Total	2,676,035	2,000,000	1,867,902		

Source: NOAA (2008). Note that the listed species represent groundfish species, except the last five ones, which are non-targeted bycatch species.

- The Tier System, which defines maximum sustainable yield (MSY) as the overfishing level (OFL). Further it defines maximum acceptable biological catch (max ABC) at a lower level than OFL to buffer uncertainty in the calculations of MSY.
- Optimum yield, where economic, social or ecological factors are taken into account to evaluate whether ABC should be reduced from maximum ABC. Maximum ABC is thus a preliminary quantity for ABC.
- The groundfish cap, which sets an upper limit for the total groundfish catches in a management area (BSAI and GOA). Consequently, in years when the ABCs for all stocks amount to a higher level than the cap, the TAC for one or more stocks must be set lower than its ABC.

The idea of implementing groundfish caps was included in the FMPs in 1984, and caps for each of the GOA and BSAI areas were developed (NPFMC, 2008a, 2008b). The underlying concept of the caps is that there is an optimum yield of groundfish for each ecosystem that will vary according to the productivity of the system.

Using an ecosystem production model, the cap range was estimated at 1.8–2.4 million metric tons for BSAI groundfish (NPFMC, 2008a). For precautionary reasons, the optimum yield was set at 85% of this range, i.e. 1.4–2.0 million metric tons. The cap range for GOA groundfish is similarly set to 116–800 thousand metric tons (NPFMC, 2008b). In practice, the cap for the GOA has never been reached and only the upper limit of the range for the BSAI has been used, with the FMP eventually implementing the single value cap of 2 million metric tons (NPFMC, 2008a).

Advisory panels for both areas are appointed by the Council and consist of industry and non-industry stakeholders. They make TAC recommendations to the Council to fit within the cap and stay below the ABC levels. The recommendations are finalised by the Council before being handed to the US Secretary of Commerce. The advisory process on TAC-setting takes about a year. During this period, science reports and further recommendations are made and are reviewed being. Public testimonies are allowed twice per year (NPFMC, 2008a, 2008b).

4.3.1 The Evolution of the Tier System

Until the mid-1970s, catches in the fisheries off Alaska were only limited by closed areas and gear restrictions. These restrictions did not prevent some stocks from declining, and the first catch limits were set in international bilateral agreements in 1973. The Council implemented catch limits in the GOA and BSAI area in 1979 and 1982 respectively (NPFMC, 2006). This change required estimates of fish stock biomass and fishing mortalities (Fs), which generated a need for scientific surveys. The developments in stock assessment modelling at the time tended towards a more statistical approach, which was readily adopted for the groundfish stocks in Alaskan waters for a number of reasons. Firstly, traditional methods such as virtual population analysis (VPA), which is the common method for TAC-setting in European waters, demand more extensive data than was available. Secondly, the assumptions required for VPA are more restrictive. Finally, estimating uncertainty was regarded as central to risk-averse management practice. As information technology evolved, the computing capacity demands for statistical approaches were no longer a barrier.

The Alaskan HCRs have become more advanced since the first version of the Tier System was adopted for the GOA and BSAI areas in 1997 (NPFMC, 2008a, 2008b). For example, HCRs are now designed to produce three different F levels (instead of two) to suit different stock conditions (Thompson, 1999).

In 1998, the U.S. Department of Commerce issued a set of national guidelines (U.S. Department of Commerce, 1998) that required further development of the HCRs (Thompson, 1999). These national guidelines include requirements for a precautionary approach stipulating that: (1) target Fs are less than limit fishing mortalities, (2) Fs at low stock sizes are lower than Fs at high stock sizes, and (3) the buffer between limit and target Fs widens as uncertainty, regarding a stock's size or productive capacity, increases. The resulting HCRs[3] are still valid so that the current Tier System is similar to the changes these amendments led to. The Tier System meets all the requirements of the national guidelines, except that the size of the uncertainty buffer does not necessarily increase with uncertainty (Thompson, 1999).

4.3.2 Description of the Tier System

The Tier System consists of six tiers of harvest control rules and defines criteria for determining the tier level for each stock (Table 4.3). Its purpose is to provide rules suitable for managing stocks for which different levels of data and biological information are available. The most data rich stocks are thus assigned to Tier 1 and the most data poor stocks to Tier 6. Each tier contains a formula or a set of formulae defining ABC and OFL, with the aim to increase the degree of caution with decreasing tier level. The three upper tiers are more advanced than the rest in that they contain three different harvest strategies depending on the assessed biomass level. They include strategies for stock recovery. This allows for relatively high Fs when the stock is considered abundant, but allows no fishing when the abundance is lower than a certain critical level. The lower tiers intend to provide relatively low Fs independent of abundance and are designed so that in order for a fishery to develop, additional information is necessary.

The stock assessment models are also more advanced for the higher tiers, with Tiers 1–3 being typically based on age-structured models. Only Tiers 1 and 2 apply the MSY concept, which requires a functional relationship between spawning stock biomass and recruitment. The other tiers are based on proxies for MSY and are less advanced and less data demanding. Some of the scientists we interviewed regarded these MSY-proxies as ad hoc, yet robust. Tier 6 only requires a reliable catch history over a certain period. The distribution of the number of stocks assigned to the tiers in the BSAI area is shown in Table 4.2. The distribution in the GOA area is similar, except that there are no stocks in Tier 1, but some in Tier 4. According to our respondents, Tier 2 has never been applied in either area.

In practice, a higher tier usually entails a higher ABC. For Eastern Bering Sea pollock, for example, Tier 1 generates a maximum ABC of 1.17 million metric tons while Tier 3 generates far less, namely 0.555 million metric tons (NOAA, 2008). Stocks can be moved from one tier to another. Most often they move upwards when data quality and/or the assessment model is improved. Occasionally a lower tier is chosen, for example when survey coverage is considered insufficient. Tier 6 is designed in such a way that it requires a certain level of knowledge and data to develop fisheries on new species. A comprehensive observer programme together with scientific survey information has made it possible to provide sufficient data to

[3]Established by Amendment 56 to the fisheries management plans in 1999.

Table 4.3 The Tier System Expressing the Information Requirements and HCRs for Each Tier

Tier	(1)	Information available: *Reliable point estimates of B and B_{MSY} and reliable pdf of F_{MSY}.*
		(1a) Stock status: $B/B_{MSY} > 1$
		$F_{OFL} = \mu_A$, the arithmetic mean of the pdf
		$F_{ABC} \le \mu_H$, the arithmetic mean of the pdf
		(1b) Stock status: $\alpha < B/B_{MSY} \le 1$
		$F_{OFL} = \mu_A \times (B/B_{MSY} - \alpha)/(1 - \alpha)$
		$F_{ABC} \le \mu_H \times (B/B_{MSY} - \alpha)/(1 - \alpha)$
		(1c) Stock status: $B/B_{MSY} \le \alpha$
		$F_{OFL} = 0$
		$F_{ABC} = 0$
	(2)	Information available: *Reliable point estimates of B, B_{MSY}, F_{MSY}, $F_{35\%}$ and $F_{40\%}$.*
		(2a) Stock status: $B/B_{MSY} > 1$
		$F_{OFL} = F_{MSY}$
		$F_{ABC} \le F_{MSY} \times (F_{40\%}/F_{35\%})$
		(2b) Stock status: $\alpha < B/B_{MSY} \le 1$
		$F_{OFL} = F_{MSY} \times (B/B_{MSY} - \alpha)/(1 - \alpha)$
		$F_{ABC} \le F_{MSY} \times (F_{40\%}/F_{35\%}) \times (B/B_{MSY} - \alpha)/(1 - \alpha)$
		(2c) Stock status: $B/B_{MSY} \le \alpha$
		$F_{OFL} = 0$
		$F_{ABC} = 0$
	(3)	Information available: *Reliable point estimates of B, $B_{40\%}$, F_{MSY}, $F_{35\%}$ and $F_{40\%}$.*
		(3a) Stock status: $B/B_{40\%} > 1$
		$F_{OFL} = F_{35\%}$
		$F_{ABC} \le F_{40\%}$
		(3b) Stock status: $\alpha < B/B_{40\%} \le 1$
		$F_{OFL} = F_{35\%} \times (B/B_{40\%} - \alpha)/(1 - \alpha)$
		$F_{ABC} \le F_{40\%} \times (B/B_{40\%} - \alpha)/(1 - \alpha)$
		(3c) Stock status: $B/B_{40\%} \le \alpha$
		$F_{OFL} = 0$
		$F_{ABC} = 0$
	(4)	Information available: *Reliable point estimates of B, $F_{35\%}$ and $F_{40\%}$.*
		$F_{OFL} = F_{35\%}$
		$F_{ABC} \le F_{40\%}$
	(5)	Information available: *Reliable point estimates of B and natural mortality rate M.*
		$F_{OFL} = M$
		$F_{ABC} \le 0.75 \times M$
	(6)	Information available: *Reliable catch history from 1978 through 1995.*
		OFL = the average catch from 1978 through 1995, unless an alternative value is established by the Scientific and Statistical Committee on the basis of the best available scientific information.
		$ABC \le 0.75 \times OFL$

Source: NOAA (2008). Note that *FX%* is defined as the harvest rate associated with an equilibrium level of spawning per recruit equal to X% of the equilibrium level in the absence of any fishing, BX% likewise.

assign non-targeted species to a higher tier. To allow targeting that stock, an amendment of the FMP(s) needs to be made, as the plans include a list of species that are allowed to be targeted.

4.3.3 Assessment of the Tier System

We assess the Tier System and the wider TAC-setting process only in terms of biological robustness and social acceptance. The economic performance of the fleets depends on the ability of the Tier System (and supporting regulations) to provide abundant stocks. We therefore assume that it suffices to assess biological robustness. We also assume that management costs are the same for most forms of scientific advice on quota management. The extent of participation in the TAC-setting process makes it more expensive than a top-down decision would be. However, since we have focused on the Tier System itself we have not investigated these costs.

4.3.3.1 The Biological Robustness of the Tier System

A comprehensive assessment of the maximum ABC levels would require an evaluation of the assessment data in combination with relevant assessment models and reference points, which is outside the scope of this study. Fishing mortalities, on the other hand, are comparable across stocks to some degree; as for example long-lived species need to be harvested with lower Fs to achieve sustainability than short-lived species.

Table 4.2 shows that the recommended fishing mortalities (F_{ABC}) for 2008 are low for most groundfish stocks in the BSAI area, except two relatively high mortalities above 0.5 (Greenland turbot, Alaska plaice). The Fs are similar for the GOA area (NOAA, 2008). As the abundance of EBS pollock is assessed to be below the desired level, the recommended F is relatively low. However, had the abundance exceeded the desired level, Tier 1 would have generated an F of 0.919 (NOAA, 2008). This level is quite high, also in a European fisheries context. This means that Tier 1 allows for a considerable fishing mortality when this stock is estimated to be above the desired abundance level. In practice, however, the average fishing mortality has been estimated to be at 0.506 since 1982. This level is basically a result of having set the TACs lower than ABC.

Table 4.2 also exemplifies that ABC can be set lower than maximum ABC, in this case for Eastern Bering Sea and Bogoslof pollock (NOAA, 2008). Furthermore, scientific advice on ABCs was followed, as no TACs were set higher than their corresponding ABC levels in 2007. On the contrary, while the ABCs added up to 2.7 million metric tons, the TACs amounted to only 2.0, which is the groundfish cap level for the BSAI area. Actual catches were even lower. The picture is similar for the GOA area. Our respondents suggested that several factors contributed to the fact that some TACs for targeted species were not taken in full: the strict bycatch regulations, socio-economic reasons, weather conditions and low catch rates.

Because of the comprehensive observer programme there is good reason to believe that the reported catches are accurate. This belief was expressed by all our respondents, including the environmental organisations. This means that, in practice, the Tier System successfully provides upper catch limits for all groundfish stocks and non-targeted species in question. The distribution of sub-tiers illustrates that most stocks in Tiers 1–3 in the BSAI area are assessed to be in a healthy condition with no stocks below the critical level (see Table 4.2).

Although our observations above suggest that the stocks are managed sustainably, our respondents identified room for improvement with regard to three aspects of the system: unsustainable features of the Tier System itself, inadequate handling of uncertainty, and experienced local stock depletion. Tier 1, as just demonstrated, may allow for rather high Fs and Tier 6 may lead to stock depletion because a fixed year range is chosen, which incidentally may represent years of overfishing. Uncertainty is accounted for by a rather fixed buffer that does not take into account that the uncertainty may vary from year to year. The environmentalists argued that uncertainty would have been better accounted for if the Council had followed the scientists' periodic recommendations on reducing ABC from maximum ABC. Finally, they expressed concern about local depletions, which had been discussed at the Council. Although the Council does split the TAC (and ABC) for some stocks, the environmentalists argued that this was not done to a sufficient degree.

Overall, our analyses suggest that the groundfish fisheries are harvested in a precautionary manner. At least it is fair to say that harvesting is cautious compared to fisheries elsewhere in the world, for example in other U.S. fisheries or in Europe. The fishing mortalities are generally set low, supported by a precautionary TAC-setting process in the Council; bycatch regulations on non-targeted species; strict bycatch enforcement and an observer programme that provides comprehensive information on what and how much is caught.

Critical voices claim that sustainability is still not ensured and that uncertainty should be better accounted for in various ways. Their concern is that today's abundance may be a coincidence and that Alaska will eventually experience the pattern seen elsewhere in the world with stock collapses and ecosystem degradation (Stump, Hocevar, Baumann & Marz, 2006; Marz & Stump, 2002). However, this same uncertainty makes it difficult to accurately determine the boundary between sustainable and unsustainable harvesting.

4.3.3.2 Stakeholders' Trust in Science and Resource Management

Our stakeholder interviews conveyed a sense of mutual trust between the scientists, the managers and the (pollock) industry. All stakeholders seemed to have a general confidence in the Tier System and the TAC setting process, except for environmental non-governmental organisations (NGOs), which were opposed to some aspects.

Industry respondents, managers and scientists all expressed the view that the industry does have trust in science and especially in the Tier System. The pollock industry in particular articulated pride in respecting the maximum ABC, supporting sustainable fisheries and contributing to an environmentally friendly fishing practice (MCA, 2007). However, this does not mean that industry support is always a given, for example when scientists recommend a reduction of the ABC from its maximum level. Yet, the scientists were content with the TAC-setting process, which implies that they did not consider past disagreements to threaten the sustainability of the fisheries. Scientists, managers and industry all confirmed that the industry had never lobbied to raise TACs above ABC levels, and that the industry did not necessarily negotiate for the highest possible catch.

There were only a few critical remarks about the Tier System as such, mostly coming from the assessment scientists themselves. These are generally to be understood as expressions of a wish to continuously improve science and scientific advice. When we asked managers and industry representatives about the scientists' reservations they expressed surprise, indicating that such improvements were not being discussed or regarded as necessary. The one exception was a suggested change in Tier 6 to avoid its perceived arbitrariness.

The environmental organisations' lower trust in the Tier System and the TAC-setting process and their view that the fisheries off Alaska are not harvested in a sufficiently precautionary manner is reflected by repeated lawsuits against the fisheries administration.[4] More concretely, the NGOs criticise the lacking consideration of uncertainties and unknown information in stock assessments and in setting ABC and TAC levels; a failure to address the needs of pollock predators in the ecosystem in setting ABC and TAC levels; and large uncertainties about stock structure and stock rebuilding (Marz & Stump, 2002). One of the scientists we interviewed indicated that he thought the environmentalists sometimes exaggerated their criticism of the management, while the environmentalists in turn questioned the scientists' credibility, pointing to the close relationship between scientists and industry in the Council and to industry-funded research.

4.4 Bering Sea Pollock Cooperatives: Economic Gains at the Cost of Social Acceptance?

The second fisheries management innovation we will analyse is the system of industry cooperatives that evolved in the Bering Sea pollock fisheries in the US (EEZ) waters off Alaska. The cooperatives represent a specific form of rights-based management: in a self-governance approach the fishing industry negotiates quota shares among themselves after access to the fishery had been limited to a defined number of participants. The cooperatives operate in three of the four major Alaskan pollock stocks, notably the Eastern Bering Sea (EBS), Aleutian Islands (AI) and Central Bering Sea/Bogoslof Island stocks, not, however in the Gulf of Alaska (GOA). The pollock cooperatives represent one of a number of rights-based management systems that have been introduced in Alaska over the past 15 years.

[4]For example, Greenpeace v. NMFS (55 F. Supp. 2d 1248, W.D. Wash. 1999); Greenpeace v. NMFS (80 F. Supp. 2d 1137, W.D. Wash. 2000); Greenpeace v. NMFS (106 F. Supp. 2d 1066, W.D. Wash. 2000); Greenpeace v. NMFS (237 F. Supp. 2d 1181, W.D. Wash. 2002); American Oceans Campaign et al. v. Daley et al. (183 F. Supp. 2d 1, D.D.C. 2000).

By way of background information, the Alaskan walleye pollock fishery is large and highly industrialised. In 2006, it accounted for 36% of total US landings (NMFS, 2007) and was the biggest single segment of the Alaskan fishing industry with 1.57 million metric tons of retained catch and an ex-vessel value of $377 million (Hiatt et al., 2007). Harvesting takes place in January/February (A season) and in fall (B season), predominantly by mid-water trawling. Products of the low-priced pollock include higher value commodities like roe and fillets as well as surimi, mince, and fish meal. Since 2005, both the BSAI and GOA pollock fisheries have been certified by the Marine Stewardship Council to be well managed and sustainable.

4.4.1 Evolution and Characteristics of the Pollock Cooperatives

Like other Alaskan fisheries, the walleye pollock fishery was historically dominated by foreign trawlers. From the early 1980s, joint ventures between US and foreign companies expanded and efforts were made to 'Americanise' the industry (NOAA, 2002; Wilen, 1998). The emerging domestic industry consisted of an offshore sector that largely operated from Washington State and included catcher/processor vessels (C/Ps) – large 'factory trawlers' both harvesting and processing fish – as well as motherships. The latter process fish delivered to them by a fleet of catcher vessels (CVs), i.e. smaller boats that only harvest fish. At the same time, the State of Alaska struggled to establish processing plants on the Alaskan coast in order to improve community infrastructures, create employment and generate landing tax revenues. Above all Asian firms, facing faze out as a result of Americanisation, started to invest in processing plants. In that way, an Alaska-based inshore sector developed, with catcher vessels now also delivering to shore-based processors.

The evolution of the cooperatives is closely tied to the emerging and quickly intensifying quota allocation conflict "Pollock Wars" between the offshore and inshore sectors in the BSAI region (Criddle, 2008). The huge capacity of the catcher/processors was seen to deprive the smaller catcher vessels, which needed to land fish to shore-based plants, of harvesting opportunities. In the late 1980s, allying with the processors and backed by powerful players such as Alaskan Senator Ted Stevens, the (overcapitalised) catcher vessel sector started to fight in the Alaska-dominated Council for a share of the TAC to be set aside for them. They achieved their goal in 1992, when 35% of the pollock TAC were allocated to the inshore sector (i.e. catcher vessels) and 65% to the offshore component (both catcher/processors and motherships). During the Pollock Wars, a 'Community Development Quota' (CDQ) programme was created, too. It served to include coastal, largely native Alaskan, communities, which received 7.5% of the pollock TAC (later set at 10%) 'upfront' (NRC, 1999).

Following the first 'Inshore-Offshore' deal, the inshore quota share was gradually raised, to the massive discontent of the (likewise overcapitalised) offshore sector (MacGregor, 2006). To make up for their losses in the allocation conflict, the catcher/processors lobbied for the creation of harvesting cooperatives, hoping these would lead to efficiency gains similar to those seen when a catcher/processor cooperative was introduced in the Pacific whiting fishery off Washington in 1997 (Sylvia, Mann, & Pugmire, 2008).[5] Once concerns about the compatibility of harvesting cooperatives with US anti-trust law had been overcome (Sullivan, 2000), the legal path seemed to be cleared for creating an analogous cooperative for Bering Sea pollock catcher/processor companies, a number of which were members of the Pacific whiting cooperative. Technically, this required dividing the offshore sector into a catcher/processor sector and a mothership sector.[6] During the third Inshore-Offshore negotiations in 1998, the Council, however, refused to consider such a formal split as called for by catcher/processor representatives (Sullivan, 2000; Holland & Ginter, 2001).

The disappointed catcher/processors turned to Congress. At the time, the American Fisheries Act (AFA) was being drafted in Washington D.C. but negotiations were at an impasse (Sullivan, 2000). In this situation, the lobbyists of the catcher/processor sector managed to convince Congress of substantial revisions to the draft bill, thus not only breaking the stalemate but facilitating the split into sector shares and, hence, the creation of cooperatives.

The AFA, as it was passed in 1998, establishes discrete and permanent allocations for the three industry components: once certain shares are deducted from the TAC of BSAI pollock, the catcher vessels receive 50%, catcher/processors 40%, and the motherships 10% of the TAC (see Fig. 4.1). The AFA also determines the eligibility criteria for vessels to participate in the fishery (based on catch history, 1995–1997), and lists the eligible vessels. Hence, no new entries are possible to the pollock fishery except when a company purchases a vessel listed in the AFA or its quota share. In addition, the AFA includes a buy-out of nine catcher/processors, thus reducing fleet capacity. It does not directly provide a legal framework governing the formation of *catcher/processor* cooperatives – these are based on contractual agreements only. However, the AFA does lay down criteria for the formation of cooperatives in the *inshore* and *mothership* sectors, which in the meantime had become interested in having cooperatives as well. To accommodate processor interests, inshore cooperatives were designed to be plant-specific: on an annual basis, they form around an affiliated shoreside processor to which they agree to deliver at least 90% of their pollock catch allocation.[7] AFA vessels may not exceed, in the aggregate, a defined amount of (non-retainable) bycatch of specific species. These 'prohibited species catch' (PSC) limits are based on the vessels' historical bycatch

[5]Cooperatives also allowed circumventing the mid-1990s US moratorium on new individual transferable quota (ITQ) systems (Matulich, Sever, & Inaba, 2001; Wyman, 2005).

[6]The other technical precondition – restricting access to the fishery – had already been met by an earlier licence limitation programme for all groundfish fisheries (Kitts & Edwards, 2003; Queirolo & Muse, 2005).

[7]The cooperatives may only form when a contract is signed by 80% of the qualified catcher vessels that delivered the majority of their pollock to the particular processor in the prior year. Note that catcher vessels may choose not to join any of the inshore cooperatives. They then have to operate in the license-limited, open-access fishery for at least one year without the benefit of a guaranteed TAC share but can freely select any processor to deliver to.

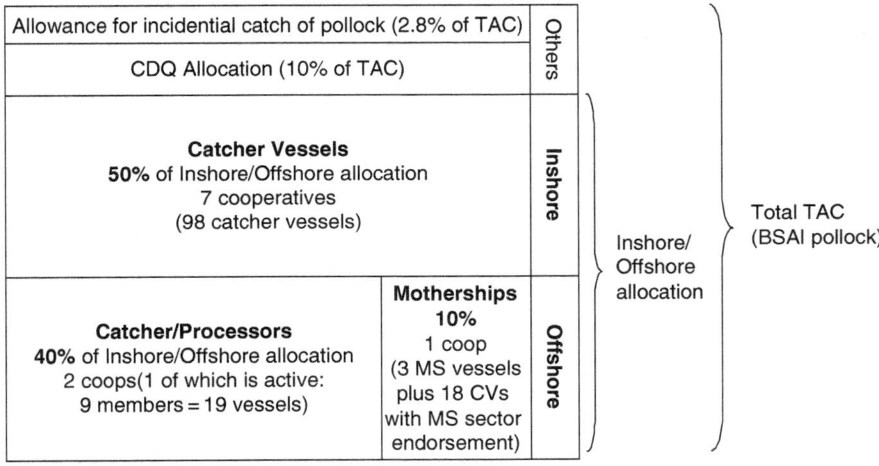

Fig. 4.1 The System of BSAI Pollock Cooperatives (2007/2008)
Source: Authors' interpretation, based on NOAA (2008); Herrmann & Haynie (2007); PCC & HSCC (2007).

of the species in question. Subsequent Council actions also established harvesting and processing restrictions (so called 'sideboards') that AFA-fishermen and processors may not exceed when targeting species other than pollock. Reflecting the AFA vessels' historical level of participation in such fisheries, sideboards are intended to prevent a 'spillover' of capacity into these fisheries and hence to protect the interests of non-AFA fishermen and processors, especially in the Gulf of Alaska.

After the AFA had been passed, the cooperatives formed and participants contractually allocated among themselves percentage shares of the sectors' total allocation. These were typically based on historical catch levels. The catcher/processor cooperatives started cooperative fishing in 1999 and the inshore and mothership cooperatives followed a year later. In 2008, ten cooperatives existed (see Fig. 4.1): two in the catcher/processor sector (only one of which is active, as one coop leased their quota shares to the other); one in the mothership sector and seven in the inshore sector. The economically strongest and politically highest profile association is the 'Pollock Conservation Cooperative' from the catcher/processor sector.

Within the confines of the AFA, the cooperatives set the modalities of their operation and decision-making. This is the realm of 'fisheries self-governance' (Townsend & Shotton, 2008). From among the various fishery participants (vessel operators, crews, processors etc.), the cooperatives are constituted by vessel operators – including native Alaskan investors in some of the companies. They shape the cooperatives' governance and take operational decisions. Shore-based processors form part of the inshore cooperatives and exert considerable influence over these. Decision-making modalities can range from unanimous to majority vote, depending on the issue and each cooperative's arrangements.

The cooperatives basically fulfil three functions: A. allocation and transfer of pollock harvest shares, of sideboards (i.e. harvesting and processing limitations on species other than pollock) and of prohibited species catch apportionments; B. bycatch reduction; and C. monitoring and enforcement.

With regard to the first function, the cooperatives contractually apportion pollock allocations, sideboard and prohibited species catch limits among their members. This happens through annual harvesting plans that are based on membership agreements. Similar to ITQs, company-specific harvesting allocations allow the firms to freely determine when and with how many vessels to catch their share, and to coordinate their efforts. Harvesting allocations can be transferred within the cooperatives and, under specific conditions, between cooperatives of the same sector. In practice, most quota trades take place at the end of the fishing season to pool and 'clean up' the remaining metric tons of TAC. With respect to sideboards, catcher/processors can trade these only within their sector, while the other sectors are allowed to trade across the sectors: an '*Inter*cooperative Agreement' was developed to govern allocation and trades of sideboard (and prohibited species catch) limits between the inshore and mothership cooperatives. These trades are handled by an inter-cooperative manager.

Bycatch management emerged as a second – and initially not foreseen – function of the cooperatives in 2001. All cooperatives concluded an intercooperative 'Salmon Bycatch Management Agreement', renewed in 2006. It contains a voluntary 'rolling hot-spot programme' to avoid bycatch of Chinook (*Oncorhynchus tshawytscha*) and chum salmon (*Oncorhynchus keta*). Salmon bycatch is a crucial issue not only because of the importance of salmon to Alaskan subsistence fisheries, but also because the pollock fishery is closed down once the salmon bycatch cap is exhausted. In the rolling hotspot programme, areas with the highest bycatch rates are identified as 'savings areas' twice a week and are closed to fishing for cooperatives with medium or poor bycatch performance. A weekly ranking of the twenty boats with the highest bycatch record creates peer pressure to reduce bycatch among the highly competitive vessel captains. When squid bycatch reached alarming levels in 2006, a similar agreement was devised for squid bycatch.

The third function of the cooperatives is the monitoring and enforcement of the above tasks. Contractual remedies exist to enforce individual allocations as well as the bycatch agreements.[8] Since enforcement presupposes monitoring of the vessels' fishing activities, the cooperatives' members have committed themselves to release vessel-specific (partly confidential) data such as landing reports and observer data to a third party, the private company 'Sea State Inc'. This enterprise is contracted and paid by the cooperatives. On the basis of vessel-specific information, NMFS Alaska Regional Office data and direct communication with boats, Sea State Inc. can determine in real time whether a vessel – and, in aggregate, a cooperative – exceeds its allocations of pollock, sideboard or prohibited species catch, or violates

[8]In the inshore coops, for instance, the penalty for exceeding one's pollock allocation amounts to 150% of the ex-vessel value of the over-harvest, and fines also exist for sideboard and prohibited species catch overharvests (§ 4b ICA 2006). Violations of salmon savings area closures under the Salmon Bycatch Agreement are penalised with $10,000 for the first, $15,000 for a second and $20,000 for a third and subsequent violations (§ 7b SBMA 2005).

a salmon savings areas closure. This private monitoring system constitutes a form of in-season management, with the aim of maximising the harvest by avoiding closures triggered by too high bycatch levels.

4.4.2 Assessment of Pollock Cooperatives

4.4.2.1 Biological Performance: Indirect Impacts of the Cooperatives on Pollock Stocks and Ecosystems

Has the establishment of cooperatives impacted the pollock stocks and implicitly the ecosystem?

Focussing on the Eastern Bering Sea (EBS) stock as the major stock harvested by the cooperatives, annual catches of the EBS pollock stock have been between 1.0 and 1.5 million metric tons in the past three decades. After more than two decades with the pollock stock estimated to be well within the desired limits, the 2008 predictions for the EBS pollock indicate a stock condition where the Tier System requires a reduction in F (NOAA, 2008). The assessed stock decline cannot be ascribed to the cooperatives, as the TACs have been set at low levels for several years the last decade compared to the maximum ABC recommendations (see Section 4.3.3).

The cooperatives have actually contributed to the setting of low TACs: coop representatives at the Council had repeatedly supported lowering the TAC (MCA, 2007), so that it was set significantly below ABC. The cooperatives' agreement to reduce the pollock TAC may have been a result of several factors. With 1.4–1.5 million metric tons, the TAC for EBS pollock was quite high and stable from 2001 to 2006, contributing to stability for the cooperatives. This means that as a result of the AFA reducing capacity and the high biomass of pollock, catches by the catcher/processor sector were far higher than in the 1990s. The cooperatives' quota shares may also have been an incentive for longer-term planning, letting the remaining fish grow and replenish. Another possible factor, however, is that the cooperatives may have expected a higher TAC to be difficult to catch anyway because of strict bycatch regulations. Furthermore, taking into consideration that the ABC for EBS pollock alone was above the groundfish cap in three of these years, the pollock TAC had to be reduced in order to allow prosecution of other important fisheries managed under the cap.

When it comes to bycatch rates, there do not seem to be any clear trends or changes in the levels of either target bycatch, non-target bycatch, or prohibited species catch in the period of 1997–2006 (NOAA, 2008). Pollock trawlers had been limited to mid-water trawling already before the cooperatives were introduced. This significantly reduced the bycatch rate to a level that makes further bycatch reductions difficult to achieve in the fishery. For some bycatch species, there are quite marked fluctuations in bycatch levels, perhaps due to fluctuating abundance of the bycatch species, changes in fishing areas or random factors. However, the doubling of bycatch levels of Chinook salmon since 2001 as well as increasing bycatch rates of other salmon species in the last few years have been a major concern (NOAA, 2008; Miller, 2008). Although the pollock fishery is considered a clean fishery, in the sense that the bycatch rates are low, bycatch in *absolute* terms can be high (AMCC, 2004). For example, the pollock fishery dominates the bycatch of the prohibited species herring and salmon (Hiatt et al., 2007) and almost half a million Chinook salmon were caught in the BSAI pollock fishery in 2006.

The fact that bycatch levels have increased in the last years points to the limits of the cooperatives' rolling hotspot programme. Although the cooperatives' self-regulation enables the pollock fishery to coordinate problem solving efforts and to react more quickly to unforeseen changes in catches and bycatches, the system did not live up to the expectations of many Council members when the fishery was confronted with conditions that exacerbated bycatch in the last years.[9] As a consequence, the Council has discussed additional bycatch reduction measures (Miller, 2008).

With regard to the impact of the pollock fishery on the ecosystems, competitive interaction between groundfish fisheries and Steller sea lion (*Eumetopias jubatus*) populations was a major issue: between the late 1960s and 2000, the endangered population segment residing in the Aleutian Islands and the central and western Gulf of Alaska declined over 80% in abundance, though non-pup populations in some sub-regions indicated slight increases in the early 2000s (NOAA, 2008). The observation that 'the pollock fishery had disproportionately high seasonal harvest rates within critical habitat that *could* lead to reduced sea lion prey densities' (NOAA, 2008) triggered a number of management measures to temporally and spatially redistribute the fishery. Further concerns on possible secondary effects of the pollock fishery include declines in northern fur seals and harbour seals, which feed on pollock (SCS, 2005); unexplained increases in jellyfish and arrowtooth flounder (ibid); and the effect of pollock mid-water trawls on bottom habitat. These trawls are estimated to be in contact with the seafloor up to 85% of the duration of the tow (Loverich, 2001, cited in Enticknap, 2002), but as the fishery areas do not consist of rocky habitats, coral reefs and sponges are not affected. The ecosystem issues mentioned are not significantly influenced by the pollock fleet's organisation in cooperatives.

In 2005, the Marine Stewardship Council certified Alaskan pollock stocks, their ecosystems and management to be sustainable. The certification was applied for and largely funded by the catcher/processors' trade association whose members at the same time constitute the Pollock Conservation Cooperative. It was motivated by the industry's search for new market opportunities. To what extent an MSC certification induces immediate 'on the water' environmental benefits in a fishery is contentious (Agnew, Grieve, Orr, Parkes & Barker, 2006).[10] However, beyond imposing concrete conditions such as measures regulating fishing in Steller sea lion critical

---

[9]These included large salmon runs and restrictions on areas where pollock fishing may at all occur (NOAA, 2008).

[10]Note that a number of NGOs contested that the fishery warranted a certification in the first place (Marz, 2004).

habitat, a certification can be assumed to create an interest in maintaining a fishery's good condition in future. Industry representatives argue that the cooperative system had enabled the successful certification in the first place: under the previous race for fish it would have been more difficult to fulfil the certification principles and criteria. Also, the different economics would have prevented the industry to deal with the conditions imposed by the certifiers.

The upshot is that the establishment of cooperatives did not particularly affect the fishery's impact on the stock and ecosystems in a *direct* way, though minor improvements were observable in bycatch management. However, *indirect* effects can be assumed: coordination through the cooperatives gives fishery participants more leeway to react to ecosystem challenges. Industry also claims that the fishery's very profitability smoothes resistance to more ecological fisheries regulations. Environmental NGOs, on the other hand, criticise the power of the cooperatives that results from their de facto exclusive rights to the profitable Bering Sea pollock fishery. They fear this powerful industry will resist environmental friendly suggestions in the future, especially should the Eastern Bering Sea pollock stock continue to decline.

4.4.2.2 Economic Performance: Stability and Different Levels of Efficiency Gains

How has formation of the cooperatives impacted on the industry's economic performance? Passage of the AFA allowed for considerable efficiency gains with regard to fleet performance, harvesting operations and above all processing yield. Employment in the fleet decreased as a result of the AFA. Data on crew income effects is not available.

Since 1999, when the first AFA cooperatives started operating, both the real ex-vessel value of the pollock catch and the gross value of pollock products have increased by 80–90%; this is some 10% more than the respective values for other groundfish species (Hiatt, Felthoven & Terry, 2002; Hiatt et al., 2007). The above-average increase of these values is likely to have resulted to some extent from eliminating the race for fish (Hiatt et al., 2007). However, market developments over this period may also have contributed to the general increase in pollock value: the waning of the Asian currency crisis, which had hit the surimi market in 1997; decline of the Russian pollock fishery as prime competitor of the US pollock industry; and an increasing demand for pollock due to the global depletion of cod and other whitefish stocks.

Fleet performance increased due to the removal of marginal vessels from the fishery that followed the AFA-buybacks and the subsequent transfer of quota allocation within cooperatives to the most efficient operators. During the first year of operation of the Pollock Conservation Cooperative, for example, only 14 of the 20 eligible catcher/processors fished, saving the operating costs of vessels, which, in the absence of the cooperative, would have fished (Wilen & Richardson, 2008). It is expected that permanent fleet reductions will be 'of the order of 30% for all three sectors of the industry' (NOAA, 2002). Felthoven (2002) estimates that cooperative fishing induced a fall in total effort (days, duration, crew) of around 30% in 1999 in the catcher/processor sector, although it slightly rose again in 2000.

Other efficiency effects are related to harvesting. These resulted from the spreading of fishing effort in time and space, which became possible through eliminating the concern of losing harvest shares and other AFA-effects (NOAA, 2002). For catcher/processors, seasons expanded from 55 days in 1997 (MacGregor, 2006) to a historical peak of 140 days in 2000 (Felthoven, 2002). For the first year of cooperative fishing within the Pollock Conservation Cooperative, the length of the spring season doubled compared to that of 1998 (Wilen & Richardson, 2008). The slower pace of fishing allowed to use smaller bags for hauls so that the fish were less damaged and of higher quality: in 1999 in this cooperative, 'Catch per haul was 27% lower, the number of hauls per day dropped by 45%' (ibid.). Furthermore, it became possible to target a more specific size range of pollock for fillet or surimi processing, to range further in order to locate higher quality catch, to better time deliveries and to serve different markets (NOAA, 2002). Finally, the TAC uptake increased due to the ability of the cooperatives to jointly 'mop up' residual TAC that would previously have been forgone to avoid TAC overruns (Sullivan, 2000).

With regard to processing operations, resources freed from the race for fish were invested in the re-tooling of boats and new on-board production technology. Many parts of fish previously treated as 'waste' were now fully processed, making possible production of additional lower-grade surimi or mince products. Utilisation rates increased by 2.3% in the inshore sector; 29% in the mothership sector; and 35% in the catcher/processor fleet between 1998 and 2000 (NPFMC, 2002; PCC & HSCC, 2007). While some of this increase resulted from 'Improved Retention/Improved Utilisation' regulations imposed by the Council in 1998, the AFA cooperatives allowed further gains. Felthoven (2002) attributes most of the perceived gains associated with the AFA to processing and the respective increases in product recovery rates and product grades, rather than to increased harvesting efficiency.

Introduction of the cooperative system in a wider sense contributed to the economic stability of the US pollock industry. Since the AFA limited access to the fishery and reduced fishing capacity, no bankruptcies have occurred. The secure harvesting rights, signalling a safe future of the industry, also made access to capital less costly as risk premiums were cut.

Looking at the development of employment and crew income, there have been lay-offs in the BSAI pollock industry. The sector most affected is the catcher/processor fleet where employment losses are estimated to amount to some 1,500 (NPFMC, 2002). These can be attributed to the buy-out of nine vessels, idling of less efficient boats and the sector's reduced harvest allocation. For the remaining sectors, including on-shore processors, employment impacts are much lower and more difficult to measure (ibid). No detailed data are available on how crew income changed as a consequence of cooperative fishing. However, with longer seasons, employment of crews throughout the year has increased. Furthermore, the fact that remuneration is share-based and depends on the value of products (which has increased) may lead us to infer that crew income has been at least stable, if not increasing.

4.4.2.3 Management Costs: Easing the Public Burden

How costly is the pollock management system as operationalised through the cooperatives? Did introduction of the cooperatives lead to a shift of costs between public and private actors? Qualitative data[11] that we gained through interviews and secondary sources suggest that, although the system of cooperatives promotes fisheries self-governance, their design, implementation, and operation required a fair amount of government expense. However, the cooperatives have freed public resources, too, particularly in the operation phase.

With regard to the design of the cooperatives as a management system, public costs were substantial. This was the case although, once the regulatory foundation was laid, setting up the cooperatives incurred private expenses only. The idea of a sector allocation and of cooperatives was first negotiated between major players of the pollock industry and with a small number of political actors in Washington D.C. (Matulich, Sever, & Inaba, 2001; Sullivan, 2000; Wyman, 2005). With the Justice Department's acquiescence and Congressional support, the public hearing process and various (environmental, regulatory impact, etc.) analysis requirements were initially circumvented (Criddle & Macinko, 2000). While this at first reduced transaction costs for the public sector, albeit at the costs of procedural standards, the Council was later required to develop the respective analyses post hoc. It was also charged with developing 'sideboards' to limit the potential for the AFA to have impacts on other sectors and regions. It is hence not clear that the initial policy process in Congress reduced expenditures in the subsequent policy process in the Council. Today, public expenses continue to accrue when regulatory adjustments to the cooperative programme become necessary and the Council carries out the respective analyses, regulatory drafting, review and revision work. Vis-à-vis industry, fisheries managers have not been able to assess cost recovery fees on the AFA cooperatives.

Costs for implementing the cooperative system encompass the buy-out of vessels, the initial quota allocation, as well as adjustments in public fisheries management. Apart from the latter, these costs were largely borne by the pollock industry. The decommissioning and scrapping of nine catcher/processor vessels was the price tag attached to enabling separate sector allocations, and hence the cooperatives, through the AFA. The buy-out was funded through a combination of $20 million in federal appropriations and $75 million in direct loan obligations to the companies in the inshore sector. With this loan, the inshore component compensated the offshore component for its loss in TAC share. The inshore sector is paying

[11] We are limited to qualitative assessments, as quantifying and delimiting the specific costs of the cooperative system from the general costs of pollock management would be extremely difficult if not impossible.

off the loan through a delivery fee of 0.6 cents per pound of pollock harvested. Negotiations of the initial quota allocation within each of the three sectors were carried out exclusively among industry participants, who bore the respective costs. The relatively low number of participants (ca. 100 companies), divided into sectors with fairly homogenous interests, implies that these costs were probably lower that the typically high up-front cost of initial allocations in the case of ITQs (Sutinen & Soboil, 2003). In the public administration, costs accrued for reprogramming catch accounting and adjusting enforcement systems.

Operational costs pertain to the running, monitoring and sanctioning of the cooperatives. Here, significant costs were shifted to industry, although the fisheries administration still meets substantial expenses, too. Major cost items for industry include the publicly prescribed observer coverage – catcher/processors are required to have two observers aboard at all times, which sample 99% of all hauls –, as well as subcontracting the private monitoring agent Sea State Inc. to safeguard the cooperatives' compliance with quota, sideboard, and bycatch limits. The cooperative self-governance system reduces public in-season management tasks related to BSAI pollock, thus freeing public resources for other purposes. Governmental agencies supervise a yearly application procedure of inshore cooperatives, are still engaged in catch accounting and remain reponsible for ultimate enforcement. However, with regard to enforcement, NOAA's Office for Law Enforcement judges the AFA cooperative system to have substantially reduced their enforcement burdens. This is because controls now relate to cooperatives rather than to individual companies. As the cooperatives' members can even out catch and bycatch rates amongst themselves the potential for violations is reduced and individual vessels' day-to-day activities become less important. In addition, the cooperatives themselves sanction offending members.

4.4.2.4 Stakeholder Acceptance: Insider Versus Outsider Attitudes

How well is the cooperative system accepted by Alaskan fishery stakeholders? To answer this question, we differentiate between groups of stakeholders that we expect to have different material interests or normative orientations. While we find that members of the pollock cooperatives themselves (i.e. the 'insiders') are highly supportive of the system, other stakeholders "outsiders" are less so: industry stakeholders are wary of the economic and political power that is feared to go along with the property rights created through the AFA, and non-industry stakeholders criticise the fleets' ecological and community impact. In the following, we describe the different stakeholders' views and actions as they reflect varying degrees of regime acceptance, but we do not assess their 'truth' or legitimacy.

Among the cooperative members (i.e., AFA vessel owners), the satisfaction with the regime is high: 'We are really firm believers in the coop system here', as a participant of the Pollock Conservation Cooperative told us. Not only did the vessel owners' political deal set an end to the cumbersome allocation conflicts between the inshore and offshore sector, but also it increased the industry's security of expectation and facilitated long-term planning. It also boosted its economic performance – albeit to differing degrees in the three industry components – in a way that over-compensated the offshore sector's loss of TAC shares as compared to the pre-1998 situation. Also, Alaskan communities, participating in the Community Development Quota (CDQ) Programme, profited from the new wealth of the industry: they leased out the – historically unique – share of pollock TAC allotted to them and some became shareholders in AFA listed companies. So far, these benefits coincided with abundant pollock stocks. Not least, according to a participant 'the beauty of the coop[erative] system is that people are managing themselves', so that the sense of ownership within the industry has increased. One of the few sore points is that catcher vessels in the inshore sector cannot easily switch cooperatives. While this provides security to shore-based processors, independent boat owners complain that it prevents them from negotiating the best price for their catch (Loy, 2000). Also, some crewmembers disapproved of longer fishing seasons with more trips. Further discontent with the coop system at crew level interestingly results from 'cultural' reasons: a number of crewmembers were said to miss the old race for fish. Competitive harvesting was seen as a fulfilment of the 'American Dream', with experienced captains taking credit for tracking down maximum amounts of fish in the shortest possible time.

However, the overall attitude of the pollock industry towards the cooperatives is positive. This is reflected first and foremost in the functioning of the system. Even in the inshore sector with its more precarious balance between harvesters and processors the system has worked out so far, as shown by the successful creation of cooperatives each year. The system's acceptance by the cooperatives' members also manifests itself in a high level of compliance with both cooperative and public rules, although other factors are likely to have contributed to this positive record as well.[12] No substantive violations have been reported with regard to catch, bycatch and sideboard limits as allocated in the cooperatives' membership agreements (see e.g. PCC & HSCC, 2007) and the Intercooperative Agreement. Infringements of the salmon bycatch programme are said to have been minor and 'confusion-based'. Since operation of the cooperatives, the fishery has never had to be closed down prematurely. This positive storyline was confirmed by public law enforcement records, which gave evidence of declining instances of infringements (NMFS, 2006). Acceptance of the cooperative system by its participants is finally reflected in a low level of conflicts within or between the cooperatives – a factor that is fostered by the small number of operators and the homogeneity of their interests.

When it comes to other industry stakeholders, key informants related a scepticism of non-pollock fishermen operating off Alaska vis-à-vis the new wealth of what is nicknamed the 'big pollock guys', created through the AFA's secure property rights to a vast resource. Many suspect that AFA vessel-owners could use their revenues or better disposable time to invest and compete in other fisheries. Such

[12]For example, the comparative 'simplicity' of the pollock fishery as a single species fishery relatively clean with regard to bycatch; its high profitability which reduces incentives to behave illegally; and the coops' own sanctions.

'Dollar spillover' can occur despite the sideboard limits. Indeed, anecdotal evidence exists of such competition, among others in parts of the open access Gulf of Alaska fisheries and in the BSAI cod fishery (NPFMC, 2002). Also, windfall profits like those of a group of catcher boats that leased most of their pollock shares to the catcher/processors are frowned upon (Loy, 2000). At the time when the AFA was being devised, a feeling of disfranchisement even prevailed among a number of (medium sized) mixed species 'Head & Gut' trawlers that had harvested minor amounts of pollock, and especially among two companies with somewhat bigger pollock catch histories. Being rather small as compared to those of the larger pollock-only catcher/processors, the vessels were not made eligible in the AFA deal and hence could not participate in the fixed allocation system. The non-eligible companies still contest this outcome of what they feel was an unfair and non-transparent backroom decision process. A subsequent offer to be grandfathered in to the system was declined by individual companies.

More generally, the pollock cooperatives' new wealth is feared to go along with increased political power, both within the institutional structure of the Council and in Congress. While the Council formally has the prime responsibility for the management of federal fisheries off Alaska, the pollock industry by addressing Congress in 1998 had gone 'forum shopping' to assert their interest in another arena. This precedent has since been repeated by other industry groups (though with less success), a move which by some is felt to potentially undermine the Council's authority and devalue its participatory management approach. Also, less well-off fishing industries feel compelled to equally invest in expensive lobbying in Washington DC. Tight relations between the pollock industry and senior Alaskan politicians have sparked some debate about conflicting interests.

Looking at the cooperative system's acceptance by non-industry stakeholders, we will focus on environmental organisations. Generally, these groups have not strongly commented on the AFA, being 'a lot more engaged in how much fish should be taken, and how and where it should be taken, than *by whom* it should be taken', as an environmentalist put it. In this vein, they have not taken any action against the cooperatives as such, though a range of aspects of pollock management have been fought, including legally and through objection procedures in the Marine Stewardship Council certifications of pollock fisheries. With the expansion of rights-based management in the fisheries off Alaska, there was however some discussion of cooperatives and ITQs among the NGOs. A split exists between those that consider rationalisation as a means to reduce fishing capacity and those that reject exclusive rights to a common pool resource, among others because it can lead to the consolidation of fleets. It was generally welcomed that the 2006 reauthorisation of the Magnuson-Stevens Act set out guidelines for 'Limited Access Privilege Programs' (MFCN, 2006), and defined these privileges as *revocable*. With regard to the pollock industry in a wider perspective, a number of marine conservation NGOs oppose large-scale industrial fishing as such, for which the Bering Sea pollock fleet stands as a stark example (Stump et al., 2006). In 1998, Greenpeace used non-violent action to prevent the fleet from leaving Seattle for Alaska. More generally, the NGOs are worried about conflicts of interest in the participatory management structure of the

Council[13] and demand a better representation of non-fishing public interests and native Alaskan communities in its voting positions (Soule, 2008; Eagle, Newkirk, & Thompson, 2003).

4.5 Conclusions

We analysed two fisheries management innovations implemented in Alaska – the Tier System as a set of harvest control rules, and the pollock cooperatives as a combination of rights-based management and fisheries self-governance. Assessing the success of these regimes in terms of biological robustness, economic performance of the fleet, management costs and stakeholder acceptance, we find that their major effects are complementary: the Tier System creates the basis for abundant fish stocks, and the cooperative system creates means for making the fishery highly profitable. We identified downsides to the regime as well, such as possible weaknesses in the Tier System's design and use, and stakeholder criticism of the pollock cooperatives. In the following, we sum up the regimes' main impacts and discuss factors that have contributed to these (positive or negative) impacts.

The study of the Tier System and the TAC-setting process revealed that its strengths lie in fostering sustainable harvesting and trust in science. Objectives for most stocks are met and levels of F are rather precautionary. In our opinion, the reason for this success is a combination of factors. The Tier System is designed to provide precautionary advice. It includes recovery strategies for depleted or declining stocks and prevents new fisheries from developing unless there is sufficient data. However, the effectiveness of any harvest control rule hinges on management decisions and compliance with the rules. In the Alaska case, a fishery is effectively closed when the TAC limit is reached, whether it is for a target or a regulated bycatch species. The real-time in-season management and the extensive observer programme prevent compliance problems and provide the data necessary to manage the fisheries effectively. Thus, in reality the Tier System is put into effect where the actual catch levels are lower than the advised levels.

Historical circumstances were a significant precondition for establishing the rather precautionary overall management system. When the Tier System, its accompanying bycatch regulations, the groundfish cap and the observer programme were first introduced, they were mostly pertaining to foreign fishermen, resulting in little resistance in the regional political processes. Our respondents also pointed to lessons learnt from the collapsed fisheries on the east coast of the USA and the early Alaska success of rebuilding a salmon stock due to science-based management. However, the Tier System and the TAC-setting process themselves must be regarded as part of the success since trust in science and faith in precautionary management have persisted. The design of the Tier System is transparent, as are the criteria for choosing

[13]For example, Section 302 (j) of the Magnuson-Stevens-Act exempts voting members of Regional Fisheries Management Councils from specific conflict-of-interest provisions.

tiers, and the system applies to all groundfish and even some non-targeted species. TAC decisions are based on a participatory process where stakeholders are represented in the advisory panel of the Council and in the Council itself. In addition, public testimonies are invited at different stages of the TAC-setting process.

While our interviews conveyed mutual trust among the industry, managers and scientists, representatives from the environmental organisations were less content with the Tier System and the TAC-setting process. They would prefer greater consideration being paid to reducing the ecosystem impacts of fishing. The groups' deviating views and their lower level of trust in the system are probably caused by underlying differences in basic values and the NGO's less significant role in the TAC-setting and decision-making process. Devoid of voting representation in the Council, they have more of an outsider role, trying to influence the decision-making process mostly through public testimony at the meetings. However, their victories in court cases against the fisheries administration show that they are a rather influential player in fisheries management. Indeed, the poor integration of green groups into the participatory management system of the regional councils and the resulting law suits were general features of US fisheries management in the 1990s and became a major force for changing the system overall.

The case study of cooperatives in the Bering Sea pollock fishery showed that in terms of biological performance the establishment of cooperatives did not, to a large extent, affect the fishery's impact on the stock and ecosystems in a *direct* way. The relatively favourable condition of the BSAI pollock stocks and their ecosystems is probably a product of conservative measures provided by public elements of the system, such as the caps, bycatch regulations and observer requirements. However, *indirect* effects can be assumed: coordination through the cooperatives gives fishery participants more leeway to react to ecosystem challenges. Industry also claims that the fishery's very profitability smoothes resistance to more ecological fisheries regulations. Environmental NGOs, on the other hand, criticise the power of the cooperatives that results from their *de facto* exclusive rights to the pollock fishery and fear resistance against sustainable management by this powerful industry should the Eastern Bering Sea pollock stock continue to decline. With regard to economic performance, the stable allocations and quota share system of the cooperatives have undoubtedly created efficiency gains, mostly but not exclusively with regard to processing. In accordance with economic theory, this success can be attributed to gains of cooperation. The cooperative system, although it promotes the self-governance of industry, has caused substantial public expense. This holds above all for the intricate policy processes related to the initial design and subsequent changes of the system. Costs of implementation and operation were to some extent shifted from the public to the private sector. The pollock industry was willing to pay this price for receiving exclusive harvesting privileges as a means to boost their profits. Note that operational costs cannot fully be rolled over to fishery participants as the ultimate responsibility to enforce the system rests with governmental agencies. The overall benefits and the high acceptance of the system by the cooperatives' members themselves are somewhat counterbalanced by a certain distrust, and maybe even a sense of disenfranchisement, of industry and non-industry stakeholders that are not cooperative participants. The scepticism is fostered by the fact that reallocating rights creates winners and losers (or at least non-winners), often in long-term 'trajectories', and involves cultural issues. In this particular case, stakeholder acceptance also suffered because the decision process lacked transparency and participation. This is where the system of pollock cooperatives could have learned from the experiences with the Tier System: it is a regime's transparency and inclusive process that warrants its long-term acceptance and functioning.

Acknowledgments We would like to thank most warmly all our respondents for kindly agreeing to be interviewed by us, and we are especially grateful for the helpful comments by David Fluharty, Jay Ginter, Sally Bibb, Glenn Merrill, Jessica Gharrett, James Ianelli and William Karp on an early version of the text. The help of our reviewers Tammo Bult, Anne McLay and Douglas Wilson is also greatly appreciated. Finally, our thanks go to Norma Schönherr for copy-editing the text.

References

Agnew, D., Grieve, C., Orr, P., Parkes, G., & Barker, N. (2006). *Environmental benefits resulting from certification against MSC's Principles and Criteria for Sustainable Fishing*. London: MRAG UK Ltd and Marine Stewardship Council.

AMCC. (2004). *By-Catch: Wasting Alaska's future* (2nd ed.). Alaska Marine Conservation Council. Retrieved September 4, 2006, from http://www.akmarine.org/publications/Bycatch-WastingAKFuture2004.pdf

Criddle, K. R., & Macinko, S. (2000). A requiem for the IFQ in US fisheries? *Marine Policy*, *24*(6), 461–469.

Criddle, K. R. (2008). The legal context of US fisheries management and the evolution of rights-based management in Alaska. In R. Townsend, R. Shotton, & H. Uchida (Eds.), *Case studies in fisheries self-governance*. FAO Fisheries Technical Paper 504 (pp. 369–381). Rome: FAO.

Eagle, J., Newkirk, S., & Thompson, B. (2003). *Taking stock of the regional fisheries management councils*. Washington, DC: Island Press.

Enticknap, B. (2002). *Understanding the effects of bottom trawl fisheries on Alaska's living seafloor*. Report on Trawling the North Pacific. Alaska Marine Conservation Council.

Felthoven, R. G. (2002). Effects of the American fisheries act on capacity, utilization and technical efficiency. *Marine Resource Economics*, *17*(3), 181–205.

Hiatt, T., Felthoven, R., & Terry, J. (2002). *Stock assessment and fishery evaluation report for the groundfish fisheries of the gulf of Alaska and Bering Sea/Aleutian Islands area: Economic status of the groundfish fisheries off Alaska, 2001*. Retrieved November 12, 2008, from http://www.afsc.noaa.gov/refm/stocks/Historic_Assess.htm

Hiatt, T., Felthoven, R., Dalton, M., Garber-Yonts, B., Haynie, A., Herrmann, K., et al. (2007). *Stock assessment and fishery evaluation report for the groundfish fisheries of the gulf of Alaska and Bering Sea/Aleutian Islands area: Economic status of the groundfish fisheries off Alaska, 2006*. Retrieved November 12, 2008, from http://www.afsc.noaa.gov/refm/stocks/Historic_Assess.htm

Holland, D. S., & Ginter, J. J. C. (2001). Common property institutions in the Alaskan groundfish fisheries. *Marine Policy*, *25*, 33–42.

Kitts, A. W., & Edwards, S. F. (2003). Cooperatives in US Fisheries: Realizing the Potential of the Fishermen's Collective Marketing Act, 27. *Marine Policy, 27*(5), 357–366.

Loy, L. (2000). Race for pollock ends: Co-op ends the competitive Bering Sea fishery, improves safety and efficiency, but not everybody is pleased with the new era. *Anchorage Daily News Sunday*, July 30.

MacGregor, P.(2006, March). *Overfishing and overcapitalization in large scale fisheries: The Bering Sea Pollock fishery 1964–2006, A Case Study.* Paper presented at Pacific Rim Fisheries Conference, Hanoi/Viet Nam.

Marz, S. (2004). Objection to final determination to certify the Eastern Bering Sea/Aleutian Islands Pollock fisheries. Submitted on behalf of Alaska Oceans Program, Greenpeace International, National Environmental Trust, Oceana. In: *Decision to allow or dismiss a further objection on the Gulf of Alaska Pollock fishery determination* (pp. 26–53). Marine Stewardship Council. Retrieved November 12, 2008, from www.msc.org/track-a-fishery/certified/pacific/gulf-of-alaska-pollock/assessment-downloads-1

Marz, S., & Stump, K. (2002). *Concerns with the Alaska Pollock Fisheries Regarding the Marine Stewardship Council Sustainability Certification Review.* Anchorage, AK: Trustees for Alaska.

Matulich, S. C., Sever, M., & Inaba, F. (2001). Fishery Cooperatives as Alternatives to ITQs: Implications of the American Fisheries Act. *Marine Resource Economics, 16*, 1–16.

MCA. (2007). MCA Supports Scientific Approach to Setting Catch Levels – Backs Council Decision to Reduce Pollock Catch during Cyclic Downturn. Marine Conservation Alliance. Retrieved September 25, 2008, from http://www.marineconservationalliance.org/press/pr20071210.pdf

MFCN. (2006). Magnuson-Stevens Act Reauthorized! *Network News*, Marine Fish Conservation Network, December 27, 2006. Retrieved November 12, 2008, from http://www.fryingpantower.com/article3546.html

Miller, S. A. (2008). Regulatory Impact Review of Measures to Reduce Chinook Salmon Bycatch in the Bering Sea and Aleutian Islands Pollock Trawl Fisheries. *Initial review draft of the Draft Environmental Impact Statement*. NPFMC.

NMFS. (2006). *Enforcement Report to the North Pacific Fishery Management Council*, January 1–December 31, 2005. National Marine Fisheries Service Alaska Enforcement Division, Juneau.

NMFS. (2007). *Fisheries of the United States 2006.* National Marine Fisheries Service. Retrieved September 25, 2008, from http://www.st.nmfs.noaa.gov/st1/fus/fus06/fus_2006.pdf

NOAA. (2002). *Final Environmental Impact Statement for American Fisheries Act Amendments 61/61/13/8, Executive Summary (ES) and Chapters I-IIX.* Retrieved September 25, 2008, from http://www.fakr.noaa.gov/sustainablefisheries/afa/afa_sf.htm

NOAA. (2008). *2007 North Pacific Groundfish Stock Assessment and Fishery Evaluation Reports for 2008.* Retrieved September 25, 2008, from http://www.afsc.noaa.gov/refm/stocks/assessments.htm

NPFMC. (2002). *Impacts of the American Fisheries Act, Report to the U.S. congress and the Secretary of Commerce.* Retrieved September 25, 2008, from http://www.fakr.noaa.gov/npfmc/summary_reports/AFACongress202.pdf

NPFMC. (2006). *Celebrating 30 years of sustainable Fisheries.* North Pacific Fishery Management Council, Anchorage, Alaska.

NPFMC. (2008a). *Fishery Management Plan for Groundfish of the Bering Sea and Aleutian Islands Management Area.* Retrieved from September 25, 2008, http://www.fakr.noaa.gov/npfmc/fmp/fmp.htm

NPFMC. (2008b). *Fishery Management Plan for Groundfish of the Gulf of Alaska.* Retrieved September 25, 2008, from http://www.fakr.noaa.gov/npfmc/fmp/fmp.htm

NRC. (1999). *The Community Development Quota Program in Alaska.* Committee to Review the Community Development Quota Program, National Research Council. Washington, DC: National Academy Press.

PCC & HSCC. (2007). *Pollock Conservation Cooperative and High Seas Catchers' Cooperative Final Joint Annual Report 2006.* Retrieved September 25, 2008, from http://www.atsea.org/images/PCC-HSCCAnnRept1-31-07.pdf

Queirolo, L. E., & Muse, B. (2005, July). *Evolving approaches to rights-based fisheries management off Alaska.* Paper presented at the Conference "People and the Sea", Amsterdam.

SCS. (2005). *MSC assessment report for the United States Bering Sea and Aleutian Islands pollock fishery* (pp. 754–844). Retrieved November 20, 2006, from http://www.msc.org/track-a-fishery/certified/pacific/bsai-pollock/assessment-downloads-1/BSAI_Pollock_Final_Rpt_15Feb05.pdf

Soule, B. (2008). Quantifying voting records in conservation decisions in the U.S. Regional Fisheries Management Councils. Presentation held at the Fisheries and Marine Ecosystems, 2008 Graduate Student Conference, Port Angeles, WA, April 11–13, 2008.

Stump, K., Hocevar, J., Baumann, B., & Marz, S. (2006). *Rethinking sustainability: A new paradigm for fisheries management.* Retrieved September 25, 2008, from www.greenpeaceusa.org/rethinking

Sullivan, J. M. (2000, July). *Harvesting Cooperatives and U.S. Antitrust Law – Recent Developments and Implications.* Paper presented at the International Institute of Fisheries Economics and Trade Conference, Corvallis, OR.

Sutinen, J. G., & Soboil, M. (2003). The performance of fisheries management systems and the ecosystem challenge. In M. Sinclair & G. Valdimarsson (Eds.), *Responsible Fisheries in the Marine Ecosystem*. Rome: FAO.

Sylvia, G., Mann, H. M., & Pugmire, C. (2008). Achievements of the Pacific whiting conservation cooperative: rational collaboration in a sea of irrational competition. In R. Townsend, R. Shotton, & H. Uchida (Eds.), *Case studies in fisheries self-governance* (pp. 425–440). FAO Fisheries Technical Paper 504. Rome: FAO.

Thompson, G. G. (1999). Optimizing harvest control rules in the presence of natural variability and parameter uncertainty. *Proceedings, 5th NMFS NSAW* (pp. 124–145). NOAA Tecn. Memo. NMFS-F/SPO-40.

Townsend, R., & Shotton, R. (2008). Fisheries self-governance: New directions in fisheries management. In R. Townsend, R. Shotton, & H. Uchida (Eds.), *Case studies in fisheries self-governance* (pp. 1–20). Rome: FAO Fisheries Technical Paper 504.

Wilen, J. E. (1998). Alaska fisheries management: A case study of power and politics. In H. N. Scheiber (Ed.), *Emerging issues in National Ocean and coastal policy* (45 pp). Berkeley: University of California.

Wilen, J. E., & Richardson, E. J. (2008). Rent generation in the Alaskan pollock conservation cooperative. In R. Townsend, R. Shotton, & H. Uchida (Eds.), *Case studies in fisheries self-governance* (pp. 361–368). Rome: FAO Fisheries Technical Paper 504.

Wyman, K. M. (2005). From fur to fish: Reconsidering the evolution of private property. *New York University Law Review, 80*(1), 117–240.

Chapter 5
The Icelandic ITQ System

Anne-Sofie Christensen, Troels Jacob Hegland, and Geir Oddsson

Abstract The fisheries sector is tremendously important for Iceland: the export of fish products accounts for a large part of the value of exported goods. Fisheries policy in Iceland is, consequently, of national importance to a degree that is not comparable to any of the EU member states. Demersal fish species (including cod (*Gadus morhua*), haddock (*Melanogrammus aeglefinus*), saithe (*Pollachius virens*), redfish (*Sebastes* spp.) and Greenland halibut (*Reinhardtius hippoglossoides*)), flatfish and shellfish constitute almost 80% of the value of landings even though around 70% of the total volume of landings is constituted by pelagic species. Cod, which is mainly caught in the Icelanders' own exclusive economic zone, is the economically most important fish.

The aim of this chapter is to evaluate the Icelandic individual transferable quota shares system with its management innovations, e.g. harvest control rule for cod, cod equivalents, temporary closed areas, community quotas and features for regulation of quota concentration. The evaluation considers four possible fisheries management objectives, namely biological robustness, cost-effectiveness of management, economic efficiency, and social robustness. In order to make this evaluation, a thorough understanding of the past and present situation on Iceland has to be established. The chapter is based on two sources of information: desk studies and a field study trip.

Keywords Cod equivalent · Harvest control rules · Individual transferable quotas · Participatory governance · Temporarily closed areas

5.1 Introduction

The fisheries sector is tremendously important for Iceland, which is – despite a modest population of little more than 300,000 – among the largest seafood-producing nations in the world measured in terms of volume of catch as well as in value. Since

A.-S. Christensen (✉)
Institute for Innovative Management (IFM) – An Aalborg University Research Centre, North Sea Centre, Willemoesvej 2, Hirtshtals, Denmark
e-mail: asc@ifm.dk

K.H. Hauge, D.C. Wilson (eds.), *Comparative Evaluations of Innovative Fisheries Management*, DOI 10.1007/978-90-481-2663-7_5,

the mid-1990s, fishing and processing have represented in the neighbourhood of 10% of the overall Icelandic gross domestic product (GDP) and the export of fish products accounts for well over half of the value of exported goods. The majority of the export of fish products goes to European Union (EU) member states and the biggest importer of Icelandic fish products is the UK (Ministry of Fisheries, 2005a). Thus, the state of the Icelandic fisheries sector strongly influences the overall state of the Icelandic economy.

Fisheries policy in Iceland is consequently of national importance to a degree which is not seen in any of the EU member states where the fisheries sectors in comparison seem insignificant: *Due to the size, scope and importance of fisheries in Iceland, policy formulation and decision-making on marine issues has far-reaching effect on the standard of living.* (Ministry for the Environment, Ministry of fisheries, & ministry for foreign affairs, 2004, p. 4). The Icelandic emphasis on national jurisdiction over fish resources has long roots and includes dramatic incidents like the so-called Cod War(s) with the UK. Iceland has never applied for membership of the EU, not least due to an unwillingness to accept the Common Fisheries Policy (CFP) of the EU, which has been perceived as severely flawed. However, also the history of disputes over access of foreign vessels to Icelandic waters is probably an explanatory factor in relation to the decision to stay outside the EU. Sharing the responsibility for managing Icelandic fish stocks with the EU member states has in any case not been considered an attractive option – no matter the performance of the CFP.

Demersal fish species (including cod (*Gadus morhua*), haddock (*Melanogrammus aeglefinus*), saithe (*Pollachius virens*), redfish (*Sebastes*spp.) and Greenland halibut (*Reinhardtius hippoglossoides*)), flatfish and shellfish constitute almost 80% of the value of landings, even though around 70% of the total volume of landings is constituted by pelagic species. Cod, which is mainly caught in the Icelanders' own exclusive economic zone (EEZ), is the economically most important species (Ministry of Fisheries, 2005a). In terms of value, most of the Icelandic fish are caught in Iceland's own, highly productive waters, but the share caught outside own EEZ has been increasing in recent years (Ministry for the Environment et al., 2004).

The aim of this chapter is to discuss the performance of the Icelandic individual transferable quota shares system – in short simply *the ITQ system* – with regard to possible fisheries management objectives, in particular biological and social robustness. In order to do this, we will initially provide an understanding of the past and present situation in Iceland in relation to fisheries management.

5.2 Research Methods

The chapter is based on two sources of information: desk studies and a field study trip. The desk studies, including review of literature and web pages, were conducted both before the field study trip to get acquainted with the general situation and after returning from the field to check up on data etc. During the field study, 19 qualitative

Table 5.1 Profiles of Interviewees

	Government	Industry	Research	Green	Total
Biologist	2	1	1	1	5
Economist/Law	4		4	1	9
Anthropologist/ social scientist			3		3
Fisheries candidate		1			1
Fisherman		1			1
Total	6	3	8	2	19

interviews were conducted with people with hands on knowledge about the Icelandic ITQ system; see Table 5.1 for profiles of the interviewees.

The interviews focused on the performance of the Icelandic fisheries management system in relation to different fisheries management objectives: economic efficiency, social and biological robustness. The individual interviews often favoured one or two of the perspectives depending on the person being interviewed. The interviews also covered the context of the ITQ system, namely the history and development of the ITQ system, the changes in costs of and benefits from fisheries management operations, the indicators used to monitor and improve on outcomes, the perceived best practices in implementing, monitoring and enforcing the system, and the resulting management measures, etc.

5.3 Perspectives on the Icelandic ITQ System

The research uncovered several debates relating to the performance and effects of the Icelandic fisheries management system. The debates seem at an overall level to be related to two different perspectives on the ITQ system in Iceland. On one side is a perspective that focuses on the positive effects of the system. The literature with this perspective looks mainly at increases in economic efficiency and to some extent at the story in relation to conservation, which seems to have turned out less negative than anticipated by some sceptics. On the other side is a perspective, which is more sceptical about the system – or at least focuses on the negative aspects. The literature representing this perspective focuses largely on the distributional effects of the ITQ-system and the issue of 'fairness' from different angles.

The existence of the two opposing perspectives can be understood with the idea of conflicting fisheries worldviews. Charles (1992, p. 379) argued that *conflict can often best be understood as rising from natural tensions between three differing fishery paradigms (or 'world views'), each based on a different set of policy objectives.* Charles (1992) identified the three paradigms to be: conservation, which focuses on the policy objective of conservation in the sense of resource maintenance; rationalisation, which focuses on economic performance in the sense of productivity;

and social/community, which focuses on community welfare in the sense of equity. Without discussing which of the two perspectives on Iceland that has 'the best case', our impression from both desk studies and interviews is that the perspective focusing on the virtues of the Icelandic ITQ system seems to have become increasingly dominant over time – perhaps as a result of the institutionalisation of the general elements of the system.

5.4 A Brief History of the Management System

The Icelandic fisheries management system, of which the cornerstone is the Fisheries Management Act no. 116/2006 (previously no. 38/1990) (Ministry of Fisheries, 2005b), is based on a system of individual transferable quota shares. Iceland extended its EEZ to 200 nautical miles (nm) in 1975 and the current ITQ system, which in essence has remained the same since the beginning of the 1990s, evolved from an initial individual vessel quota (IVQ) system, which took effect for the first time in 1984.

Since 1984 the most important Icelandic ground fish fisheries have been managed by means of IVQs and subsequently ITQs. The IVQs were distributed on the basis of historical catches in the period from 1981 to 1983. Initially the IVQ system was only adopted for one year, 1984; however, the system was subsequently reinstated for one year (with minor changes in allocations between vessel categories) and afterwards extended for two more – still time limited – periods, which resulted in the system running throughout 1990. In connection with the last extension, quota transferability was furthermore increased. In parallel it was decided after the first year of the IVQ system to create an alternative, optional system of effort quotas. This system persisted more or less unchanged together with the IVQ system until 1990 with the adoption of the Fisheries Management Act (Gudmundsson et al., 2004; Ministry of Fisheries, 2005b).

The Fisheries Management Act entered into force 1 January 1990 for the fishing year 1990/91 (the fishing year starts 1 September and finishes 31 August the next year.). The act extended the IVQ system without time limits and made quota shares divisible and transferable, although with certain restrictions – effectively converting the IVQ system into one of the most comprehensive ITQ systems in the world. Of more notable exemptions it can be mentioned that the smallest vessels, less than 10 gross registered tonnes (GRT), from the outset had the possibility to stay outside the IVQ/ITQ system. Moreover, the management set-up for the smaller vessels was originally not very restrictive, which in the 1980s resulted in a massive build-up of capacity of smaller vessels (see Figs. 5.3 and 5.4). As a consequence, the management set-up for the smallest vessels was over the years made increasingly restrictive by setting effort limits and limiting the types of gear to be used if the vessel wished to remain outside the ITQ system etc. Moreover, over time more and more of the smaller vessels have, as a result of changes in the management set-up, been included in the ITQ system (Arnason, 1996; Gudmundsson et al., 2004).

Table 5.2 Main Icelandic Fisheries Policy Developments (Gudmundsson et al., 2004; Ministry of Fisheries, 2005b)

1975	The Icelandic EEZ is finally extended to 200 nm after a series of expansions of the EEZ in the previous decades. IVQs are established in the herring fisheries.
1984	A system of IVQs is applied for the cod fisheries in 1984. The system is in the following years changed and expanded.
1991	The Fisheries Management Act (no. 38/1990) enters into force: Catch quotas without time limitations become divisible and fully transferable as from 1 January 1991 – effectively introducing an ITQ system.
1995	A harvest control rule for the Icelandic cod stock is adopted.
2004	A resource fee on quota holders is introduced.

The ITQ system entails in short that the Ministry of Fisheries sets a total allowable catch (TAC) for individual species after having received advice from the Icelandic Marine Research Institute. The TAC for each species is subsequently divided among those holding rights to catch a share of the species in question. For most species the Minister is not obliged to follow the advice from the Marine Research Institute, although in reality this is often the case. Notably, for cod, the most important species to the Icelanders, a harvest control rule (HCR) was introduced in 1996, which means that the TAC for cod is in principle a direct calculation based on the advice from the Marine Research Institute. A further addition to the system is a moderate resource fee (9.5% of the gross profit when fully implemented) imposed on quota holders from September 2004 to collect from the fisheries (Ministry of Fisheries, 2005b) (Table 5.2).

5.5 Biological Robustness/State of Stocks

An overview of the general trends for the commercially important species in Icelandic waters shows that there have been significant changes in abundance over the last few years. The following species have experienced a positive development: haddock, saithe, ling (*Molva molva*), tusk (*Brosme brosme*), flatfish stocks, monkfish (*Squatina squatina*), wolffish (*Alepisaurus ferox*), nephrops, herring (*Clupea harengus harengus*), capelin (*Mallotus villosus*), and redfish. The following species have experienced a negative development: cod, halibut (*Hippoglossus hippoglossus*), Greenland halibut, shrimp, and scallop. The development has to some extent been attributed to environmental changes, mostly higher ocean temperatures, in the waters around Iceland (personal communication with a representative from the Marine Research Institute).

The TACs in Fig. 5.1 reflect a quite divergent trend for four important commercial species in Icelandic waters. Haddock has been increasing greatly over the last few years and offshore shrimp has declined by almost 90% from the high in 1997/98. Cod has been declining more or less continuously from 1987, while herring has been continuously increasing over the same time period.

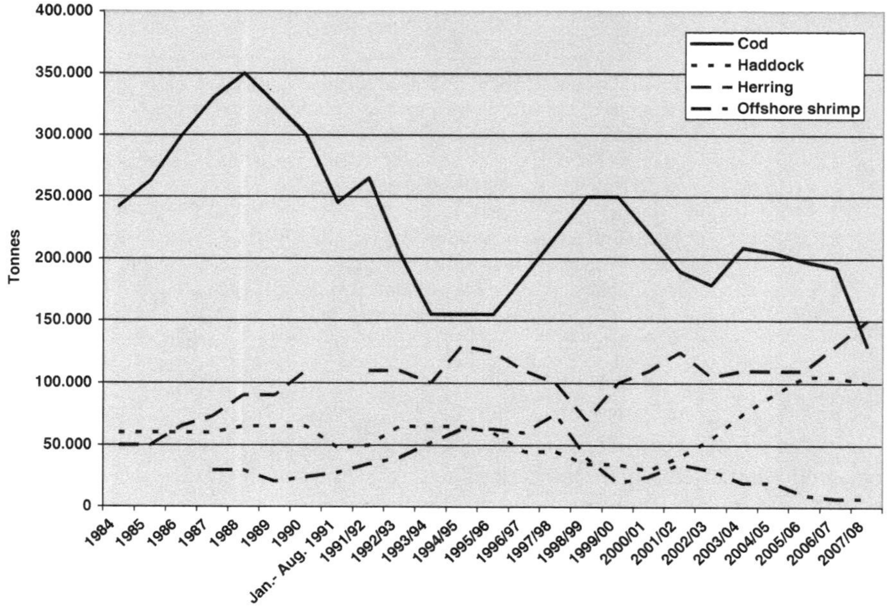

Fig. 5.1 TACs for Cod, Haddock, Herring and Offshore Shrimp from 1984 to 2007/08 (Data Obtained from Directorate of Fisheries Through Personal Correspondence)

The change in the fisheries management regime into the current ITQ system, first reflected in the Fisheries Management Act of 1983, was largely a response to the declining recruitment of cod in Icelandic waters and the inability of the existing effort limitation management system to address the decline. The declining recruitment resulted in a great decline in the fishable biomass of cod and highly diminished catches. The stock (4 years and older) had grown from 844,000 tonnes in 1973 to 1,500,000 tonnes in 1980 (e.g. Marine Research Institute, 2007). In 1983, however, the stock was down below 800,000 tonnes and drastic measures were deemed necessary to arrest this development. Despite great hope for the rebuilding of the cod stock and some apparent success in the 1980s the cod stock was down to 550,000 tonnes in 1995. The status of the stock was reflected in catches, which went from being 392,000 tonnes of cod in 1987 to 187,000 tonnes in 1994/5 (Marine Research Institute, 2004; 2005; 2006; 2007, see Fig. 5.1).

The State of Marine Stocks in Icelandic Waters 2006/2007 report by the Marine Research Institute recommended a serious reduction in cod TACs for the fishing year 2007/2008, but at the same time a considerable increase in haddock TACs. The reduction in the cod TAC was instigated by the continued poor recruitment of the cod stock and the increased risk of collapse of this most important stock in Icelandic waters. The Ministry of Fisheries followed the advice and cut down the cod quota to 135,000 tonnes, the lowest level ever. It remains to be seen if this action by the Ministry of Fisheries is sufficient to ensure the continued viability of the cod stock, but at least it is a sign that there is currently a political will to work within the boundaries of the fisheries management system in times of crisis.

However, although the Icelandic fisheries management is most famous for its ITQ features the system is not solely based on ITQs; these are backed up by several complementary measures to ensure the biological robustness of the system: a comprehensive system for gathering data on the fisheries, a harvest control rule for cod, a system of cod equivalents, and temporarily closed areas. We will in the following section look at these features of the Icelandic fisheries management system.

5.5.1 Landing Statistics

In Iceland a comprehensive system for gathering landing statistics is in place. The system primarily facilitates control purposes, but also lays the foundation for scientific work related to monitoring the stake of fish resources and eventually the setting of the TACs.

All catches landed in Icelandic harbours are weighed by a licensed weight-master at accredited harbour scales upon landing (Regulation no. 224/2006 replacing earlier regulations from 1996; Ministry of Fisheries, 2005b). All the 60 accredited harbour scales in Iceland are connected to a Directorate of Fisheries database. The Directorate of Fisheries publishes landings per boat per species and the resulting changes in quota status per species per boat every day (see their homepage where it is possible to access information on the quota status of individual vessels by species and by year: http://www.fiskistofa.is/aflastodulisti.php). Special provisions are made for ice landed with iced fish and for gutted versus un-gutted fish. For catch processed at sea there are processing efficiency indexes accounting for the loss in the processing process.

Catch statistics are collected on a continuous basis in the Icelandic fisheries. Every licensed vessel is mandated to report catches electronically through an electronic logbook system (latest regulation no. 557/2007, Ministry of Fisheries, 2005b) to the Directorate of Fisheries. When combined with a satellite vessel monitoring system (VMS) this provides very reliable information to the Directorate of Fisheries for enforcement purposes and reliable statistics to the Marine Research Institute for stock assessment purposes. The information reported in the electronic logbook are: (1) name of ship, registration number and call code, (2) fishing gear, kind and size, (3) latitude and longitude of start of fishing, (4) catch by weight and species composition, (5) date, and (6) landing harbour.

The collection of fisheries related data in Iceland is accomplished through what is probably the most advanced data collection system currently in operation for a whole sector. This system presents the opportunity to monitor in near real-time the harvesting sub-sector both regarding individual species and particular vessels. Consequently, the opportunity to manage the fisheries by for example adaptive, regional, species-specific criteria as a complement to the ITQs exists. Such measures might detract from the economic efficiency of the current system, but could address some of the emerging and pressing biological and social issues facing the system.

Fish stock assessment in Icelandic waters is conducted through a number of surveys and research programmes. The most important ones are probably the annual surveys for the major commercial species. These include the Icelandic Groundfish Survey, conducted in March, and the Autumn Groundfish Survey, conducted in October (Björnsson et al., 2007), the acoustic surveys for herring, capelin and blue whiting (e.g. Marine Research Institute, 2008) and surveys for shrimp (Skúladóttir, 2001) and lobster (Pálsson & Kristinsson, 2005), to mention a few. The demersal stock assessment surveys are conducted in close cooperation with the fishing industry. The Icelandic Groundfish Survey has been ongoing since 1985, and the Autumn Groundfish Survey has been ongoing since 1996 (Björnsson et al., 2007). In the Icelandic Groundfish Survey half of the 600 original trawling stations were selected by random stratification and half by captains of fishing vessels. Furthermore the Icelandic Groundfish Survey is conducted onboard commercial vessels using commercial gear, for comparison purposes (Björnsson et al., 2007). The acoustic surveys for capelin and herring are also partly carried out by commercial vessels (Th. Sigurdsson, Marine Research Institute, personal communication).

5.5.2 *Harvest Control Rule*

The Minister of Fisheries, in response to concerns over the declining status of stocks, formed a working group in 1992 that was intended to provide recommendations for the long term sustainable utilisation of fish stocks in Icelandic waters (Anonymus, 1994). The main conclusion of that work was the formation of a harvest rule for the cod stock in Icelandic waters. The proposal for the harvest rule was that the TAC for cod would be 22% of the average of fishable biomass at the beginning of the year and the quota allocation of the previous year (Agnarsson, Haraldsson, Jóhannsdóttir, & Arnason, 2007). The Marine Research Institute recommended a harvest rule that would allow a catch of 22–25% of the average of the fishable stock of the current year and the stock estimate for the coming year.

In the end, the Ministry of Fisheries decided to enforce a harvest rule that allowed a catch of 25% of the fishable stock, but never less than 155,000 tonnes (Agnarsson et al., 2007). This was first in force the fishing year 1995/6. In the year 2000, it became apparent that the size of the cod stock had been overestimated for a number of years. As a result the harvest rule was changed in such a way that the minimum of 155,000 tonnes was abandoned, instead a buffer of a 30,000 tonne maximum change in either direction was implemented (Agnarsson et al., 2007).

In 2001, the Minister of Fisheries formed a committee to look into the result of the harvest rule set in 1995 (Anonymus, 2004). The committee found that despite the discrepancies between the recommendations of the working group from 1994 and the final version of the harvest rule implemented by the Ministry of Fisheries, the harvest rule had had a positive effect on the cod stock (Anonymus, 2004). The committee proposed that the recommendations of the working group from 1994 should be implemented, i.e. a harvest rule of 22% of the average of the fishable stock of the current year and the stock estimate for the coming year (Anonymus, 2004). This was not implemented by the Ministry of Fisheries. The last changes to the harvest rule were made in 2006, the fraction still being 25%, but now of the average as proposed by the 2004 committee and the max buffer is no longer in effect (Agnarsson et al., 2007).

5.5.3 Cod Equivalents

All commercial fish species in Icelandic waters that fall under the ITQ system (25 species in all) are allocated annually according to so-called cod equivalents, where cod is assigned a value of 1. According to the Fisheries Management Act no. 116/2006 (originally no. 38/1990) Article 19, cod equivalents are calculated as the value of individual species in proportion to the value of gutted cod (Ministry of Fisheries, 2005b). The basis of the calculation is statistics for the total catch and total value of these species from the Directorate of Fisheries. For example, assuming that the value of 1 kg of gutted cod is 150 ISK, the value of one kg of redfish is 75 ISK and the value of one kg of lobster tails is 750 ISK, then 1 tonne of redfish is half a cod equivalent tonne but one tonne of lobster equals five cod equivalent tonnes. The cod equivalent index would then be 0.5 for redfish and 5.0 for lobster. The Directorate of Fisheries publishes cod equivalent tables annually (see Table 5.3). The cod equivalents fluctuate considerably between years, e.g. capelin between 2005/06 and 2006/07 by +67%, redfish between the same years by +28% and lobster between 2006/07 and 2007/08 by –27%. This is mostly due to changes in market prices.

The cod equivalents enable trading of quota shares in multi-species fisheries, such as the Icelandic, to be done according to a unit of measurement. This makes it easier to trade one species for another on the quota share market.

The system of cod equivalents contributes potentially to the biological robustness of the important cod stock. The system enables fishermen to catch other species and withdraw the catches from the cod quota, but not the other way around. A fisherman can catch all other species, without owning quota for it or having to lease it, and

Table 5.3 Cod Equivalent Indexes for Selected Commercial Species from 2000/01 to 2007/08 (Website of Directorate of Fisheries)

	2007/08	2006/07	2005/06	2004/05	2003/04	2002/03	2001/02
Cod	1	1	1	1	1	1	1
Haddock	0.82	0.81	0.75	0.68	0.94	1.2	1.2
Saithe	0.43	0.42	0.37	0.36	0.43	0.48	0.45
Readfish	0.6	0.69	0.54	0.47	0.5	0.54	0.55
Herring	0.1	0.13	0.14	0.09	0.1	0.16	0.06
Capelin	0.09	0.1	0.06	0.06	0.05	0.05	0.04
Lobster (tails)	5.05	6.42	6.45	6.52	6.74	7.15	6.95

have the catches deducted from his cod quota. This increases the flexibility of the system and should also help prevent discarding.

5.5.4 Temporarily Closed Areas

The Marine Research Institute has for decades had a legal mandate to temporarily close areas for fishing. These closures are generally based on information received from fisheries inspectors monitoring catch composition of vessels at sea (Law no. 79/1997, article 10). The main criterion for closure is that the proportion of undersized (juvenile or immature) fish reaches a certain level. The Marine Research Institute sends out a closure notice to the fleet. Such closures can last for up to two weeks. If longer closures are deemed to be necessary the Ministry of Fisheries can publish a closure regulation stipulating the details of the closure (e.g. fishing gear, area and time). Other seasonal or permanent area closures or protections are also in force as stipulated in the Law on Fishing in Icelandic Waters (no. 79/1997 with later changes).

5.6 Economic Efficiency

The implementation of the Icelandic ITQ system was justified in several ways. With the transferability of quotas, the system would, it was argued, be flexible and capital could be used efficiently. What the national economy would loose by giving away the resource, economists argued, would be regained through resource rents and sector efficiency (Eythórsson, 2000).

One of the main anticipated benefits of the ITQ system was a reduction in fleet capacity and a resulting increased efficiency in the industry. After almost 20 years of IVQ/ITQ, Gudmundsson et al. (2004) indicate that the Icelandic trawler fleet is becoming increasingly efficient, as the following example from the cod fishery describes:

> *[I]n 1990 three fishing companies owned 10 vessels measuring a total of 6,850 GRT. The three companies held quota of about 20,000 metric tons in cod equivalent values or 5.6% of the overall TAC measured in cod equivalent values. By 2004 these three companies had merged into one. The new company controlled about 20,000 metric tons of cod equivalent value (5% of the overall TAC), but it now used only five vessels to harvest this quota. These five vessels measured 3,850 GRT. Three new vessels were bought instead of the eight vessels that the company either sold domestically or abroad or scrapped* (Gudmundsson et al., 2004, p. 12).

In the trawler fleet, which catches more than half of Iceland's demersal catch by volume (Ministry of Fisheries, 2005a), there seems, consequently, to be an ongoing development where fewer vessels are needed to catch the same volume of quotas – yet the number of vessels has not decreased in Iceland as dramatically as seen in other countries after introducing tradability of fishing rights. The companies are increasingly taking advantage of economies of scale – as an adjustment to the incentives in the ITQ management system (Gudmundsson et al., 2004).

In a relatively recent article, Arnason (2005) reviews the Icelandic fisheries management system and presents the economic benefits, which, he argues, have been provided by the ITQ system. Arnason concludes that there has been a significant, voluntary decrease in the number of vessels licensed to fish in Icelandic waters. This has happened in response to the incentives provided by the ITQ system. This development leads Arnason to conclude:

> *1. Since 1984, under the ITQ fisheries management system, the efficiency of the fisheries has increased dramatically. 2. Currently, the economic rents generated by the fisheries, as measured by the quota price evaluation, constitutes a substantial fraction of the average landed value* (Arnason, 2005, p. 259).

The Icelandic ITQ system has, it seems, shown to be successful in minimising the overcapacity (and increasing efficiency) of the fisheries industry – by many considered the biggest challenge of the fisheries sector worldwide. Moreover, when fishing capacity decreases and fishermen sell their vessels; the ITQ system may facilitate and smooth this transaction because earnings from the sale of quotas compensate fishermen leaving the industry. However, from the opposing perspective it has been argued that people in local communities who are not fishermen, but nevertheless depend on the fishing sector (e.g. people in the processing or gear industry), are not compensated for their lost earnings, which creates unfair imbalances between those privileged with free quota shares and those not (Eythórsson, 2000). This will be discussed further beneath.

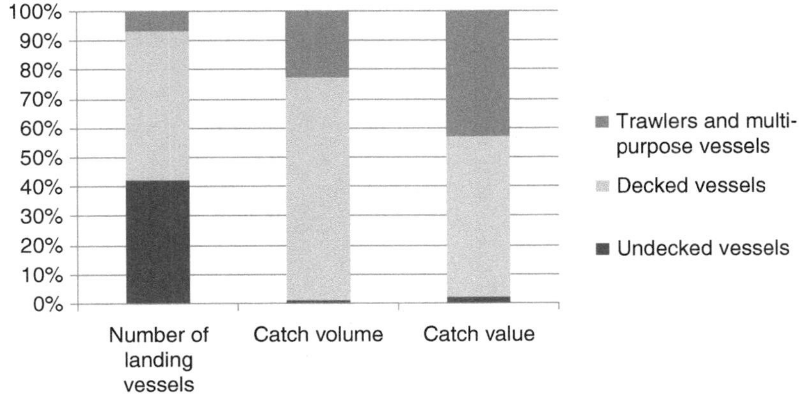

Fig. 5.2 Vessel Types and Their Share of the (1) Fleet in Numbers, (2) Catch Volume and (3) Catch Value in 2005 (Data from Website of Icelandic Ministry of Fisheries)

5.6.1 Fleet Composition and Development

Figure 5.2 shows an overall distribution of vessels and their shares of the catches in both tonnes and value for the year 2005. It shows that a small part of the fleet, the 63 trawlers and multi purpose vessels (6% of the fleet in terms of numbers) catch more than 40% of the harvest in value and more than 20% in weight. It also shows that about 40% of the Icelandic fleet (small vessels) catches less than 1% in terms of tonnes and less than 4% in terms of value.

Trends in the fishing fleet composition, both in numbers and displacement fit with the expectations of how an ITQ system works. Fewer, larger, more efficient vessels catch a majority of the fish both in value and weight. This consolidation is seen clearly in the declining number of fishing permits. At the same time, both newcomers and established fishermen are exploiting all possible loopholes in the system. This is well expressed in the historical trends in the small vessel fleet.

After the implementation of the IVQ system in 1984, it was expected that there would be a reduction in the Icelandic fleet. However, the fleet increased in both numbers and GRT (see Figs. 5.3 and 5.4). The number of boats reached a maximum in 1990, the year before the new Fisheries Act was implemented. That year there were 2,321 vessels registered, an increase of 666 vessels from the implementation of the initial quota system in 1984 (see Fig. 5.3). Most of this increase can be attributed to a great increase in open boats, from 825 in 1984 to 1,325 in 1990, an increase of 61%. Decked vessels remained fairly even

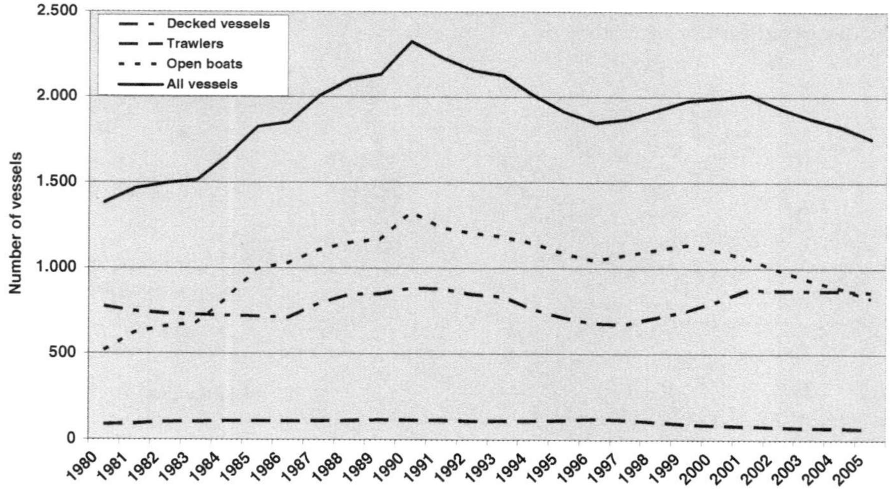

Fig. 5.3 The Icelandic Fishing Fleet in Numbers by Type of Vessel from 1980 to 2005 (Data Obtained Through Personal Correspondence; Original Data for 1980 to 1997 is from the Icelandic Fishing Society (Fiskifélag Íslands) and Data from 1999 to 2005 is from Hagstofa – Bureau of Statistics Iceland; 1998 is Estimated)

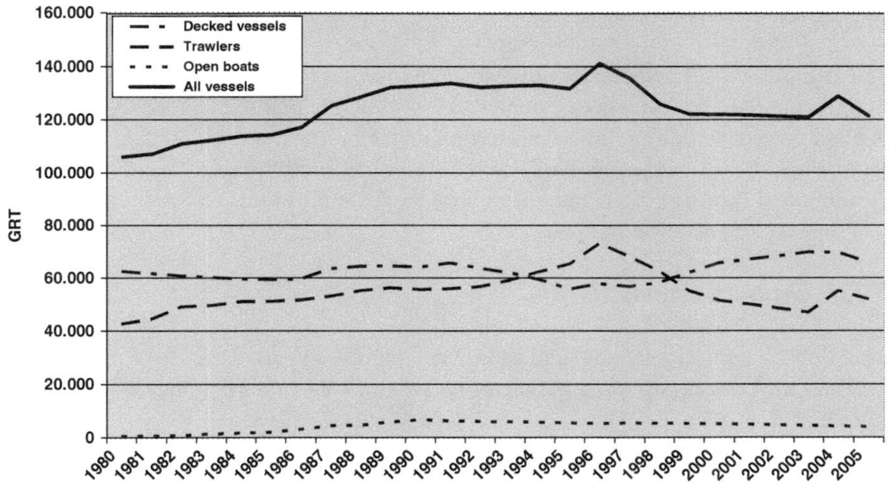

Fig. 5.4 The Icelandic Fishing Fleet in GRT by Type of Vessel from 1980 to 2005 (Data Obtained Through Personal Correspondence; Original Data for 1980 to 1997 is from the Icelandic Fishing Society (Fiskifélag Íslands) and Data from 1999 to 2005 is from Hagstofa – Bureau of Statistics Iceland; 1998 is Estimated)

during this period of time, reaching a maximum of 883 in 1990 and a minimum of 675 in 1997. The number of trawlers increased slowly from 107 in 1984 to a maximum of 121 in 1996. From the high in 1990 the overall number of vessels has decreased dramatically, with the number of fishing vessels in the Icelandic fleet in 2006 being 1692, a decrease of 28%. Most considerable is the decrease in number of open boats, by 548 vessels from a high of 883 in 1990, a 70% decrease, and in trawlers by 58 vessels from a high of 121 in 1996, a decrease of 52%.

The trends for trawlers and open boats have different explanations. It is inherent in ITQ systems that effort can be concentrated on larger more efficient vessels. This is what happened in the Icelandic fisheries. Larger vessels were brought into the fisheries, and quota was moved from less efficient vessels to more efficient vessels within the same fishing company. That is a definite trend towards a fleet consisting of fewer larger trawlers, exactly what one would expect.

The trend for open boats is quite different. Initially open boats were excluded from the IVQ system. This loophole resulted in an explosive growth in the number of small vessels (the initial cut off was at 10 GRT). In 1980, there were 518 small vessels. In 1984, there were 825 and in 1991 after the advent of the ITQ system, the number peaked at 1,325. The share of catch, especially in cod, caught by the small vessels increased commensurately, from less than 5% in 1983 to more than 20% in the 1990s. The small vessels have now been incorporated into the ITQ system, although with some special provisions, and the observed decline in the number of small vessels continues unabated.

5.7 Social Robustness

The debates over the social robustness of the Icelandic ITQ system have focussed mainly on two issues: quota concentration and the marginalisation of fisheries dependent coastal communities. The critics of the system have highlighted the problems while the advocates of the system have argued that the negative effects of the system are minor and often a result of broader societal forces.

5.7.1 Quota Concentration

The debate over quota concentration has mainly focussed on two interrelated aspects: the concentration of quotas in larger firms and the regional concentration of quotas.

The ITQ system has, according to a number of critics, to some extent had negative effects on the remote and smaller fishing communities. These problems have mostly been associated with perceptions of unfair distribution of opportunities, concentration of wealth and structural changes removing fisheries related activities away from small communities traditionally specialised in fisheries. The core of this ITQ critique is nicely summed up by Copes (cited in Copes & Pálsson, 2000):

> *[The] gratis quota allocation gives windfall gains to the privileged few. Capitalization of quota rights at high values encourages their accumulation in the hands of corporations and wealthy investors. This facilitates financial and geographical concentration of fishing operations, with substitution of capital for labour, causing irrational excessive job loss. High quota costs deprive crewmembers of the traditional opportunity to become independent owner-operators, as they can no longer afford to purchase vessel with quota privileges. Communities historically dependent on adjacent fish stocks, find their economic viability – and sometimes their very existence – threatened when their resources are alienated to outsiders. Members of the public are scandalised by the gifting of access rights to public resources, privileged an emerging class of 'armchair fishermen' who become retires, living off the avails of quota leasing.*

As the following numbers illustrate, the introduction of the ITQ system has indeed and as expected resulted in a significant concentration of quotas (fisheries access rights). In 1999, the 20 biggest quota owners held 57% of the total quotas in comparison to 1991 where the 22 largest quota owners held only 26% of the total quotas (Eythórsson, 2000). Similarly, there was a dramatic reduction in the number of quota holders from 535 in 1984 to 391 in 1994, a reduction of 144 quota-holders – or 27%. Moreover, the quotas have been concentrated among large-scale quota-holders, whereas the number of people owning small amounts of quotas has diminished (Pálsson & Helgason, 1995; Pálsson, 1998).

One result of the ITQ system has according to Eythórsson (2000) been accelerated marginalisation of some fishing communities. This has especially been the case for the smallest communities under 500 inhabitants. These small communities have lost to the larger communities in a competition for quotas. The ITQ system has supposedly not only led to a consolidation in terms of larger companies but also a relative concentration of activity in larger fishing communities leaving the smaller communities with fewer sources of income as the processing plants loose their source of raw material when fleets move elsewhere.

For many of the smaller fishing communities of Iceland one of the fundamental difficulties has been that many of the people do not have the means to obtain fishing rights, and when excluded from the fishery, they find that alternative livelihood activities are scarce (Eythórsson, 2000; Orebech, 2005). Where financial resources used to circulate in the local communities, critics argue, they are increasingly transferred to non-local institutions in larger urban centres or worldwide (Orebech, 2005). Not only do the local communities have few alternative employment opportunities outside the fisheries sector, the contract fishing, which earlier was an employment option for the people of the small communities, has also diminished. Moreover, overall the land-based fish processing has also declined as processing increasingly occurs on the vessels (Eythórsson, 2000).

It has on the other hand been argued, e.g. by Agnarsson et al. (2007) and Hall, Jónsson, and Agnarson (2002), that the ITQ system has become a synonym for everything negative that happened in the development of small coastal communities around Iceland without taking into consideration that much of the experienced development is a result of circumstances and changes in the economy and society in general. As we discuss beneath, it should also be mentioned that some management instruments are in place that should help the communities that are affected most by this development.

5.7.2 Community Quotas and Coastal Communities

The community quotas (byggðakvótar) were introduced in the fishing year 2002/03 (regulation no. 909/2002) to address some of the criticism of the ITQ system, specifically the reputed effect with quota consolidation in larger communities resulting in movement of people away from smaller communities.

The term 'community quotas' refers to a small part (currently around 4,000 tonnes of cod equivalent) of the Icelandic quota allowance that is given to small communities. Introducing the regional quotas was a highly political decision, which caused legal problems. The distribution of the regional quotas is based on a formula of employment, how much the communities are dependent on fisheries, if quotas have been transferred away from the areas and so on. This has no basis in any existing legislation; the Minister himself decides on the distribution of these quotas annually by a special regulation (no. 909/2002, no. 596/2003, no. 960/2004, no. 722/2005 and no. 439/2007). Though contentious, there seems to have been established a consensus on the broad lines of the process and most people have accepted the system of distribution under the new regulation.

Nonetheless, the regulation on community quotas was reviewed before the fishing year 2007/08. Until then the municipalities had decided for themselves who in the community should get the quota. This has caused some problems within the local communities, where everybody knows everybody; and everybody wants the quotas. The changes to be made are to ensure that these processes are running smoother. In some cases the local communities have failed to distribute the quotas in proper ways. The community quotas are supposed to support communities, not the vessel owners.

In addition to the community quotas, another measure has been introduced to improve the situation of the coastal communities: 'Línuívilnun', which means that the quota for the longliners having their lines prepared on shore is reduced by 16% less than for other vessels. Hence, this measure enhances both the use of longliners, which are considered to be biologically sustainable, and preserves jobs on shore.

It is still highly contentious if indeed trading quota shares away from (or towards) a community has any effect either way. Looking at the trends in proportional employment in the fisheries sector in Iceland the overall trend is clearly moving towards a diminished importance of fisheries as a source of employment as evidenced by Table 5.4. Official estimates indicate that the number of persons employed has dropped from approximately 7,000 to under 5,000 in the period from 1992 to 2004 (Ministry of Fisheries, 2005a).

The most recent published study on the effect of quota trading on community development by Hall et al. (2002) does not find any patterns or trends. Agnarsson has recently presented a further analysis on the subject, showing no significant relationship between landings (local quota share ownership) and population change in Iceland (Agnarsson, 2006). However, there is, not surprisingly, a weak relationship between fish processed in the local community and population change.

People stay in the communities if the work available is commensurate to their expectations of income and services. The changes in suitable employment opportunities in the fishing communities around Iceland is therefore of great interest when examining migration patterns.

Table 5.4 Employment in the Fishing Industry as Percentage of Total Employment (Website of Statistics Iceland under 'Fish and Agriculture')

	1998	2002	2005
Reykjavik area	2.7	2.0	1.7
Sudurnes	23.8	18.1	15.3
Vesturland	19.3	17.0	15.5
Vestfirdir	37.7	32.2	28.6
Nordurland west	16.8	13.7	11.2
Nordurland east	17.8	14.7	13.4
Austurland	28.3	26.2	17.5
Sudurland	14.9	11.8	9.9
All Iceland	10,1	7.9	6.6

5.7.3 Legality and Fairness

The Icelandic ITQ system has led to much debate over definitions of 'property and access rights' (see Chapter 10), not least because the Icelandic government continues to argue that even under the current ITQ system the Icelandic fisheries resources can be characterised as a common property, albeit under the supervision of the state on behalf of the public. However, arguments such as the *transactions of uncaught fish violate the rule of capture and the common property nature of the fishing stock* (Pálsson, 1998, p. 283) have been numerous in the debate. While quota-holders under the ITQ system have private property rights in the sense that their share of the right to utilise the resource can be traded on a free market, their utilisation of their particular share of the resource is submitted to rules entirely defined by the state.

When the IVQ system was introduced in 1984, fishing vessels were allocated unequal shares of quotas based on their catches the previous three years. This, as well as the semi-perpetual awarding of ITQ shares by the Fisheries Management Act, has since led to many disputes over the fairness and legality of the process by which a public property resource was handed over to individual fishermen. Eythórsson (2000) argues that a number of court cases indicate that the legislation relating to the ITQ system was not sufficiently well designed from the outset and that the Parliament would probably have been reluctant to award the ITQ shares for free without time-limits if the implications had been realised in 1990. However, the Fisheries Management Act represents *de facto* a 'point of no return'. On the other hand Helgi Áss Gétarsson, specialist at the Law Institute of the University of Iceland, has pointed out firstly that almost all of the current quota shares have changed hands since the initial allocation, i.e. have been traded, and secondly that the allocation in 1990 was based on catch records/history going back at least to 1980. The argument of fairness or lack thereof, has to be viewed in light of this fact.

According to Eythórsson (2000) one of the main legal problems has been the need to uphold a paradoxical status of quota shares, which means that they are by law public property, but for all practical purposes function as private property. Because of this paradox, the sector experienced several court cases in the late 1990s. An important example is the 1998 case (Kvótadómur) regarding a fisher who was denied a fishing licence and a catch quota because he had not been an active fisherman in the 1980s when quotas were allocated. Eythórsson (2000, p. 490) describes the outcome of the case in this way:

> *Considering the Icelandic constitution, which claims equal employment rights for every citizen, and the Fisheries Management Act of 1990, which defines the fish resources as public property, the majority of the Court found [. . .] that by introducing the ITQ system the government had given away exclusive rights to the publicly owned Icelandic fish resources. These had been given away as perpetual rights to a group of people who happened to be the owners of active fishing vessels at a certain point in time. Such an act could not be justified by the need to preserve the resources or by the best public interest.*

In short, the implementation of the IVQ/ITQ system was declared unconstitutional due to the *de facto* perpetual character of the allocation of ITQ shares. The High Court was not unanimous, a minority opinion pointed out that the statement made by the majority of the Court about giving away perpetual rights to an exclusive group of people in 1983 is simply not right since the allocation in 1990 had been based on much wider criteria. The research by Grétarsson and others, mentioned earlier, seems to support this minority opinion. However, the ruling resulted in an amendment to the Fisheries Act 38/1990 giving Icelandic citizens a general right to obtaining a fishing license on demand. This did, however, not change the fact that newcomers still had to either permanently buy or temporarily lease a costly quota to be able to fish. The need for a quota was later challenged in the courts. However, the fisher, who had fished without quota, lost the case in the High Court after having won in a lower court. In this case, the majority of the High Court found that the quotas were not formally defined as private property, and that they were justified on conservation grounds. It is notable that none of the rulings were unanimous (Copes & Pálsson, 2000; Eythórsson, 2000).

One major concern has been that it may be very difficult to reverse the ITQ system. Many citizens feel that they have lost, definitively, what used to belong to them. The discussion about 'resource rent fees' to compensate society became increasingly central in the ITQ debate. While the public largely supported the idea in the light of the large resources having been handed over to the quota owners, the industry saw resource rent fees as yet another tax that would diminish their competitiveness. Another alternative that was put forward in the Icelandic fisheries debate was the idea that the fishermen should annually, instead of paying resource rent fees, return a small percentage of their quotas to the state, which would then be auctioned at an open auction (on auctions in quota systems see e.g. Anderson & Holland, 2006). Consequently, the public would, in the long run, regain their ownership over fish resources (Eythórsson, 2000). The outcome of this debate was the resource fee, which is described earlier.

Regarding the issue of fairness, Orebech (2005, p. 166) adds the argument that *technically, the privatisation policy of ITQs involves robbing the excluded fishermen of their assets (open access) without the payment of compensation.* The excluded fishermen refer to coming generations of fishermen or newcomers into the sector as underprivileged because they must buy quotas, which the first generation of quota holders got for free. Consequently, in the transition period between first and second generation of quota holders, the relative competitiveness of the second generation compared to the first generation is imbalanced (Orebech, 2005). It has therefore been increasingly difficult for newcomers to enter into the fishery, since the price of the right to fish is too high to make their activities profitable. On the other hand, it can be argued that larger enterprises often have a larger group of shareholders, and therefore there are probably many more owners/participants in the Icelandic fisheries sector right now than at any point in time.

The conclusions that Orebech draws are possibly not specifically applicable to the Icelandic case. Both in the initial 1983 allocation and in the 1990 allocation, the quota shares went to boat owners (in most cases not fishermen as such) that had history in the fisheries. This implicates that the individuals/enterprises that were allocated the quota got it because they had made a considerable investment into the harvesting sector. This raises the question of the fairness of newcomers getting access and quota shares without similarly making an initial investment. This has been pointed out numerous times in the literature (e.g. Arnason, 1996; 1999; 2005; Hall et al., 2002).

The latest development in the debate of equality and fairness in the ITQ system is a decision by the United Nations Human Rights Committee (HRC), made public in December 2007 (UN Human Rights Committee, 2007). The decision regards two Icelandic fishermen that submitted a communication to the HRC regarding the compatibility of the fisheries management system with the non-discrimination principle as stated in Article 26 of the International Covenant on Civil and Political Rights. The HRC was in favour of the fishermen, against the state of Iceland. The main findings of the HRC were that while the quota system constitutes a legitimate means of protecting the limited fish resources *the property entitlement privilege accorded permanently to the original quota owners, to the detriment of the authors, is not based on reasonable grounds*. This constitutes, in the opinion of the HRC, a violation of Article 26 of the Covenant. The HRC concludes that the State of Iceland is under an obligation *to provide the authors with an effective remedy, including adequate compensation and review of its fisheries management system.*

This decision by the UN-HRC understandably caused uproar in the Icelandic society, and the different camps in the ITQ debate have been discussing in the Icelandic media. It is even debated whether the Government of Iceland is under an obligation to react to the views of the HRC. The Government of Iceland will have to review the ITQ part of the fisheries management system according to the HRC. What changes will result from that review is unclear, but at least the issue of access and equality will have to be addressed by some actions.

5.8 Conclusions

The Icelandic fisheries were among the first in the world to be subject to a comprehensive system of individual quotas. Initially the quota was on a vessel basis, non transferable and an option along with effort management for a part of the fleet, albeit the most important part of the fleet. Over the last 23 years several changes have been made to the quota system and currently all commercial fisheries in Iceland are subject to ITQ management. It is important to realise that the ITQ system is only a part of the fisheries management system in Iceland. The ITQ system is as described supported by other elements. On the most basic level, the Marine Research Institute recommends an annual TAC for all commercial species. This recommendation derives from catch statistics, independent survey data and other scientific information available. Based on the advice and discussions with stakeholder representatives, the Minister of Fisheries decides on the TACs for the coming fishing year. The biological performance of the fisheries management system is thus to a large degree a product of a scientific process that precedes the setting of TACs and the ITQ mechanism of distributing annual catch allowances.

The ITQ part of the fisheries management system has contributed to the fairly good economic performance of the fisheries sector in Iceland. However, the resulting consolidation of quota shares and processing capacity has been hard for many small fishing communities in Iceland. A large part of the Icelandic public relates the ITQ system directly to the changes in population and employment patterns that have occurred over the last two decades. This is certainly an oversimplification of causes and effects, but there may nonetheless be some truth in it – particularly in relation to the pace of the process by which societal changes have happened.

The evolution of the Icelandic fisheries management regime over the last two decades contains a number of lessons both of success and mistakes. These lessons are useful to bear in mind when considering future development of fisheries management.

The first lessons we would like to emphasise are those related to the *initial allocation of quotas*. The ITQ systems of Iceland and New Zealand (see Chapter 2) have many similarities, but they handled the initial allocation of the quota quite differently. In New Zealand they made the initial allocation in fixed tonnages, which caused severe problems afterwards. Fixed amount quotas potentially put managers in a difficult position when having to reduce the TAC, as they will then have to buy back quota at market price. New Zealand therefore changed the system into allocation in percentages of the quota. In Iceland, they made the initial allocation in percentages, and hence avoided the issues they had in this respect in New Zealand. So based on these experiences the initial allocation should be made in percentages of the quota (although making it in fixed amounts allow the state to sell the extra amounts if the quota goes up, which is not the case when shares are distributed). Another related lesson is that ITQs should not be given out in perpetuity. The ITQs in Iceland are not given out in perpetuity, but the Icelandic system gives the same benefits, as one would expect from a rights-based system where quotas are handed out for good. Hence there is no need to hand out the quota shares in perpetuity as in New Zealand, since the Icelandic system marketwise performs just as well as New Zealand's system. Giving out quota in perpetuity just makes it even harder to change the system fundamentally if there should ever be a need for that.

Flexibility of the ITQ system is a vital feature as all quota systems face the challenge of fishermen not being able to completely control the composition of their catches. Hence, all quota systems have to deal with the differences between the allowed catches and the unintended catches. In Iceland, discard is banned. The Icelandic fisheries management system has developed a number of flexibilities in the quota system to minimise discard. These flexibilities are essential for the success in order for the quotas to match the fluctuations of nature and unpredictable fisheries, and the following lessons can be extracted: flexibility of quota over both previous and next year is essential if introducing a discard ban, flexibility in buying/leasing quota is essential if introducing a discard ban, and flexibility in buying/leasing quota between fishermen is essential for an ITQ system to work.

The Icelandic case also points to the central role that *the enforcement system* plays. One lesson from the case is the need for a strong enforcement framework accompanied by an accurate system of monitoring. The enforcement system in Iceland is supported by a real time, online catch reporting system, which is coordinated with the amounts of quota of the vessel. Everybody has access to the information of this system. Moreover a lesson from the Icelandic system is the need for a flexible enforcement system that is reviewed regularly. In Iceland, this has for instance been done by giving the biologists the authority to decide on the timing of area closures up to two weeks without political amendment. This ensures a short response time from the observation of undersize fish in an area to the closure.

Also when it comes to *the participation of fishermen* the Icelandic case offers useful lessons. In Iceland, fishermen formally play a small role in the management system, but in practise they have easy and direct access to the Minister, who has the final say in most matters. In relation to the setting of TACs in Iceland, a HCR on cod has been introduced. This makes the setting of TACs more robust to both economic and biological changes. The fishermen were part of the formulation process for HCR.

References

Agnarsson, S. (2006). The ties that (do not) bind: The ITQ system and concentration in the Icelandic fish processing industry 1987–2004. Presented at IIFET conference in Portsmouth, July 2006.

Agnarsson, S., Haraldsson, G., Jóhannsdóttir, K. B., & Arnason, R. (2007). *Þjóðhagsleg áhrif aflareglu* (Macroeconomic effects of the harvest rule). C07:09. Institute of Economic Studies, University of Iceland, Reykjavik, Iceland.

Anderson, C. M., & Holland, D. S. (2006). Auctions for initial Sale of Annual Catch Entitlement. *Land Economics,* 82(3), 333–352.

Anonymus. (1994). *Hagkvæm nýting fiskistofna. Lokaskýrsla vinnuhóps um nýtingu fiskistofna.* (Efficient utilization of fish stocks. Final report of the working group on utilization of fish stocks). Ministry of Fisheries.

Anonymus. (2004). *Aflaregla fyrir orskveiðar á Íslandsmiðum. Skýrsla nefndar um langtímanýtingu fiskistofna.* (Harvest rule for codfishing in Icelandic waters. Report of the committee on long term utilization of fish stocks). Ministry of Fisheries. Available at www.sjavarutvegsraduneyti.is/ frettir/nr/825.

Arnason, R. (1996). On the ITQ fisheries management system in Iceland. *Reviews in Fish Biology and Fisheries,* 6, 63–90.

Arnason, R. (1999). Advances in ITQ Fisheries Management. In R. Arnason & H. H. Gissurarson (Eds.), *Individual Transferable Quotas Theory and Practice*(pp. 31–42). Reykjavik, Iceland: The University of Iceland Press.

Arnason, R. (2005). Property rights in Iceland: Iceland's experience with ITQs. *Reviews in Fish Biology and Fisheries,* 15, 243–264.

Björnsson, H., Sólmundsson, J., Kristinsson, K., Steinarsson, B. Æ., Hjörleifsson, E., Jónsson, E. et al. (2007). The Icelandic groundfish surveys in March 1985–2006 and in October 1996–2006. Marine Research Institute, Report 131: 220.

Charles, A. T. (1992). Fishery conflicts: A unified framework. *Marine Policy, 16*(5), 379–393.

Copes, P., & Pálsson, G. (2000). *Challenging ITQs: Legal and political action in Iceland, Canada and Latin America.* A preliminary overview. IIFET 2000 Proceedings. Corvallis, Oregon, USA: 1–6.

Eythórsson, E. (2000). A decade of ITQ-management in Icelandic fisheries: consolidation without consensus. *Marine Policy, 24,* 483–492.

Gudmundsson, E., Bergsson, A. B., et al. (2004). *Development of fishing effort and fishing fleet capacity in the icelandic cod fishery.* IIFET 2004, Japan.

Hall, A., Jónsson, A., & Agnarson, S. (2002). Byggðir og búseta (Communities and settlement). B02:01. Institute of Economic Studies, University of Iceland, Reykjavik, Iceland.

Marine Research Institute. (2004). State of marine stocks in Icelandic waters 2003/2004 – prospects for the quota year 2004/2005. Reykjavik, Marine Research Institute.
Marine Research Institute. (2005). State of marine stocks in Icelandic waters 2004/2005 – prospects for the quota year 2005/2006. Reykjavik, Marine Research Institute.
Marine Research Institute. (2006). State of marine stocks in Icelandic waters 2005/2006 – prospects for the quota year 2006/2007. Reykjavik, Marine Research Institute.
Marine Research Institute. (2007). State of marine stocks in Icelandic waters 2006/2007 – prospects for the quota year 2007/2008. Reykjavik, Marine Research Institute.
Marine Research Institute. (2008). State of marine stocks in Icelandic waters 2007/2008 – prospects for the quota year 2008/2009. Reykjavik, Marine Research Institute.
Ministry of Fisheries. (2005a). Icelandic fisheries in figures. Reykjavik, Ministry of Fisheries.
Ministry of Fisheries. (2005b). Fisheries management act and supporting legislation. Fisheries management Act no. 116/2006. Retrieved 20 June, 2006, from http://www.fisheries.is/managem/legisl.htm.
Ministry for the Environment, Ministry of fisheries, & ministry for foreign affairs. (2004). The ocean – Iceland's policy. Reykjavik, Ministry of the Environment, Ministry of Fisheries and Ministry of Foreign Affairs.
Orebech, P. (2005). What restoration schemes can do? Or, getting it right without fisheries transferable quotas. *Ocean Development and International Law,* 36(2), 159–178.
Pálsson, G. (1998). The virtual aquarium: Commodity fiction and cod fishing. *Ecological Economics,*24, 275–288.
Pálsson, G., & Helgason, A. (1995). Figuring fish and measuring men: the individual transferable quota system in the Icelandic cod fishery. *Ocean & Coastal Management,* 28(1–3), 117–146.
Pálsson, J., & Kristinsson, K. (2005). *Flatfiskar í humarleiðangri 1995–2003* (Flatfish in lobster surveys 1995–2003). Marine Research Institute, Report 112. 90 pp.
Skúladóttir, U. (2001). Stofnmæling rækju (Stock Assessment of Shrimp). *Hafrannsóknir,*56, 34–37.
UN Human Rights Committee. (2007). Communication N. 1306/2004. CCPR/C/91/D/1306/2004.

Personal communication with representatives from

Fiskifélag Íslands
Hagstofa – Bureau of Statistics Iceland
Marine Research Institute

Legal documents

Fisheries Management Act no. 116/2006 (originally no. 38/1990)
Law no. 79/1997, article 10
Law on Fishing in Icelandic Waters (no. 79/1997 with later changes)
Regulation no. 909/2002 on allocation on 2.000 tons of cod to seaside settlements
Regulation no. 596/2003 on allocation on 1.500 tons of cod to support seaside settlements
Regulation no. 960/2004 on allocation on 3.200 tons of cod to support inhabited area
Regulation no. 722/2005 om úthlutun byggðakvóta fiskveiðiárið 2005/2006
Regulation no. 224/2006 on Weighing and Recording of Catch
Regulation no. 439/2007 um úthlutun byggðakvóta til fiskiskipa á fiskveiðiárinu 2006/2007 samkvæmt 2. tl. 1. mgr. 10. gr. laga nr. 116/2006, um stjórn fiskveiða með síðari breytingum.
Regulation no. 557/2007 on logbooks

Websites

Directorate of Fisheries: http://www.fiskistofa.is/
Icelandic Ministry of Fisheries
http://www.fisheries.is/
Statistics Iceland
Fish and agriculture: http://www.statice.is/Statistics/Fisheries-and-agriculture/Fish-processing
Fishing vessels: http://www.statice.is/Statistics/Fisheries-and-agriculture/Fishing-vessels

Chapter 6
Evaluating Biological Robustness of Innovative Management Alternatives

Francois Bastardie, Alan Baudron, Richard Bilocca, Jesper Boje, Tammo P. Bult, Dorleta Garcia, Niels T. Hintzen, J. Rasmus Nielsen, Gudrun Petursdottir, Sonia Sanchez, and Clara Ulrich

Abstract The influence of innovative management alternatives (participatory governance, effort management, decision rules) on biological robustness (BR) in various fisheries relevant to the EU (Baltic, Western Shelf, Faroe Islands, North Sea), was investigated with a numerical simulation model developed in the EU projects EFIMAS (2004–2008) and COMMIT (2004–2007). The index for BR was set as the percentage of years in which standard biological reference points (B_{pa}, F_{pa}) were met. The results suggest that new information obtained through participatory governance may affect BR by reducing bias rather than increasing precision, implying that participatory governance should rather focus on potential sources of bias than on (perceived) low sampling efforts. Further analyses suggest that effort-based regimes combined with catch quota restrictions improve BR. However, the relative effect of catch quotas versus effort management on BR varies with circumstances, implying that careful and case-specific analyses are needed to weigh one against the other. This requires more detailed data than generally available at present, including electronic surveillance, detailed catch data, environmental/productivity data, recruitment and misreporting. Finally we analysed a decision rule consisting of a two-step management system, which allows TAC adjustment according to the state of the stock monitored during the fisheries season. Such measures may improve the BR.

Keywords Biological robustness · Decision rules · Evaluation · Harvest control rules · Management strategy evaluation · Reference points · Simulation

---

F. Bastardie (✉)
Technical University of Denmark, National Institute of Aquatic Resources, Charlottenlund Slot, DK-2920 Charlottenlund, Denmark
e-mail: fba@aqua.dtu.dk

K.H. Hauge, D.C. Wilson (eds.), *Comparative Evaluations of Innovative Fisheries Management*, DOI 10.1007/978-90-481-2663-7_6,

6.1 Introduction

6.1.1 Fisheries Management Innovations

Fisheries worldwide are under increasing pressure to design and apply management measures that aim for sustainability. The urgency is obvious as various fisheries seem to be threatened – according to the FAO only 3% of the world fish stocks are underexploited on a biological sustainability scale, 20% are moderately exploited, 52% are fully exploited, 17% are over-exploited, 7% are depleted, and 1% is recovering (FAO, 2005). In the light of these continuing trends it remains important to consider the impact of fishing on marine ecosystems in a wide context (UN, 1972) and consequently to come up with biologically robust management strategies (Degnbol & Jarre, 2004).

A range of management innovations is evolving worldwide, aimed at improving the current state of fisheries. These are mainly based on the principles of participatory governance, rights-based approaches, effort control, and decision rule systems. In this chapter, the relationship of some of these management innovations to biological robustness is explored through a small set of case studies, acknowledging the need for further studies.

6.1.2 Biological Robustness

Biological robustness (BR) has been defined in various ways, including: *robustness is a property that allows a system to maintain its functions against internal and external perturbations* (Kitano, 2004), and *the ability to maintain performance in the face of perturbations and uncertainty, is a long-recognized key property of living systems* (Stelling, Sauer, Szallasi, Doyle, & Doyle, 2004). In this study, an operational definition is used when referring to the biological robustness of a management strategy: *the ability of a management strategy* (i.e. 'innovation') *to account for uncertainty and error in the biological knowledge thereby reducing the impact of uncertainties on the sustainability of the resources.*

In order to quantify BR one may use standard biological indicators, like spawning stock biomass (SSB) and fishing mortality (F): B_{lim} indicates an SSB below which recruitment is thought to be impaired or the dynamics of the stock are unknown; F_{lim} indicates the Fishing Pressure that drives the stock towards B_{lim}. Because of uncertainty in the annual estimations of F and SSB, ICES has defined operational reference points: B_{pa} (higher than B_{lim}), and F_{pa} (lower than F_{lim}), where the subscript pa stands for precautionary approach. A stock estimated to be at B_{pa} should with high probability be above B_{lim}. At F_{pa} the probability that F is higher than F_{lim} should be low.

In this study, the indicator for BR was set as the percentage of years in which these standard biological reference points (B_{pa}, F_{pa}) were met. This implies a narrow definition of BR, in the sense that it focuses only on the demographic stock dynamics of a single species. Robustness may indeed not only be reflected in the properties of individual elements, but in the dynamic feedback between the interacting elements, for example in an ecosystem (Lenski, Barrick, & Ofria, 2006). For example, the biological robustness of management could be expressed in terms of changes in the life history parameters of individual species or within an entire complex of species. This would take into account the rising concerns about the evolutionary changes that fishing may bring to fish stocks (Law, 2000). For the purpose of this study, however, these system indicators were not taken into account, mainly for practical reasons and also because the BR indices chosen were thought satisfactory as a first step in these exploratory analyses.

6.1.3 General Approach

The influence of innovative management alternatives (participatory governance, effort management, decision rules) on biological robustness (BR) in various fisheries relevant to the EU (Baltic, Western Shelf, Faroes, North Sea), was estimated using a numerical simulation model developed in the EU projects EFIMAS (2004–2008) and COMMIT (2004–2007). The analyses were structured around the three general hypotheses in Table 6.1.

This chapter offers a general description of our approach, main findings and conclusions concerning the innovative management alternatives and biological robustness.

Table 6.1 A List of Tested Hypotheses Indicating Which Case Studies were Used

H1	Improved information, assumed to result from participatory governance, increases biological robustness.	North Sea, Western Shelf, Faroe Islands
H2	Management through effort restrictions leads to higher biological robustness than management based on TACs.	North Sea, Baltic, Western Shelf, Faroe Islands
H3	Decision rule systems that include recent information lead to higher biological robustness.	Western Shelf

6.2 Case Studies – Evaluating Innovative Management Regimes

6.2.1 North SEA – Evaluating Participatory Governance and Effort Management

Two hypotheses, H1 and H2, were tested for the North Sea case, focusing on participatory governance and effort management (Table 6.1). The aim was to analyse whether BR may be affected by new information obtained through participatory governance, as well as analysing the effects of shifting from TAC to TAE management in the North Sea demersal fisheries. H1 was specified to answer whether decreased variability of discard estimates, as an (assumed) result of participatory governance, results in increased biological robustness of the North Sea stocks of plaice (*Pleuronectes platessa*) and sole (*Solea solea*).

6.2.1.1 Hypothesis 1

Background

Biologically robust management strategies are able to account for uncertainties and errors in biological information and therefore reduce their impact on the biological sustainability of a resource. Hence it is essential to understand possible sources of uncertainty and errors.

In the context of fisheries management, uncertainties in fish stock assessment and management can be categorized as follows (Rosenberg, 1994; Francis, 1997), as summarized in Kell et al. (2007):

- process error — caused by disregarding temporal and spatial variability in dynamic population and fisheries processes;
- observation error — sampling error and measurement error;
- estimation error — arising when estimating parameters of the models used in the assessment procedure;
- model error — related to the ability of the model structure to capture the core of the system dynamics;
- implementation error—where the effects of management actions may differ from those intended (including uncertainty about fleet adaptations).

Evaluation of the BR of management strategies must take into account both variance and bias regarding data, models, and implementation. Robustness in relation to uncertainty regarding the future state of nature, including regime shifts, is tested with special focus on the sensitivity to changes in productivity at low stock sizes. Hence, one of the requirements to be able to measure BR is determining reference points and the performance of the system relative to these reference points. Managing a resource with a biologically robust strategy will contribute to its sustainability both in the long and short term, ensuring that future generations can meet their needs concerning goods and services provided by the resource.

A particular problem concerning management of the North Sea plaice is that the main plaice fishery, the beam trawl fishery (Rijnsdorp & Millner, 1996), also targets sole. The minimum mesh size of 80 mm in the beam trawl fishery selects sole at its minimum landing size. However, this mesh size generates high numbers of discards of undersized plaice (Daan, 1997) from 17 cm to its minimum landing size of 27 cm.

Recent estimates indicate discards around 50% of the total catch weight. In order to account for this discarding in the fisheries management, the stock assessment takes discard data into account (ICES 2007c) based on onboard observer programs with low sampling effort, because of the high costs per sample. The low sampling results in uncertain estimates of SSB and fishing mortality (ICES, 2007c). Hence, scientists and the fishing industry debate on discard levels; the fishing sector claims that due to bias and small sample sizes scientific discard estimates are too high.

Participatory governance and cooperative research may improve information on discards from the fisheries (Johnson & van Densen, 2007), especially because the fisheries can generate high sampling levels at low costs. Therefore, an extensive self-sampling programme was put in place in the Netherlands in 2004, aimed at proving that scientific estimates were indeed too high. Dutch scientists agree that their sample sizes are low, but question the quality of the discard estimates of the self-sampling program. They argue that the discrepancy between research and sector estimates is primarily due to differences in sampling procedures and lack of standardization (c.f. Liggins, Bradley, & Kennelly, 1998). In order to clarify the issue, a cooperative research program was organized. Its aim is to ascertain the degree to which differences in discard estimates are due to bias or sample size, to identify sources of bias and to develop procedures to deal with these. This hopefully leads to more precise discard estimates and to better management decisions and improved acceptance if all stakeholders are involved in the process.

In a study executed separately from the cooperative field programme mentioned above, we investigated the effects of both bias and precision in discard estimates on the management and BR of the plaice stock, both as separate components of a cost-benefit analysis.

Approach

A simulation model is used to analyse the effects of bias and uncertainty in discard estimates on the BR of a management strategy. Here we assume biological robustness to increase when SSB increases. In this model, the beam trawl fishery and population dynamics of the age structured plaice stock in the North Sea are coherently simulated. The model separately describes the plaice population, where the mature part of the population determines next year's recruitment based upon the Beverton & Holt relationship, and the behaviour of the fishing fleet that exploits the plaice stock resulting in a yearly fishing mortality, while an XSA stock assessment is executed to estimate the plaice abundance in the North Sea. Finally, the management advice is based on a simple harvest control rule, loosely based on the ICES precautionary approach, trying to maintain the stock at B_{pa}. While fishing, the fleet generates discards which, if over- or underestimated, may introduce an artificial measurement error into the model. By varying the precision of the measurement error we simulate the number of observations available, where error is assumed to decline with increasing observations. Data from ICES WGNSSK have been used to set the conditions of the model (ICES, 2007c).

Findings

Increased sampling (through participatory governance) does not necessarily lead to higher BR: When sampling addresses possible sources of bias, BR may increase; but when a larger sample size serves only to improve the precision, no substantial effects on BR are found. Stock assessments provide information on the status of the SSB, as long as unbiased discard estimates are available, even if these estimates are imprecise. Management decisions based on biased information lead to decreased BR. This implies that research should be directed towards understanding possible sources of bias, rather than increasing the sampling size per se.

6.2.1.2 Hypothesis 2

Background

In the North Sea the Dutch beam-trawl fisheries are mixed fisheries. For such fisheries TACs may be set to achieve single species target levels (F, SSB). On the other hand, effort control regimes may lead to a situation where target levels of one species are met at the cost of over- or under fishing others. For single species fisheries, such concerns are of less importance, as target levels may be achieved by either TAC or effort control.

Approach

We analysed the beam-trawl fisheries using the modelling approach of hypothesis 1, and simulating three scenarios for TAC and effort control:

(1) TAC: based on the EU long-term management plan for plaice and sole, assuming that plaice and sole can be fished independently
(2) Maximum effort constraint: the effort was set such that one of the species reached target levels, whilst overexploiting the other,
(3) Minimum effort constraint: the effort was set such that one of the species reached target levels, whilst under exploiting the other.

Findings

The results suggest that in the scenarios studied management through effort restrictions give a higher BR than management based on TACs: BR and total catches were highest in the long term for the third scenario (Minimum effort constraint) and lowest for the second scenario (Maximum effort constraint), with the first scenario (TAC) in between. In the short term, the second scenario (Maximum effort constraint) provided the highest catches.

These results imply that in a mixed fishery, BR may be achieved both through TAC and effort regulations, and that effort regulation may achieve higher BR, depending on the effort levels chosen. BR was highest in the third scenario (Minimum effort constraint), which aims at achieving target levels for one species, resulting in lower fishing pressure for other species and higher catches in the long run.

6.2.2 Baltic System – Evaluating Effort Management

The Spawning Stock Biomass (SSB) of the Eastern Baltic cod stock is far below the biological precautionary limit and target reference points set by the EU and ICES. The Baltic cod fishery has been managed with a TAC (total allowable catch) system, which has resulted in high levels of unreported and illegal landings. The management bodies need evaluations to test the cost and benefit of switching from a TAC-based regulation system to a direct effort control system by setting a total allowable effort (TAE) for this fishery. The argument for this is that TAE is easier to control and enforce in order to reduce misreporting. In this case study, we test hypothesis H2 by several approaches.

6.2.2.1 F-Adaptive Management

Background

In 2008 the EU Commission introduced a multi-annual management plan with an adaptive regulation system to gradually reduce the fishing mortality rate (F) of the Eastern Baltic cod (*Gadus morhua*) stock by 10% per year until the biological precautionary limits are reached (EU Commission, 2007). In the regulation this 10% reduction in F is translated directly into an overall 10% reduction in effort (TAE). Under the Baltic cod recovery plan, the EU Commission maintains seasonal and area-based closures as well, to protect the cod spawning zones from fishing activities.

In this case study, we compare the relative performance of both direct effort control (TAE), through F-adaptive management and of indirect effort control through spatio-temporal closure designs, to a traditional EU TAC regulation system. The overall purpose was to test the biological robustness (BR) of all the above management innovations and strategies to different environmental pre-conditions for cod recruitment, different levels of misreporting, and diverse fleet adaptation (capacity) levels ('as circumstantial indicators'). The environmental pre-conditions include either frequent or no major inflow of Atlantic water into the Baltic Sea system, which improves the survival conditions for cod eggs and fry (recruitment). The evaluations of the innovations include different reactions of the fleets resulting in different partial fishing mortalities and effort allocation patterns ('as innovation indicators'). The performance of the innovations were measured as the resulting fishing mortality (F), SSB precautionary limits and how fast these were reached ('as performance indicators'). The evaluations were performed on a 15-year time horizon.

Approach

A spatially and seasonally determined stock- and fleet-based bio-economic model using FLR (http://www.flr-project.org; http://www.efimas.org; Kell et al., 2007) was established as described in detail in Bastardie, Nielsen, and Kraus (2009). This monthly model based on ICES rectangles was used to model multi-fleet dynamics and simulate the performance and sensitivity of the adaptive approach under both TAE and TAC systems. The main characteristics and differences between the two systems lay in the decision rules used according to the cod recovery management plan (EU Commission, 2007). The TAC restrictions for the coming year corresponds to a 10% reduction in F from the assessed F level in the previous year. Meanwhile, the TAE restriction corresponds to a 10% reduction in the total fishing effort in the previous year (if the assessed $F > 0.3$), assuming a linear relationship between F and E, i.e. constant catchability.

Given these decision rules, the TAC system is characterized by the high impact of uncertainty in the assessed F when the TAC is set according to the ICES short-term forecast. In a TAC system, a short-term forecast is performed to estimate the stock abundance, which requires assumptions (qualified guesses) on fish growth (i.e. weights-at-age), stock recruitment, and adoption of previous year(s) exploitation pattern for the forecast year. On this basis, TAC is distributed among countries following the relative stability principle. The TAE system is characterized by only using the assessed F as a signal to decide on small-scale adjustments of the long term TAE for the coming year. The TAE therefore does not rely on assumptions made on stock development in the future, as is the case in the TAC using a short term forecast, but rather takes into account a short history of TAE values and relies on these values to decide on a TAE for the upcoming year.

Another specific characteristic of the TAC regulation compared with the TAE control is the different response of the fleet to the regulations (Nielsen, Sparre, Hovgaard, Frost, & Tserpes, 2006). If the TAC of each country is exhausted the fleets have the choice of continuing the fishery (non-compliance) or discard all age groups of a given species, and/or to direct their effort to other areas and other species. With effort control regulation the total effort is assumed to be exactly the TAE, and the share of each fleet segment is assumed to remain constant over the years if it was initially decided on the basis of effort allocation patterns that led to relative stability. The effort regulation system is assumed to induce fishermen to increase the fishing efficiency either through improved fishing (catching) power or improved spatio-temporal allocation of the fishing effort towards the target species (Nielsen et al., 2006).

Findings

The effect of environmental pre-conditions on the BR of the regulatory systems: The environmental context proves to be a dominant factor, as both TAE and TAC regulations failed to drive the biomass to the target reference point (B_{pa}) under adverse environmental conditions. The F target of 0.3, however, was reached under the TAE system. Under adverse environmental conditions, the recruitment success was too low to rebuild the stock within the given time horizon. In addition, the TAC management enhanced this failure by computing too optimistic short-term forecasts of the biomass level from the high historic recruitments. This caused a higher TAC being set than actually needed to catch the amount corresponding to the targeted fishing mortality at F_{y+1}. As a cascade effect, F could be set too high for several years for several years. This happened as the targeted F for the coming years was continuously set higher by the HCR from the dynamically assessed F_{y-1}. This overshadowed the effect of the 10% step-by-step decrease in F, advocated by the adaptive decision rule. In case of steady favourable environmental pre-conditions (or at least when a few Atlantic water inflows occurred to the Baltic), the stock recovered within the timeframe set with both regulation systems.

The effect of uncertainty due to misreporting on the two regulation systems: The TAC system is much more sensitive to misreported landings than the TAE system. This is due to the fact that the step-by-step effort reduction is not affected by lack of catch compliance and almost independent of the effect of misreporting on the catch-at-age matrix used in the stock assessment model. By contrast, misreporting regarding the effort (evaluated e.g. by simulating a systematic underestimation of the actual effort) leads to a dramatic change in the performance of the effort control system, which will not reach SSB targets and exceed F targets.

Robustness of the TAE system to changing fleet capacity: Change in fleet capacity as included, either (i) in terms of catching power, where we simulated a yearly continuous increase in fishing efficiency (e.g. continuous technical improvements), or (ii) in terms of the number of vessels when new vessels enter the fishery depending on increasing past profits. It turned out that the ability of the effort control system to reach the management target was reduced by these changes in capacity. In both cases, the assumption of constant catchability when deciding on the next period TAE is violated, and accordingly the allowed effort will systematically be set too high. Using a scenario with a dynamically changing number of vessels (Hoff & Frost, 2006) led to the collapse of certain fleet segments and a large depletion of the stock. Longer time was required to reach targets under the effort control system if the fleets chose to reallocate their fishing effort by area proportionally to the catchability (CPUE) in the fishing areas the year before.

6.2.2.2 Closures

Background

A spatio-temporal management regime for the Baltic Sea has been enforced by the EU Commission since 1995. In January 2005, an MPA (marine protected areas) network consisting of three large closures, which banned fishing in the main spawning areas of the Eastern Baltic cod, were enforced to restore the cod stock. Two closure scenarios were tested: (i) the potential effect of the EU closure proposal to protect spawning zones from all fishing activities, and (ii) a seasonal closure of all fishing activities in the ICES subdivisions 25, 26 and 27 from June 1st to September 31st.

Approach

Spatially- and temporally-explicit regulations are simulated, using the same framework as in the previous section, specifying seasons, areas, years, and fleets affected by the regulations. When the regulation occurs in a given time step, i.e. when some areas are closed in this time step, the involved fleets will move from the closed areas to other possible fishing areas (i.e. spatial effort displacement). In all cases, the possible change in the fishing mortality pattern due to the different effort re-allocation responses is embedded in the model as stock availability per area.

Findings

Spatio-temporal closure effect alone: For both closure designs tested, reduced landings in comparison to the 'F status quo' scenario were neither balanced out during the closure when the fishing effort was equally re-allocated to the remaining areas, nor if re-allocation was proportional to the catchability (i.e. CPUE) in each area. It is assumed that no additional effort was added to compensate for the losses.

In the open periods landings increased with higher CPUE for cod, since fish had been protected in certain periods and areas, but these increased landings were not sufficient to balance out the losses during the closed periods. Temporal redistribution of effort from closure to inter-closure periods is not assumed to take place.

Decreased fish mortality from the first year of closure was mainly due to drastic re-distribution of fishing effort out of the areas where the eastern Baltic cod is generally found in other areas, mainly the western Baltic Sea.

Both closure designs tested, temporal closures and the protection of spawning fish, resulted in increased SSB in the second year and enhanced recruitment in the third year and onwards. Closures, however, do not suffice alone to enable indicators to reach the management targets. The indirect effect of spatio-temporal closures was also mitigated by a possible change in the spatial pattern of effort re-allocation.

Robustness of the current TAC system with spatio-temporal closure: In relation to the F-adaptive approach the TAC system alone was neither able to reach the management targets when environmental conditions were adverse, nor with the spatio-temporal closures alone. However, combining the TAC system with spatio-temporal closures increased the robustness of the system, as shown by the fact that the management objective was reached for both the F and the SSB. Certain fleet-segments were not able to exhaust their specific quotas. As a result, the total catches are closer to the exact TAC, i.e. the total catch restriction needed to get the targeted level of F at year y, and the adaptive F-approach behave better than the TAC system alone. This statement remains valid even if the fleets choose to focus their effort on the higher catchability areas in response to the closure.

6.2.2.3 Conclusions

F-Adaptive Management

Environmental pre-conditions seem to play a dominant role in the overall performance of the management options. Accordingly, it is necessary to provide good indicators for environmental drivers in the fishery system and decide how information on the environment can be obtained most efficiently in order to include it into the advisory process. The biological robustness of the TAC system is also highly influenced by misreported landings, while the TAE system seems more robust in this regard. The TAE system is not so dependent on or sensitive to high quality information on actual landings, as no short-term forecast is performed here. Consequently, the TAE system is not so sensitive to assumptions on landings and F in the terminal assessment year. However, the performance of the effort regulation system can be greatly reduced if vessels misreport their effort. In general, it is necessary to improve the information on actual landings and obtain robust indicators for misreporting to include this information into the fishery advice. Furthermore, it would help to clarify which parts of the fishery system provides the best information in this regard.

Closures

The indirect effort regulation through spatio-temporal closures improves the biological robustness of both the TAC system and the TAE system. The total fishing effort exerted on the stock is reduced by moving the effort to areas where the particular stock is not found. This increases the SSB, leading to enhancement of the recruitment in the following years. It is assumed that the fleets do not respond to the closure by reallocating effort to other seasons. Regarding the TAE and spatio-temporal closures, robustness seems exclusively dependent on how effort is re-allocated between fleet-segments, areas, and seasons as well as on the assumed fleet specific (constant) catchability of cod. Information on effort allocation by vessel and fleet behaviour needs to be detailed and disaggregated in order to keep this system robust.

6.2.3 Western Shelf Hake (Merluccius merluccius) *– Evaluating Participatory Governance and Effort Management*

6.2.3.1 Background

For the Northern hake fishery in Western Shelf, hypotheses H1 and H2, and a combination of these, were tested, centred on participatory governance and effort management (Table 6.1). The objective of analysing these was to explore possible effects on BR of alternative management schemes that place more emphasis on effort restrictions (TAE) versus catch restrictions (TAC), and assuming that participatory governance will result in a reduction of unreported catches and better discard information.

Since the establishment of a minimum landing size of 27 cm in 1998 for Northern hake, the catches of the younger ages are almost completely discarded. According to ICES assessment working group of this species (ICES, 2008b) age 0 catches are completely discarded, age 1 fish are discarded at 85% and age 2 fish at 5% of the total catch. Older fish are not discarded. These discard rates may vary among years, areas and fleets.

Recently, more discard information has become available, as a result of an increase in sampling efforts by on-board surveys. However, this program does not yet cover all fleets that target hake and reliable discard information on these fisheries is still lacking. From this, discards are not taken into account in the assessment of this species. The question is whether these results lead to improper management decisions, given the fact that discard rates are high.

Another important issue in the management of the Northern hake stock is the level of unreported catches: Although there is no data available to quantify the level of underreporting in the Northern Hake fishery, it is believed that levels could be high. Unreported catches are not part of the current assessment process and predictions are made based on reported landings only. Doing a maximization of the private benefit function a level of 30% of underreporting was estimated in COBECOS EU-project.

An option that may reduce the level of underreporting and increase the accuracy of discard data is to increase the level of participation of fishers in the data collection and policy making. The idea behind this is that such involvement may lead to better compliance, especially if fishing practises and management actions become more compatible as a result of participation in the management process, costs of non-compliance for fishermen increase and fishers feel that the new situation indeed results in better management decisions and healthier fish stocks. In addition, if fishermen realized the importance of including reliable discard information into the assessment they may provide better information on discards.

Another option aimed at reducing the level of underreporting is the introduction of an effort control system (in our example implemented as days at sea), instead of, or on top of, a TAC approach: In such a system, the incentive to not report catches may be removed if indeed all catches can be landed legally (Shepherd, 2003) and as long as the fleet behaviour and landings are largely driven by the TAE instead of TAC. An additional advantage of such a TAE driven approach is that this is easier to control than a TAC system based on reported catches: All vessels involved in the hake fishery are currently monitored by satellite Vessel Monitoring System and it would be relatively easy to use this system for control purposes. Under TAE-based management the vessels would increase the catchability of hake directing their effort to this species in detriment of other less valuable species or through technological improvements. One solution to avoid this increment is to set a TAC together with the TAE.

6.2.3.2 Approach

The analyses were done using a simulation model written in FLR (Kell et al., 2007) developed within the COMMIT (2004–2007) and EFIMAS (2004–2008) EU-projects. The algorithm was described in detail in García, Santurtún, Iriondo, & Quincoces (2008) and was used to analyse possible long-term management plans for this stock (STECF, 2007).

Simulations involved both the population and fleet dynamics as well as the management process. The parameterization of the model was based on ICES data available to the working group. An initial random population of 2007 was projected to 2040 under different management options. Each year of projection a management process was run from which a management advice was obtained for next year. This management advice was then assimilated by the fleets. The management advice was determined by the HCR that aimed at reaching a fixed target fishing mortality. Each year the observed fishing mortality was compared with the target and the management advice was adjusted in order to be able to reach the target in the next year. This HCR was the same used in the STECF group on long-term management plans for Northern hake (STECF, 2008). Depending on the scenario, this advice and further actions relate to either TAC and/or TAE. In scenarios that combine TAC and TAE, the fleet behaviour was dominated by the TAE: When catches resulting from a TAE advice were smaller than those determined by TAC, it was assumed that the fleets will stop fishing before exhausting the catch limit and will report the total catch. In contrary, when potential catches of a TAE advice are larger than those of TAC, it was assumed that fishers would continue fishing until they exhaust the TAE and do not report the extra catch. This behaviour was a consequence of the fact that effort can be controlled easier than catch.

A total of five scenarios were simulated to analyse the biological robustness of different management regimes:

1. Actual TAC-based management: underreporting (catch of legal size fishes) = 30% and reliable discard estimates (undersized fish) are not available
2. TAC-based management regime with reliable discard estimates (as an assumed result of participatory governance), underreporting = 30%
3. TAC-based management regime with gradually decreasing levels of underreporting (from 30% to 5% over 5 years, as an assumed result of participatory governance), reliable discard estimates are available.
4. TAE-based management regime: Underreporting is a result of possible mismatches between TAE and TAC, reliable discard estimates are not available.

    a. TAE-based management regime combined with participatory governance: Underreporting is a result of possible mismatches between TAE and TAC, reliable discard estimates (undersized fish) are available.

6.2.3.3 Findings

The inclusion of discard information in the assessment process (scenarios 2, 3 and 5), did not lead to major differences in BR, neither in TAC-based management regimes nor in effort-based management regimes. The level of SSB, catches and fishing mortality obtained in scenarios 2 and 5, were very similar to those obtained in scenarios 1 and 4 respectively. This can be explained by the fact that fishing mortality of younger ages is not used to calculate the reference fishing mortality, and they are immature so they do not contribute to the SSB (note that an improvement in BR can only be achieved by increasing the accuracy of the data and not by reducing the discard level).

The simulations further suggest that a reduction of underreporting as a result of either effort-based management or participatory governance, substantially improves the biological robustness of the system:

In the effort-based management regime of scenarios 4 and 5, the levels of underreporting decreased substantially from 30% down to 8% and SSB stayed at levels above safe biological limits. That is, BR was positively influenced as a result of a decrease in underreporting, resulting from effort control measures. In contrast the TAC management regimes of scenarios 1 and 2 led to SSBs well below safe biological limits (in roughly 15% of the years in the long term, until 2040), high fishing mortalities and lower catches. Fishing mortality targets could not be achieved. This suggests that, indeed, management through effort restrictions results in higher biological robustness than management based on TACs.

A reduction of underreporting as an assumed result of participatory governance (scenario 3), may also substantially increase BR: The probability of the system being below safe biological limits was 0 in all the years of the simulation. The fishing mortality was around or well below the target fishing mortality. Catches of this scenario were lower in initial years but highest in the long run. This implies that participatory governance that substantially reduce underreporting may be interesting to consider in situations where underreporting is an issue.

6.2.4 *Western Shelf Anchovy*(**Engraulis Encrasicolus**)*: Evaluating Decision Rules*

For the Western Shelf anchovy case the hypothesis H3 (Table 6.1) was tested, centred upon decision rules. The goal of testing this hypothesis was to explore the possible effects of harvest control rules (HCR) on biological robustness (BR).

6.2.4.1 Background

The Western Shelf anchovy fisheries have been closed since the second half of 2005, except for 1750 tons captured during a reopening in the first half of 2006 and 136 tons captured by the French fleet in 2007 in the framework of an experimental-commercial fishery with observers onboard. The closure was a reaction to the poor condition of the spawning stock, assessed to be below B_{lim} (21,000 t) in 2005 and to remain below B_{pa} (33,000 t) from 2006 to 2008. Until the closure, the anchovy fishery was managed by a fixed annual total allowable catch (TAC) that was split into country quotas, with associated technical measures, such as minimum landing size and minimum mesh size in some areas and for some gears (Council Regulation No 850/98). Historically, the TAC has been set around 30.000–33.000 tons without taking variations in recruitment into account.

Currently two research surveys are conducted in the spring each year to assess the anchovy stock: an ichthyoplankton and adult survey for the implementation of the daily egg production method and an acoustic survey. Since 2003 a juvenile acoustic survey is conducted in the autumn each year (ICES, 2008a). Currently this survey is under development and being tested, but will hopefully provide yearly estimates of next year's recruitment that would assist management advice.

Both scientific working groups and the fishing industry agree that the current management of this stock is not adequate and needs to be reviewed. Decision rules are looked upon as a potential innovative option for the anchovy fisheries, as an adaptive management tool suggested by the ICES working group on the assessment of mackerel, horse mackerel, sardine and anchovy (ICES, 2004). In addition, the STECF states in 2008 that 'complementary management measures to output control (TAC) need to be further investigated to maintain the longer-term viability of the stock (closed seasons, closed areas, minimum size, etc.).' (STECF, 2008).

Currently the European Commission intends to propose a long-term management plan for the anchovy, to be implemented as soon as the stock has recovered. To provide the necessary scientific basis, the biological and economical performance of two basic harvest control rules (HCRs) has been evaluated (STECF, 2008). In these management strategy evaluations, the traditional January to December management calendar has been replaced by a management cycle from July to June, in which the HCRs catch options are based solely on the results of the just completed spring survey. With this approach the major uncertainty arises from the recruitment level and the fisheries during the first half of the year, which are unknown at the time of setting the TAC. Therefore the HCR has to be robust, not only regarding the observation errors of the spring surveys, but also regarding the uncertainty about the next recruitment. With juvenile acoustic surveys in the autumn, a management calendar from January to December with a possible mid-year revision after the spring surveys could be reconsidered.

Decision rule systems are management systems that intend to transfer the decision power from politicians to a system that gives 'automatic' responses. There are two kinds of decision rule systems: harvest control rule systems that aim at reducing the reliance on political processes in management decision-making, and non-predictive adaptive systems, which furthermore aim at reducing the traditional reliance on specific predictions about stock dynamics. In this study we focus on the latter by examining the potential benefits of a harvest control rule that includes new and additional stock information (two step management) versus a management system that does not include such information (one step management).

6.2.4.2 Approach

Given that the current official management calendar still goes from January to December, and that there might be difficulties with changing to a July to June calendar, we have tested the performance of HCRs for the same recruitment scenarios as adopted by STECF, in which the original TAC can be revised in the middle of the year. The aim is to see if using new information to correct the TAC in the middle of the year reduces the biological risks for the stock, based on the assumed recruitment in the beginning of the year.

These analyses were done using the management strategy evaluation algorithm implemented within the FLR framework (Kell et al., 2007) to compare the biological robustness of the TAC management regimes in different scenarios. The model simulates the population dynamics (operating model), the fleet dynamics, the surveys (observation model) and the management process. The operating model consists of a single age-structured stock exploited by a single fleet with harvest rates (ratio of catch to total biomass) in seasonal time steps (half year basis). The parameters of the operating model were based on the results from the seasonal Integrated Catch-at-Age Analysis (ICA) of the relevant ICES Working Group (ICES, 2006; STECF, 2008). The 2007 population was projected to 2017 with the different scenarios defined. The HCR is based on the biomass, the age one recruitment and the TAC in the previous year.

Firstly, regarding the recruitment levels used to set every year's TAC in January, three different scenarios are considered:

(1) recruitment based on a survey estimate (from the autumn juvenile acoustic survey), (2) precautionary recruitment level (at the historical 25 percentile), and (3) non-precautionary recruitment level (at the historical median recruitment level).

Secondly, two scenarios are considered, with the TAC set in either yearly or half-yearly steps. For these, two sets of HCR are considered: (a) a management scheme based on a fixed yearly TAC determined in January and (b) a two-step management scheme where the January TAC is revised in the middle of the year, using SSB estimates from the spring surveys. In both sets of harvest control rules, the TAC is set as a percentage (given by the harvest rate) of the estimated SSB (either foreseen -at the beginning of the year- or observed with the spring surveys -at the middle of the year-).

This combination of scenarios for recruitment assumptions and TAC updating dynamics leads to 6 scenarios in total. In these, the biological robustness has been evaluated in terms of the following performance statistics: median SSB, average catch, inter-annual variability in TAC, probability of the population being below B_{lim}, average number of years to get the population above B_{lim} and differences between the initial and the revised TACs.

6.2.4.3 Conclusions

Our results suggest that indeed, a harvest control rule that includes additional and most up to date (through the two step management) may be more biologically robust than a management regime that does not include such information (one step management), especially at high harvest rates.

We found that a two-step management approach, with a TAC update in June, performs better than a one step management approach because at similar harvest rates it allows either higher catches at smaller or similar risks (cases of precautionary and non precautionary recruitment scenarios) or similar catches at smaller risks (case of recruitment input coming from a survey); leading to an overall higher BR. A partial exception to this conclusion occurs at harvest rates higher than 0.5 (TAC = 50% of estimated SSB) for the scenario assuming a precautionary level of incoming recruitment for which a revision at the middle of the year implies similar or slightly higher biological risks than maintaining the TAC set in January; this being due to the higher catches then allowed (by about 40% in comparison with single step approach) and to the observation errors in the spring surveys.

In addition, we found that for the two step management approach, setting the original TAC in January, based on a recruitment survey in the previous autumn, improves the performance of management by allowing similar catches at smaller risks than assuming a recruitment level. For the single step approach this is also true for the comparison with the assumption of non-precautionary recruitment. However, the precautionary recruitment at high harvest rates implies smaller risks than using a recruitment survey's estimate due to the strong limitation of the allowable catches.

6.2.5 Faroe Islands – Evaluating Effort Management

6.2.5.1 Background and Hypotheses

In 1996, an effort regulation system consisting of individual transferable quotas in terms of fishing days within fleet categories was introduced to manage the mixed demersal fisheries of the Faroe Plateau, after the previous TAC system had been rejected (Jákupsstovu, Cruz, Maguire, & Reinert, 2007; Nielsen et al., 2006). Not all the main objectives of the new system have been achieved. For instance the average yearly fishing mortality value of cod (*Gadus morhua*), haddock (*Melanogrammus aeglefinus*) and saithe (*Pollachius virens*) was set at 0.45, a target which has several times been exceeded for cod (ICES, 2006). Furthermore, the system was based on the underlying assumption that fishermen will switch target species according to the relative availability of the stocks. This assumption has not been verified (Jákupsstovu et al., 2007; ICES, 2006). Therefore, effort management without stock specific measures may not be appropriate for the mixed fisheries in the Faroe Plateau. In this study the bio-economic consequences of the effort regulation system are compared to a traditional TAC system, based on some main assumptions of the two systems.

The Faroese case study focused on H2. It was examined in several ways in order to assess the efficiency of the effort-based management (TAE) by comparing it with a management based on total allowable catch (TAC). We operationalize this relationship in several ways, which included both the consideration of other management measures and of environmental conditions.

6.2.5.2 Approach

A management strategy evaluation model was developed within the FLR framework (Kell et al., 2007) to compare the Faroese effort management system with a TAC system currently applied in EU fisheries, both on a single and multi-stock basis. Standard stock assessment data from ICES was used (ICES, 2006). Fleet data was obtained from the Faroese Fisheries Laboratory. The main differences between the TAC and effort regulation evaluated here lie in the different rules employed to regulate fisheries in the coming year. In both the Effort and TAC systems, the target fish mortality in the year following the stock assessment (F_{y+1}) is given by the harvest control rule (HCR) as follows:

- $F_{y+1} = F_{low}$ (0.2) if $SSB_y < B_{lim}$,
- $F_{y+1} = F_{target}$ (0.45) if $SSB_y > B_{pa}$
- $F_{y+1} = F_{low} + (F_{pa} - F_{low}) * (SSB_y - B_{lim})/(B_{pa} - B_{lim})$ if $B_{lim} < SSB_y < B_{pa}$

In a TAC system, a short-term forecast is performed to estimate the stock abundance in year *y+1*, which requires assumptions (traditionally an average over the most recent years) concerning fish growth (i.e. weights-at-age), stock recruitment, and fishing mortality (fishing pattern) for the forecast year. The target fishing mortality to be applied is then converted into a TAC using the traditional Baranov catch equation. In an effort regulation system, the effort level (total number of fishing days allowed in the fishery) is directly deducted from the target fishing mortality, simply assuming that the catchability is constant.

In the case of mixed fisheries where one single effort is applied on several species, this effort level is set at the minimum across effort levels corresponding to the fishing mortality defined by the HCR for each individual species. Furthermore, no discards were assumed in the TAC simulations, and constant catchability and effort-share between fleets were assumed in the TAE simulations.

The current Faroese effort system has little flexibility, with only minor adjustments of the amount of days at sea allowed, in spite of scientific recommendations for large reductions in fishing mortality. Indeed, the total nominal effort has only been reduced by 15% in ten years (Jákupsstovu et al., 2007). Therefore, the BR of both systems with regards to their flexibility to follow drastic changes in scientific advice were tested under three different levels of year-to-year variation ('bounds') in management decisions: 1% (as a proxy of the current rigid Faroese system), 15% (as a proxy for current EU long-term management plans), and without bound (full flexibility). This model included several sources of uncertainty (on abundance indices, recruitment levels, weight-at-age estimates, and fleet selectivity patterns) to mimic both environmental influence and variability induced in the system in order to assess the biological robustness of the effort regulation model. Observations were made by manipulating the overall fishing activity and investigating its effect on stock trends.

6.2.5.3 Results

Effort-based regulation simulated with a low level of management flexibility as currently observed in the Faroes did not appear sustainable with regards to the stock biomass. To be efficient, the Faroese effort regulation system would need to allow more effort variation from one year to the other. Current measures (e.g. areas closed to trawling) could prevent excessive fishing mortality, which suggests that the effort-based regulations currently applied in the Faroese demersal gadoid fisheries system are not sustainable (precautionary) as shown by the current levels of fishing mortality and the potentially serious impact of environmental fluctuations on recruitment and fish growth.

One possibility we examined was that an effort-based HCR is more biologically robust and economically efficient than catch-based HCR. This may be the case as it is less dependent on uncertainty in growth, recruitment and stock assessment. On the other hand, effort-based HCR is more dependent on uncertainty in the relationships between effort and stock, i.e. in the catchability parameter, which can vary dramatically. Hence, provided there is enough year-to-year flexibility the TAE appears to be a better management strategy than TAC-based management because it results in more sustainable stock biomasses given the same uncertainty about biological parameters. These assumptions, in turn, are based on the absence of discards when implementing the TAC regulation. Our results confirmed only partly these possibilities. Given the modelling assumptions, the TAE system in a single-species context may provide on average higher SSB and revenues than the TAC system. But the results depend strongly on the level of flexibility of the system (reactivity to changes in scientific advice), the state and the dynamics of the stock, and the level of uncertainty. In particular, the variability in catchability, assumed to be partly driven by the environmental fluctuations for some fleets (ICES, 2007a), may induce as much uncertainty about future states as when using a traditional TAC forecast.

In a mixed-fisheries context, modelling showed that single-species objectives could not be met simultaneously, because of differences in the dynamics and initial state of the various stocks. One single precautionary effort applied to all fleets would ensure biological sustainability and low risks but would result in economic losses and underexploitation of some valuable species in the short-term. On the contrary, one high effort level would jeopardize the sustainability of the most depleted stocks. Setting the fleet level at an intermediate level of effort but with additional measures to protect the depleted stock would help prevent under-exploitation of certain stocks and contribute to sustainable and profitable fisheries, provided that there is some spatio-temporal separation of stocks on some fishing grounds. The Faroese also use an advanced system of technical measures ensuring clear spatial separation between gears, but these should also be designed in order to ensure the best exploitation of the various stocks.

6.2.5.4 Conclusions

In a single-species context, and in the absence of discards, a TAE system does not necessarily perform better than a TAC system. It all depends on the state and the dynamics of the stock, the level of uncertainty and the level of flexibility of the system (reactivity to changes in scientific advice).

But in a mixed-fisheries context, TAE would appear to be a more robust management strategy than TAC-based management, considering the large fluctuations in management derived from the single-species HCR and the large amount of expected discards this would create.

The effort-based regulations currently applied in the Faroese demersal gadoid fisheries system are not sustainable (precautionary) as shown by the current levels of fishing mortality and the serious impact of environmental fluctuations on recruitment and fish growth. However, the main issue may not be the effort-management in itself, but rather its lack of reactivity to adjust to scientific recommendations.

6.3 Conclusions

In this chapter, several detailed case studies have been evaluated to explore the potential effects of innovative management on Biological Robustness, as indicated by the percentage of years in which standard biological reference points (B_{pa}, F_{pa}) were met. The case studies differed considerably from one another and will be published in detail elsewhere, where the focus will be on the individual case-studies, rather than the generic mechanisms. The advantage of using detailed case studies is that they may reveal pitfalls and strengths of the innovations in real-life. The drawback, however, is that the differences between the cases makes it difficult to draw generic conclusions about the outcomes. Still, some conclusions about the different innovations can be drawn from the collection of case studies:

6.3.1 Participatory Governance

In both the North Sea case and the Western shelf case a narrow definition of participatory governance was used, referring only to the collection and exchange of information and assuming that this leads to better estimates of landings and discards for the stock assessment that the fish stock management is based on. In both cases it was studied how BR is affected by new and additional discard information, assumed to be the result of participatory governance.

In the case of the North Sea the effects of precision and bias in discard estimates were studied separately, concluding that bias has important implications for BR. It turns out to be more important to identify and address possible sources of bias than to increase sample sizes, which emphasizes the need for uniform sampling procedures and understanding the factors that influence discard levels.

In the Western shelf case the effect of including discard information was studied and regarded as reducing bias in catch data on the younger age groups. Contrary to the findings in the North Sea case, reducing this bias did not affect the BR.

A possible explanation for this difference could be that the discard estimates in the North Sea case comprise both mature and immature individuals. Mature individuals are part of the SSB, and if their discard estimates are biased it is bound to affect SSB estimates. In the Western shelf case, however, the discards consist only of immature individuals and their discard bias would not affect SSB estimates and management decisions based on these. This does not imply that discarding does not affect SSB.

This possible explanation is supported by the fact that in the Western Shelf hake case, BR is affected by changes in the underreporting of landings. Avoiding such underreporting can also be seen as reducing bias regarding the catch data that are used in the assessment. However, in this case the bias concerns mature fish that indeed contributes to the SSB. This makes the situation similar to the North Sea discards case, and the findings seem to correspond.

It is concluded that in participatory governance the collection and exchange of data can indeed lead to improved BR providing that such information is important to the assessment process. However, care must be taken to avoid bias, especially in the data that affects the SSB if it is used as a management reference point.

Similar conclusions were drawn by Dickey-Collas, Pastoors, and van Keeken (2007). They, however, only studied the effect on the SSB estimation and did not include a full feedback model to study the effect of management on the BR as was done in our analyses.

6.3.2 Effort Management

The effects of effort management on the biological robustness were tested for several mixed fisheries, where effort management is thought to be a potential innovation (Nielsen et al., 2006, Rijnsdorp, Daan, Dekker, Poos, & Van Densen, 2007). One of the obvious results from the simulations performed in the different case studies is that the BR depends on the level of effort set annually. As implemented in the simulation studies, the effort management allowed fishing efforts to be set so as to prevent over-quota discarding of one or more species.

Another main conclusion is that a pre-requisite for effort management is that the link between fishing effort and fishing mortality is at least predictable or controllable. This link is influenced by changes in the dynamics of the species (Paloheimo & Dickie, 1964; Horwood & Millner, 1998), the dynamics of the fleets (Marchal, Nielsen, Hovgård, & Lassen, 2001; Ulrich, Pascoe, Sparre, de Wilde, & Marchal, 2002) and environmental factors (Bastardie, Nielsen, & Kraus, 2009). Therefore the management regime should account for such influences, e.g. by including the possibility to allocate effort in certain seasons and/or areas. The low biological robustness of the Faroese effort management of mixed fisheries can be traced back to lacking catchability control combined with high fish mortality.

With enough flexibility (in the year-to-year variation in allowed effort) effort-based management appears to be more biologically robust TAC-based management, as it provides better results with a given uncertainty in biological parameters. This assumption is based on no discards in the TAC regulated fisheries.

Our evaluations suggest that biological robustness in a TAC system is strongly influenced by misreported landings, which is not the case in effort regulation systems. This is because the effort regulation system is not as sensitive to and dependent on high quality landing information and assumptions in relation to short-term forecasts in the terminal assessment year. With respect to misreporting it is furthermore thought to be easier to control effort restricted fisheries than catch restricted fisheries.

Spatio-temporal closures act as indirect effort regulations that displace effort from one area to another. This is shown to improve the BR of both TAC and direct effort control systems by reducing the fishing effort exerted on the stock(s). An SSB surplus is created leading to enhanced recruitment in the following years. However, the BR is closely linked to how the effort is re-allocated to fleet-segments, areas, and seasons, as well as being sensitive to fleet-specific catchability. In an effort regulation system, it is therefore necessary to obtain high resolution information on the behaviour of the fleet as well as actual effort allocation by individual vessel.

6.3.3 Harvest Control Rules

The hypotheses we analysed referring to harvest control rules explicitly (cf. Western Shelf case), focus on differences between harvest control rules, instead of comparing a situation with or without a harvest control rule. In addition, although not said explicitly, most of the other hypotheses considered in the different cases studies are tested simulating harvest control rules.

In the Western Shelf anchovy case, we conclude that in the case of a short-lived species with high fluctuations in abundance, the in-year updates of the harvest control rules, according to new information about the state of the system (SSB for anchovy), improves the biological robustness of the harvest control rule itself. In other words, in-season adaptive management (two step management) overcomes the single step application of the harvest control rule on annual basis. Another finding is that, for this type of short-living species, a neat improvement of the management is achieved through incorporation of a recruitment survey in the observation of the system.

6.3.4 Innovations Work, Provided

While we use the term 'innovations' to indicate that the relevant approaches to management have not been used extensively in Europe, these are not new or untested ideas – all of them have been incorporated in various forms into modern fisheries management regimes in developed countries. Moreover, being 'ideal types', these innovations can only be explored in their pure form on paper. In practice these innovations take many different forms, also leading to different analytical definitions in the present context, and can therefore not be examined in their pure form. In the present study the innovations were tested as they appear in the cases covered – in combination with other management strategies and circumstances.

This study demonstrates that the innovative approaches to management may perform better than those currently used, both with respect to BR as well as resulting in larger catches. This suggest that if such innovations and harvest rules are adopted, substantial improvements may be possible compared to current fisheries practices, especially if correct, relevant and recent information on the state of the system is available, combined with a precautionary approach regarding fishing pressure.

References

Bastardie, F., Nielsen, J. R., & Kraus, G. (2009). A Management Strategy Evaluation framework for the Eastern Baltic cod fishery to test the robustness of regulations against environmental variation and fleet response. (Accepted. ICES Journal of Marine Science).

COMMIT. (2004–2007). Creation of Multiannual Management Plans for Commitment. Specific-Targeted-Research-Project-EU-6th-Framework Programme-SSP8-CT-2003-502289.

Daan, N. (1997). TAC management in North Sea flatfish fisheries. *Journal of Sea Research, 37*(3–4), 321–341.

Degnbol, P., & Jarre, A. (2004). Review of indicators in fisheries management – A development perspective. *African Journal of Marine Science, 26*(1), 303–326.

Dickey-Collas, M., Pastoors, M. A., & van Keeken, O. A. (2007). "Precisely wrong or vaguely right: simulations of noisy discard data and trends in fishing effort being included in the stock assessment of North Sea plaice."*ICES Journal of Marine Science, 64*(9), 1641–1649.

EFIMAS. (2004–2008). Operational Evaluation Tools for Fisheries Management Options. Specific-Targeted-Research-Project-EU-6th-Framework Programme-SSP8-CT-2003-502516. *WWW.efimas.org*

EU Commission 2007. Council Regulation (EC) No 1098/2007 establishing a multi-annual plan for the cod stocks in the Baltic Sea and the fisheries exploiting those stocks, amending Regulation (ECC) No 2847/93 and repealing Regulation (EC) No 779/97

FAO. (2005). *Review of the state of marine fishery resources*. FAO Fisheries Technical Paper, 457.

Francis, R. (1997). "Risk" in fisheries management: A review. *Canadian Journal of Fisheries and Aquatic Sciences, 54*(8), 1699.

Hoff, A., &. Frost, H. (2006). *Economic response to harvest and effort control in fishery*, Fødevareøkonomisk Institut: 43.

Horwood, J. W., & Millner, R. S. (1998). "Cold induced abnormal catches of Sole." *Journal of the Marine Biological Association of the United Kingdom, 78*, 345–347.

García, D., Santurtún M., Iriondo, A., & Quincoces I. (2008). Northern Hake long term management plan evaluation. *Working document, ICES WGHMM*, Copenhagen, May 2008.

ICES. (2004). Report of the working group on the assessment of mackerel, horse mackerel, sardine and anchovy (WGMHSA), 9–18 September 2003. ICES CM 2004/ACFM:08.

ICES. (2006). Report of the north-western working group, 25 April–4 May 2006. ICES Document CM 2006/ACFM: 26.

ICES. (2007a). Report of the north-western working group (NWWG), 24 April–3 May 2007. ICES CM 2007/ACFM:17.

ICES. (2007b). Report of the working group on the assessment of mackerel, Horse Mackerel, Sardine and Anchovy (WGMHSA), 4–13 September 2007. ICES CM 2007/ACFM:31.

ICES. (2007c). Report of the working group on the assessment of demersal stocks in the North Sea and Skagerrak – Combined Spring and Autumn (WGNSSK), 1–8 May 2007, ICES CM 2007\ACFM:18 and 30.

ICES. (2008a). Report of the working group on the Anchovy, ICES Headquarters, 13–16 June 2008. ICES CM 2008/ACOM:04.

ICES. (2008b). Report of the working group on the assessment of southern shelf stocks of Hake, Monk and Megrim. ICES Document CM 2008/ACFM: 02.

Jákupsstovu, S. H., Cruz, L.R., Maguire, J-J., & Reinert, J. (2007). Effort regulation of the demersal fisheries at the Faroe Islands: a 10-year appraisal. – *ICES Journal of Marine Science, 64*, 730–737.

Johnson, T. R., & van Densen, W. L. T. (2007). Benefits and organization of cooperative research for fisheries management. *ICES Journal of Marine Science, 64*(4), 834–840.

Kell, L. T., Mosqueira, I., Grosjean, P., Fromentin, J.-M., Hillary, R., Jardim, E., et al. (2007). FLR: an open-source framework for the evaluation and development of management strategies. *ICES Journal of Marine Science, 64*, 640–646.

Kitano, H. (2004). Biological robustness. *Nature Reviews Genetics, 5*(11), 826–837.

Law, R. (2000). Fishing, selection, and phenotypic evolution. *ICES Journal of Marine Science, 57*, 659–669.
Lenski, E. L., Barrick, J. E., & Ofria, C. (2006). Balancing Robustness and Evolvability. *Plos Biology, 4*(12), 2190–2192.
Liggins, G. W., Bradley M. J., & Kennelly S. J. (1998). Detection of bias in observer-based estimates of retained and discarded catches from a multi species trawl fishery. *Fisheries Research, 32*(2), 133–147.
Marchal, P., Nielsen, J. R., Hovgård, H., & Lassen, H. (2001). Time changes in fishing power in Danish cod fisheries of the Baltic Sea. *ICES Journal of Marine Science, 58*, 298–310.
Nielsen, J. R., Sparre, P. J., Hovgaard, H., Frost, H., & Tserpes, G. (2006). Effort and Capacity Based Fisheries Management. In L. Motos & D. Wilson (Eds.), The Knowledge Base for Fisheries Management. *Developments in Aquaculture and Fisheries Sciences Series* 36 (pp. 163–216). Amsterdam: Elsevier.
Rijnsdorp, A. D., & Millner R. S. (1996). Trends in population dynamics and exploitation of North Sea plaice (*Pleuronectes platessa* L) since the late 1800s. *ICES Journal of Marine Science, 53*(6), 1170–1184.
Rijnsdorp, A. D., Daan, N., Dekker, W., Poos, J. J., & Van Densen, W. L. T. (2007). Sustainable use of flatfish resources: Addressing the credibility crisis in mixed fisheries management. *Journal of Sea Research, 57*(2–3), 114–125.
Rosenberg, A. A. (1994). Uncertainty and risk evaluation in stock assessment advice for U. S. marine fisheries. *Canadian Journal of Fisheries and Aquatic Sciences, 51*(12), 2715.
Paloheimo, J. E. & Dickie, L. M. (1964). Abundance and fishing success. *Rapports et Proc's-Verbaux des Réunions Conseil International pour l'Exploration de la Mer, 155*, 152–163.
Shepherd, J. G. (2003). Fishing effort control: Could it work under the common fisheries policy? *Fisheries Research, 63*, 149–153.
STECF. (2007). Northern hake long-term management plans. Report of the sub-group on balance between resources and their exploitation (SGBRE-07-03) of the STECF. *Commission Staff Working Paper.*
STECF. (2008). Report of the Meeting to inform the Commission on a long-term management plan for the stock of anchovy in the Bay of Biscay (ICES Sub-area VIII). STECF/SGBRE 0801.*Commission Staff Working Paper:* 119 pp + annexes
Stelling, J. Sauer, U., Szallasi, Z., Doyle III, F. J., & Doyle, J.(2004). Robustness of Cellular Functions, *Cell Volume, 118*(6), 675–685.
Ulrich, C., Pascoe, S., Sparre, P. J., de Wilde, J.-W., & Marchal, P. (2002). Influence of trends in fishing power on bioeconomics in the North Sea flatfish fishery regulated by catches or by effort quotas. *Canadian Journal of Fisheries and Aquatic Science, 59*, 829–843.
United Nations. (1972). *The United Nations Conference on the Human Environment*. Stockholm, UN.

Websites

COBECOS EU-FP6-044153-Project: Costs and Benefits of Control Strategies s. https://maritimeaffairs.jrc.ec.europa.eu/web/cobecos/1
FLR. Fisheries Library in R. http://flr-project.org
EFIMAS EU-FP6-502516-Project: Operational Evaluation Tools for Fisheries Management Options. http://www.efimas.org and http://wiki.difres.dk/efimas/doku.php/

Chapter 7
Evaluating Economic Efficiency of Innovative Management Regimes

Erik Buisman, Hans Frost, Ayoe Hoff, Arantza Murillas, and Jeffrey P. Powell

Abstract The objective of this chapter is to estimate the likely implications for economic efficiency from the introduction of various innovative management systems in a number of European fisheries. Bio-economic models are developed to simulate and evaluate the impact of the management systems as they are applied to the following fisheries: Baltic cod (*Gadus morhua*), North Sea flatfish, Spanish northern hake (*Merluccius merluccius*), Faroe cod, haddock (*Melanogrammus aeglefinus*) and saithe (*Pollachius virens*).

For each fishery, a number of scenarios are evaluated and ranked according to economic performance measures. The analyses suggest that economic performance is significantly influenced by the type of management system implemented.

With respect to participatory governance, it is shown that in the part of the northern hake fishery exploited by Spain, the introduction of participatory governance provides better economic performance in the long-run when compared to the more traditional total allowable catch (TAC) system. With respect to effort-based control, it is shown in both the Faroe and the North Sea cases that rights-based effort control can have both positive and negative implications for economic performance when compared to TAC systems. In the North Sea case, the impact on economic performance is shown to depend on how effort restrictions are set. Finally, it is shown that marine protected areas may influence the economic performance of fisheries negatively in the short and medium term, even though this effort-based control system has a positive influence on recovery of fish stocks.

Keywords Economic efficiency · Economic performance · Management systems · Net present value · Simulation models

E. Buisman (✉)
LEI, Alexanderveld 5, 2585 DB Den haag, The Netherlands
e-mail: erik.buisman@wur.nl

K.H. Hauge, D.C. Wilson (eds.), *Comparative Evaluations of Innovative Fisheries Management*, DOI 10.1007/978-90-481-2663-7_7,

7.1 Introduction

When considering the current initiatives within the EU towards fish resource conservation, it is critical to assess the impact of changes in fish stocks on the economic efficiency of fisheries. Indeed, different management tools will have different implications for stocks and thereby for the economic efficiency of fisheries targeting those stocks. The objective of this chapter is to estimate the likely implications for economic efficiency relative to the introduction of various innovative fishery management systems.

From an economic perspective, a fishing fleet is said to be economically efficient, or more correctly, Pareto efficient, if it is impossible to reallocate goods and production factors between agents involved in the fishery in such a way that somebody is made better off without making somebody else worse off (Varian, 1999). This is defined as the *Pareto condition*. The *Kaldor Hicks* condition says that if gains by one agent, groups of agents, etc., are sufficiently high to compensate the losses of another agent, groups of agents etc., then the solution is an improvement in economic efficiency. The Kaldor Hicks criterion does not require that compensation is actually paid, which means that a more efficient outcome can leave some people worse off. In practice, it is difficult, if not impossible, to implement changes that result in Pareto improvement, whereas it is much easier to identify solutions which can result in outcomes that will satisfy the Kaldor-Hicks criterion for improvement in efficiency.

Often heard phrases when dealing with fisheries, are: 'One vessel is more economically efficient than another', or 'the vessel has become more economically efficient'. These are not wrong from a linguistic perspective, but such phrases conflict with the definition of economic efficiency stated above as they describe states of the system that, given the heavily exploited nature of European fisheries, will be economically *inefficient* in most cases, even though some agents may have increased their individual economic performance.

In principle, a fishery is Pareto efficient if the resource rent is maximised (Frost, Buisman, Hoff, & Murillas, 2008). The resource rent is the amount left over after all costs have been deducted from the landings value, and is a key variable with respect to assessing the conditions for economic efficiency of a fishery. The costs include both direct exploitation costs and so-called opportunity costs, i.e., how much could have been earned if the capital and labour had been used in the best alternative to fishing. At vessel level, landings values and costs can be extracted from recorded statistics. For the whole industry, the exploitation costs will depend on both the total fishing effort and the level of exploited stocks, as one vessel's catches will decrease the stocks in the future and hence increase the fishing costs. This external effect is not accounted for in the costs and earnings statistics. As such, economic efficiency cannot be assessed by solely using available economic data, but must be estimated using bio-economic models that include information about the costs and earnings of fishing vessels as well as the size of the fish stocks. In such models, the spawning stock biomass should optimally be considered a capital input in the same way as vessel capital, and return (remuneration) on the fish stocks should be included in a similar manner. The return or rent of fish stocks can be viewed as compensation to society for the common property of the fish stock, and this should be added to the effort costs.

The value of the resource rent will, as such, depend on the level of effort used, as well as the state of the fish stocks. The latter is implicitly a function of the former, and, as such, some effort level will exist for which the resource rent is maximised for a given fishery. When only one stock is exploited, the maximum resource rent of the fishing fleet will coincide with the maximum economic yield of the stock (Gordon, 1954), i.e. with the optimal sustainable exploitation level seen from the fishermen's point of view. Most fisheries, however, do not exhibit this ideal one-stock situation, as several stocks and fleets usually compose a fishery. That makes it difficult or impossible to arrive at the optimal solution in which the resource rent is maximised at the same time as the maximum economic yield is reached for all stocks. In this case, the resource rent is still the total catch value minus all costs, but the maximisation of this is likely to coincide with over-exploitation of some species and under-exploitation of others.

However, achieving economic efficiency in fisheries, as defined above, is in many cases not possible as most fisheries are subject to embodied externalities. Externalities are any external effects caused by individual fishermen, but not included in their accounting system or behaviour (Seijo, Defeo, & Salas, 1998). In fisheries, externalities are most often negative, i.e., the actions of one fisherman cause the rest of the fishery to be worse off. Three types of negative externalities can be identified for most fisheries, stock, crowding, and technological externalities. The stock externality refers to the fact that a fish stock is reduced every time a fisherman harvests it, thus increasing harvest costs for subsequent fishermen. Crowding externalities occurs when vessel aggregation on fishing grounds increases the marginal harvest costs. Technological externalities occur when fishing gear technologies change the population dynamics of fish stocks, e.g., through targeting certain age groups, or through bycatch. From above, it should be clear that in most practical situations a fisherman increases his own welfare at the expense of the welfare of other fishermen because of one or more of the above mentioned externalities. Thus, his welfare actually increases the economic inefficiency of the entire fishery unless he is able to compensate other fishermen, which is generally not the case in fisheries subject to overfishing and/or open access. Thus, it seems more correct to monitor economic development of fishing fleets using economic performance measures, including measures of resource rent.

Typical indicators for economic performance are gross revenue, intermediate consumption, gross cash flow, and net profit. In a dynamic assessment over a time period these will each form a path in time. An additional indicator that measures the economic performance over the entire simulation period is the net present value (NPV), which, in this chapter, is used to compare different innovative management regimes. NPV is defined as the cumulated net profit (total revenue minus total costs) over a period of time, taking into account the time preference of profits by using a discount rate. The discount rate depends, among other considerations, on the time horizon considered. From the point of view of most fishermen it would be appropriate to take a high discount rate (20–30%), see Hillis and Whelan (1992) and Harrison, Lau, and Williams (2002), but from the viewpoint of policy makers, who are interested in sustaining the fish resource, a lower discount rate (3–5%) would be more appropriate. Fishermen apply a shorter time horizon than society, and weigh short run losses higher than the long-term gains compared to society, i.e. fishermen's discount rate is higher than society's discount rate. See e.g. HM Treasury (2003) for discount rates proposed by the UK Treasury. As this study evaluates the economic effects of management innovations, it naturally takes the point of view of the policy makers. NPV is used in this chapter to assess the economic impact of a change in regime brought about by (i) participatory governance and (ii) effort-based control including marine protected areas. To investigate these systems, bio-economic models are developed to simulate and evaluate the impact of the management tools as they are applied in a number of European fisheries.

Participatory governance can take many forms. In this chapter, focus is on participatory governance facilitated through access to better quality information including aid with the research and enforcement of fisheries. The effects on the economic performance of the Basque Baka and pair trawl fleets from including participatory governance in the fishery for northern hake (*Merluccius merluccius*) on the western shelf are assessed. Participatory governance has just been introduced in the management of northern hake alongside other well-known instruments. The participatory governance is organised through the development of the North and South Western Waters Regional Advisory Council (NWW RAC and SWW RAC) where fishing industry groups are included in the assessment and decision process. The RAC system was established as part of the Common Fisheries Policy (CFP) at the revision from 2003 with the objective to include stakeholders' (fishermen's associations, non-governmental organisations, etc.) knowledge in the decision process and enhance their responsibility for the CFP.

Effort-based control comprises several aspects, among these, marine protected areas (MPAs) and transferable fishing days. Both kinds of effort controls have gained attention in Europe as they are seen by some people as working very well in neighbouring areas outside of the CFP, especially in the Faroe Islands, where the demersal fisheries are governed by individual fishing day allocations that are transferable within fleet segments (Løkkegaard, Andersen, Boje, Frost, & Hovgård, 2007). The economic performance of the Faroe Islands' demersal fisheries for cod (*Gadus morhua*), haddock (*Melanogrammus aeglefinus*) and saithe (*Pollachius virens*) is therefore assessed in the present context with respect to effort control, together with the Dutch beam trawl flatfish fishery in the North Sea. The latter is also subject to days at sea allocations, based on fishing area and type of gear. The days at sea can in the North Sea be transferred as horsepower-days. In both cases the economic performance under effort control is compared to economic performance under a more traditional TAC regime.

MPAs range from highly protected nature reserves to large multi-use areas with modest limitations on specific types of human activities. As a fisheries management tool, MPA's have gained increasing attention over the last couple of decades, and some consider their establishment as a necessary condition for successful fisheries management (Kjærsgaard & Frost, 2008). MPAs are expected to reduce fishing on spawning stocks and recruits, to increase fish abundance within the protected area, and to promote spillover of increased fish abundance into neighbouring areas where it may lead to improved catches. The impact of MPAs on economic performance is, in the present context, assessed for the Danish fishing fleet operating in the Baltic, where large spawning areas for cod are closed for fishing in the spawning season or longer.

7.2 Management Systems in Selected EU Fisheries

As discussed above, this chapter focuses on two types of management regime innovations that could potentially result in positive economic impacts for EU fisheries, namely, effort-based management and participatory management regimes. Likely effects of potential innovations are analysed for the following four fisheries: The Danish fleet fishing cod in the Baltic Sea, The Dutch beam trawl fishery for plaice (*Pleuronectes platessa*) and sole (*Solea solea*) in the North Sea, the fishery for cod, haddock and saithe by Pair trawlers and Long liners at the Faroe Islands, and the Northern hake fishery by Spanish Basque Baka and pair trawlers on the western shelf.

In the Baltic Sea, effort management in the form of closed areas and seasons was introduced in 1995 (EC, 2006) in addition to the TAC management. Closure was introduced to selectively mitigate the fishing pressure in spawning areas for cod because the cod stock in the eastern Baltic had been below safe biological limits since 1990. The presumed effects of extended closures are analysed in Sections 7.3 and 7.4. It is expected that in the short run closures will decrease the economic performance of the fishery as they limit fishing possibilities. At the same time, however, closures will especially benefit Eastern Baltic cod, a benefit that will also influence the areas in which fishing is allowed, through spillover from the closed areas. In the long run it is expected that the economic performance of the fishery will improve as a result of the closures. The total influence is, therefore, composed by short run economic losses that should be counted against long run economic gains. In that respect, it should be noted that fishermen apply a shorter time horizon than society, and weigh short run losses higher than the long-term gains compared to society.

In the North Sea flatfish fishery another type of effort management is analysed. This fishery is presently managed through a combination of TAC and effort restrictions. In this case, the impact of introducing a system of days at sea restrictions without complementing TAC constraints will be considered with an emphasis on the problems associated with using single species TAC's in a mixed fishery. It has been theorised, and in some cases documented, that single species TAC's under certain conditions provide incentives for misreporting, discarding of over-quota catches, and high-grading, i.e., the discarding of low valued fish to make room in the quota for higher valued fish (OECD, 2004). The hypothesis is that effort-based management without TAC constraints has a positive influence on fish stocks and on economic results in the long run. Whether this really is the case is discussed in

Section 7.4. The same hypothesis is tested for the Faroese case, where a quota system was replaced by a system of individually transferable effort quotas in 1996. One of the reasons for the change was that the quota system was causing considerable discarding and misreporting problems. In both the case of the North Sea flatfish fishery and the Faroese fishery, scenarios of effort and TAC management will be compared using simulation models. The results of these simulations are discussed in Section 7.4.

The northern hake fishery is presently managed through a traditional TAC system alongside other complementary measures (technical measures, an emergency plan, and a total allowable effort system). In this case, the possible impact of the introduction of participatory management is considered. Participatory management here means that not only information from scientists is used for estimating catch and discards data, but also information from fishermen. The hypothesis is that participatory management will improve economic results because of its expected positive effects on compliance, e.g. reducing misreporting, and on the quality of input data for stock assessments (observed discards). An example of this is the Dutch co-management system in which a different form of participatory management was introduced in 1993 as a complement to the ITQ system, which is widely believed to have increased compliance in the beam trawl fishery. Under the assumption that this will also be the result in the Spanish case, the impacts of participatory management on economic performance in the northern hake fishery are compared, using a simulation model, to economic results under traditional TAC management.

Table 7.1 gives an overview of the hypotheses discussed above.

Table 7.1 Hypotheses Connected to Case Studies Assessing Economic Efficiency

	North sea	Baltic	Western shelf – hake	Faroe islands
Technical measures and MPA's are likely to reduce economic efficiency (in the short run).		X		
Participatory management will in the long run increase economic efficiency (by increasing biological robustness[6]).			X	
A system of Effort Control leads to higher economic efficiency than a system of TAC's[7]	X			X

[6] As defined in Chapter 6.

[7] Notice that this hypothesis is a combination of 'Effort-based management is more easy to enforce than resource based and will therefore on average be more economic efficient' and 'A system of Effort Control leads to a higher biological robustness and economic efficiency than a system of TACs'.

7.3 Assessment of Economic Performance Under Innovative Management Systems

In order to investigate the impact of introducing the innovations on economic performance, bioeconomic models are developed for each case study to simulate the fisheries in different innovation scenarios. Each model builds on the same foundation as illustrated in Fig. 7.1, i.e., each model includes an economic and a biological simulation module that interacts via stock and catches of the focus species, and via effort. Each model includes a management decision procedure, which in the present context comprises effort management, marine protected areas (MPA's) and participatory governance. The management decisions are in all the case studies based on the production (implementation) of knowledge, based on observed (estimated) stock and catches.

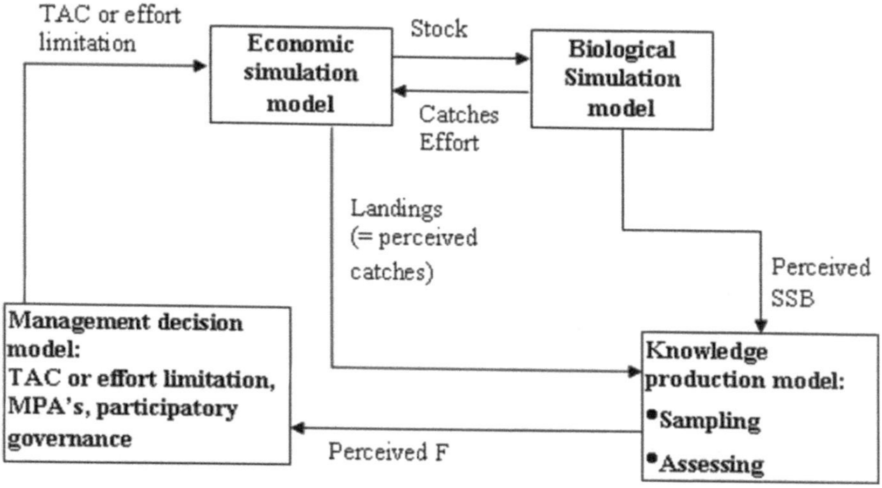

Fig. 7.1 Basic Structure of the Bio-economic Models Used to Assess Economic Performance

7.3.1 The North Sea Case Study

The North Sea flatfish case study compares the economic performance of the Dutch beam trawl fishery under TAC management and under effort management (days at sea). The comparison is made using a bio-economic simulation model for the North Sea flatfish fishery. The model has been developed in the EU project EFIMAS (see Oostenbrugge et al., 2008; EFIMAS ECONOWS report, 2008).

The core of the model (see Fig. 7.1) is a yearly simulation of the biology of sole and plaice in the North Sea in which, among other variables, SSB is calculated on the basis of recruitment, natural mortality, and catches. The model further includes a management module, a harvest control rule, and an economic sub-model.

The economic sub-model essentially imports SSB data from the biological model and then returns calculated fishing effort and catches back into the biological model. Costs and revenues of the fishery are calculated based on effort, landings and price flexibilities for sole and plaice.

A basic assumption of the economic sub-model is that for every restriction placed on the fishery, fishermen respond by dropping the trips with the lowest net revenues per unit of the restricted factor, based on their experiences in previous years. This essentially means that under effort management, the trips with the lowest net revenues per day at sea will be dropped first, while under TAC management, trips with the lowest net revenue per unit of TAC used will be dropped. This optimising behaviour will cause the production functions to differ according to the applied management regime. All else equal, a TAC reduction for a species causes fishing activities to be less directed to that species, while the impact of an effort reduction will cause fishermen to drop the trips in which net revenue per day are lowest. The trips dropped under an effort management system are not necessarily the same trips that are dropped under a TAC management system; it's this difference that accounts for the different *economic* impacts of the two systems.

The core of the economic sub-model consists of a catch equation,[1] where catches of sole and plaice depend on fishing effort and biomass. The catch equations are based on analyses of historical catch and effort data per trip taken from official logbook data gathered by the Dutch ministry of Agriculture, Nature and Food Quality. An important feature of the methodology is that catches increase with fishing effort but the rate of increase in catches decreases with increasing effort, in other words, the marginal return from an additional expenditure of effort results in fewer fish being caught. This is reflected in our assumption that if the fishery is restricted, the least efficient trips, those with fewer catches for a given amount of effort, will be dropped first. A direct consequence of using this type of production function is that catchability varies with the amount of fishing effort, implying that optimising behaviour by fishermen will affect the productivity of fishing effort. Most importantly for current purposes, optimising behaviour by fishermen will be influenced by the nature of the management constraint (effort versus TAC) that is applied to the fishery.

7.3.1.1 Management Scenarios

Three management scenarios are simulated; one scenario for TAC management, and two scenarios for effort management. The scenarios are based on the present management plan for North Sea plaice and sole,[2] which contains both effort and TAC constraints. The management plan aims at reducing the fishing mortality rate for plaice and *sole* by 10% each year, until safe biological limits are reached for

[1]The catch equation has the form of a Cobb-Douglas production function.

[2]EC (2007).

both stocks.[3] When these limits are reached, fishing mortality for both species will be kept at a level corresponding to the MSY, which is set at 0.3 for plaice and 0.2 for *sole*. TAC and effort restrictions are set annually, and correspond to this targeted level of fishing mortality with the additional stipulation that the maximum TAC variation is limited to 15% per year.

In the TAC management scenario, the procedure for setting TAC's for *sole* and plaice is exactly the same as in the management plan described above. A central assumption in this scenario is that *sole* is the main target species that drives the effort of the beam trawl fleet. Plaice catches are derived from the effort that is needed to deplete the sole, and as a consequence, plaice catches can be lower or higher than the plaice quota. In the latter case they are assumed to be discarded. The sole catches and landings are assumed to be always equal to the fleet's share of the TAC.

In the effort scenarios, the procedure for determining the total allowable catch of sole and plaice is the same as in the TAC scenario. In this case, however, catch restrictions are not imposed on the fishery. On the contrary, it is assumed that the authorities impose an effort constraint at the level that would be needed to take the Dutch share of both the plaice and sole TAC. Two scenarios for effort management were simulated. In the 'Effort Max' scenario, the effort constraint is annually adjusted to the *maximum* of the effort needed to take the TAC for either sole or plaice. In this scenario, one of the species will be overexploited each year. In the 'Effort Min' scenario, the effort constraint is set at the *minimum* of the effort needed to deplete the TAC for either sole or plaice. Consequently, one of the stocks will be underexploited each year.

The simulations are performed over periods of one, three, seven, ten, twenty, and thirty years. The reason for including the short as well as long run is that impacts on economic performance work through stock adjustments, which takes a long time.

7.3.2 The Faroe Islands Case Study: Model and Scenario Description

The Faroe Islands case study investigates how the economic performance of Faroese long liners and pair trawlers is influenced by the varying degrees of the innovative effort-control systems introduced in 1996. These results are compared with the situation in which the Faroese fishery is assumed to be managed by a more traditional quota control. The analysis of economic performance is carried out in two steps. In the first step, a model developed by Baudron (Baudron, 2007) for the evaluation of the biological impacts of effort-based management in the Faroe Plateau cod fishery is simulated to obtain catch and effort distributions for the two fleet segments. These outputs are then used as inputs in the second step to evaluate economic performance given the evaluated effort-based management system.

---

[3] Ibid.

The biological module evaluates the effort for a given year based on stock data from the two previous years. This corresponds to the usual practice when implementing TAC and effort systems for most EU fisheries, where the working groups take data from the previous year to set quotas and effort in the following year. The effort is set on the basis of a target fishing mortality, based on a harvest control rule (HCR) aiming at keeping the stock above safe biological limits. In the Faroese fisheries the target fishing mortality is 0.45 for the most important species cod, haddock, and saithe. The model thus first finds a target effort for each of the three species in relation to this management rule, but then decides which of the three efforts will be used to catch all three species, as it is assumed that these species are caught in mixed fisheries so that only one effort can be applied for all three species. The final effort could be the minimum of the three target efforts, resulting in under-exploitation of two stocks or the maximum resulting in over-exploitation of two stocks. The chosen effort thus depends on the strategy used by fishermen, either by their own choices or through management limitations. The output from the Baudron model is (i) applied effort by the two fleet segments, (ii) resulting catches of cod, haddock and saithe, (iii) spawning stock biomass of cod, haddock and saithe, all over a predefined assessment period.

The economic module uses the output from the biological module to evaluate economic performance indicators for the two fleet segments each year of the assessment period. The output, which is based on price and cost information (Rasmussen and Wiehe, see below), includes catch value for each species, crew share as a fraction of the catch value, other variable costs as a function of the effort and, finally, yearly cash flow for each fleet segment equal to catch value minus crew share and variable costs. Present values of the cash flow are then summed for each fleet segment to get a proxy for the resource rent. Market adjustments, i.e. price fluctuations caused by changes in fish supplies, are not considered in the Faroese model.

The biological module of the Faroese model builds on data provided by ICES North-Western Working Group (NWWG), and includes stock data for Faroe plateau cod, haddock, and saithe, for the period 1998–2005. Economic data used in the model has been provided by the Faroese auditing company Rasmussen and Wiehe, and covers prices and cost data (crew share and other variable costs per sea day) for the Faroese fleets in the period 1999–2005.

7.3.2.1 Management Scenarios

To evaluate the economic performance of the two fleet segments in an effort-control management system, five scenarios are considered. It is considered important to assess the economic performance of the Faroese fishing fleets (i) when the permitted effort level is chosen to be the lowest of the target efforts for the three species, and (ii) when the effort level is chosen to be the maximum target effort. These two effort decisions yield the outer points of possible effort scenarios, and as such give measures of the extremities of the economic outcomes of an effort-management scenario in the Faroese fishery. To assess the sensitivity of these results, scenarios are tested in which the effort has been allowed to vary with ±10% and ±20% per year. Thus, the first four scenarios cover the cases where the minimum respectively maximum target effort are chosen, and where the effort has been allowed to vary with 10 and 20% per year.

Finally, the TAC situation is considered in scenario 5, i.e., a TAC is set for each species according to the above mentioned HCR. An effort corresponding to each TAC is then evaluated for each species. Again, the minimum or the maximum of these efforts can be chosen, but evaluations have shown that the maximum effort scenario leads to quick extinctions, especially so in the case of the cod stock, so this scenario is left out. Thus, the minimum of the three efforts is used to calculate the final TAC/landings for each species. The species TAC's are allowed to vary with ±15% each year.

All simulations are run over a period of 10 years, which is considered an adequate time-span for capturing possible effects of the different scenarios considered. Discards are assumed equal to zero in the model.

7.3.3 The Baltic Sea Case Study

The Baltic Sea case study investigates how the economic performance of the Danish fishing fleets is affected by the introduction of MPAs in the eastern Baltic Sea, which aims to protect Baltic cod. The model[4] evaluates the economically optimal allocation of effort in the Danish fleet given a certain MPA scenario. In other words, the model finds the number of days at sea that should be used by each fleet segment in the Danish fleet each year in a pre-defined assessment period in order to get the highest total profit over the assessment period for the Danish fleet.

The basic feature of the model is that the fishing area (the Baltic Sea) is disaggregated into ICES-squares. This makes it possible to open and close areas (squares) for the cod fishery in the model and thereby create artificial MPAs in the Baltic on a very detailed level. Each square is equipped with a basic catch per unit effort (CPUE) value for each target species (cod, herring, sprat and other species), where the effort is measured in number of days at sea. The CPUE values are further disaggregated to fleet segment and seasonal levels, thus making it possible to make seasonal closures. When a square is closed for e.g. the cod fishery the CPUE for cod is set equal to zero in this square. The initial CPUE values are based on historical observations for the Danish fleet, and these are then scaled throughout the assessment period depending on the size of the stocks.

The model consists of biological and economic operational modules (cf. Fig. 7.1). The biological module is focused on the development of cod stocks in the Baltic Sea as MPAs are introduced to aid the recovery of, especially, the eastern Baltic cod. The biological module thus evaluates the eastern and western Baltic cod

[4]The model used for the Baltic case is BEMCOM (BioEconomic Model to evaluate the Consequences of Marine protected areas), also used in the EU 6FP PROTECT (Marine Protected Areas as a tool for Ecosystem Conservation and Fisheries Management).

stocks each year, based on stock and catch values from the previous year, together with natural mortality. The economic module evaluates (i) Catches, divided into landings and discards, in each ICES square as CPUE multiplied by effort (days at sea) (this is done on the fleet segment level), (ii) Catch value equal to catch weight multiplied by prices, (iii) Variable costs, as a function of effort and catch, (iv) Fixed costs, depending on the number of vessels, (v) price fluctuations as a function of changes in fish supplies as the stock changes and finally (vi) Total fleet profit equal to catch value minus all costs and summed over vessel segments and assessment years. The model evaluates these values for each year of the assessment period and feeds them back from one year into the next, as the stock in a given year depends on the catch the year before. For herring (*Clupea harengus membras*), sprat (*Sprattus sprattus balticus*), and other species, the stock development is not assessed, but the catch of these species is included to give the full catch and thus profit for the fleet.

The biological module of the Baltic model builds on stock data for Baltic cod provided by ICES Baltic Fisheries Assessment Working Group (WGBFAS). The model is initiated with stock data for 2005. The economic module uses cost data for the Danish fleet obtained from the Danish fisheries account statistics (FOI, 2005). Fish prices are disaggregated down to fleet segment level and are based on historical landings data from 2005.

7.3.3.1 Management Scenarios

To evaluate the effect of MPAs on the economic performance of the Danish fishing fleet three scenarios are assessed. The first scenario is the base case, which is basically illustrating the 'status quo' situation, where the management situation in 2005 (cf. Section 7.2), including closed areas and seasons is assumed to continue for the analysed period. In this scenario, the model determines the economically optimal distribution of catches, and thus allocation of days at sea, disaggregated to ICES-square for the Sound, Western and Eastern Baltic Sea and for the main areas outside the Baltic.

Historical data shows that extensive fishing activity takes place on the boundaries of the currently closed areas in the Eastern Baltic Sea. This, of course, has a significant influence on the possibilities for the cod to go in and out of the spawning areas without getting caught. Therefore, a scenario is analysed in which the closures around the Bornholm and Gotland deeps are expanded.

Finally, the area in and around the Bornholm deep is recognised as the most important spawning area for eastern Baltic cod. As such, it can be argued that an even larger area around the Bornholm deep should be permanently closed to cod fishery. Therefore, the last scenario closes most of the ICES area 25 around the Bornholm deep together with the extended closure around the Gotland deep used in the previous scenario.

All three scenarios are run for the period 2006–2015, i.e. a 10-year period, which is sufficient to illustrate the effects of the MPA assumptions on the fishing fleet as well as on cod stocks.

7.3.4 The Western Shelf Case Study

Taking into account the specific characteristics of this fishery (described in Section 7.2), the economic performance of the Baka and pair trawler fleets is tested including participatory governance through access to better quality information alongside TAC or effort management. With this aim, a management procedure is assessed with an operational bio-economic model that has been implemented (developed) to simulate the real world, that is, real stock population and other key variables. The operational model consists of biological and economic sub-models (see Fig. 7.1). The biological sub-model is described in Garcia (2008), and is based on an age-structured population model for hake and three different fleets, harvesting the resource: the Baka fleet, the pair trawlers and a third category that covers other fleets. The northern hake stock is estimated each year of the simulation using an age-based assessment model.[5] The initial stock is also estimated using assumptions about discards and different levels of underreporting. Finally, management advice is produced for northern hake using a harvest control rule based on a target fishing mortality. The management advice can either be a proposed TAC or effort. For the remaining stocks fished especially by the Baka fleet, i.e. megrim (*Lepidorhombus whiffiagonis*), monkfish (*Lophius* spp.), horse mackerel (*Trachurus trachurus*), pout (*Trisopterus luscus*), and squids (*Logilo* spp.), similar advice is not produced by the model. Megrim and monkfish catches are simulated from the historical biomass estimates and using a production function, and for the rest of them it is assumed that they are caught in fixed proportions equal to the historical mean, even though this may not always correspond to the correct advice for these species. Thus, the effect on economic performance is directly linked to changes in the northern hake population and the assumed benefits from participatory governance, while the effects for the other species are considered as being constant.

The economic sub-model depends on the fishing mortality target specified in the biological sub-model, and assumes the fishing mortality to be linearly dependent on effort. The effort is calculated in each year and iteration of the algorithm, using this effort and fishing mortality relationship and the Baranov catch equation to link the fleet with the biological population, i.e. to link the fishing mortality with the catch. The economic sub-model includes price functions for the target species, hake, megrim and monkfish according to the EIAA model (ECONOWS, 2008). For the rest of the species, prices are taken to be constant. Additionally, cost indicators (variable and fixed) are used to evaluate historical economic performance of the fishery (from 1992 to 2006) and to develop future economic performance projections (from 2007 to 2040). In particular, economic performance has been measured for the fleets in the short term and in the long term by means of the following indicators: hake

[5]That is an XSA model chosen by the ICES Working Group on the Assessment of Southern Shelf Stocks of hake, monk and megrim (WGHMM) to assess the history of the Northern Stock dynamics.

catch revenue, total catch revenue, gross cash flow, gross surplus, net surplus, financial profitability, full equity profit, net present value, and the break even revenue.

7.3.4.1 Management Scenarios

Five different management scenarios are analysed depending on the hypothesis related to the introduction of the participatory governance regime. First, a base case is evaluated under which the fishery is managed by means of a TAC without including participatory governance.

Secondly, a scenario is evaluated in which it is assumed that the introduction of participatory governance will permit observing 100% of the discard data at ages 0, 1 and 2 (not included in the traditional assessment based on TAC). Thirdly, a scenario is considered in which it is assumed that the participatory governance regime introduced relates not only to the discards that are observed, but also includes the assumption that the level of underreporting will be reduced.

Finally, two additional scenarios are analysed to check the effect of an effort management and including both discard observation and underreporting. Latter cases will allow for comparison with the introduction of participatory governance in the more traditional TAC system.

As discussed later in this chapter, discard observation and underreporting reduction will, in general, contribute to an increase in the differences between the real world and the observed one in terms of the economic performance of the involved fleets in the medium and long run. Notice that in the observed world, catches are equal to the TAC while within the real world catches could be above the TAC given the underreporting.

All scenarios are run for four time periods: (i) 2008–2010, (ii) 2008–2014, (iii) 2008–2028, and (iv) 2008–2040, going from a short-run to a long-run perspective. The same procedure was used in the STECF assessment of Northern Hake long-term management plans, where the long-run simulations were included to test whether the Northern Hake population would stabilise.

7.4 Economic Effects of the Innovations

As described above, two types of effort-based management innovations are considered in relation to economic performance: days at sea control and marine protected areas. The former has been tested in three cases: the North Sea flatfish fishery conducted by Dutch beam trawlers, the Faroese fishery and the Western Shelf fishery. The MPA setting has been tested for the Danish fleet fishing in the Baltic Sea. Furthermore, participatory governance has been tested in the Western Shelf case study. Two types of effects of participatory governance regimes are considered, firstly 100% reporting of discards, and secondly a reduced level of underreporting besides the discard reporting, as described above. The effects are tested in a TAC as well as in an effort-control setting.

The basic hypothesis in the case of days at sea control is that effort management based on control of fishing time will result in better economic performance than TAC management. For the MPA innovation the basic hypothesis is that MPA's are likely to reduce economic efficiency in the short run. In case of participatory governance the hypothesis is that this will in the long run increase the economic performance of the fisheries.

In the North Sea case, three management scenarios are tested, two scenarios (Effort Min; Effort Max) in which the allowed effort is restricted, and one in which the total allowable catch (TAC) is restricted. An outline of how the scenarios are constructed is given above in the description of the North Sea case. In the Faroese case, five management scenarios are tested, four in which effort is the limiting factor (Effort Max, 10%; Effort Min, 10%, Effort Max, 20%; Effort Min, 20%) and one TAC scenario. In the Baltic Sea case, three scenarios with different assumptions regarding the extent of the MPA's are analysed (Scenario 0, Scenario 1, and Scenario 2). In the Western Shelf case study, 5 scenarios are tested: (i) a base case TAC regulation (TAC base case), (ii) a TAC system with observed discards (TAC, observed discards), (iii) a TAC system with observed discards and the level of underreporting reduced (TAC, observed discard, reduced underreporting), (iv) an effort-based case management without participatory governance (EFF base case), and (v) a participatory governance based on effort (EFF, observed discard, reduced underreporting) with observed discards and reduced underreporting. All scenarios are described in Section 7.3. Economic performance has been analysed using net present value.

7.4.1 North Sea Case Results

Using a discount rate of 4%, the NPV for the North Sea case over the years 2007–2037 is 484 Million Euros for the TAC scenario, 671 Million Euros for the minimum effort scenario, and 328 Million Euros for the maximum effort scenario. From this we may conclude that over the whole simulation period, the minimum effort scenario performs better than the TAC scenario followed by the maximum effort scenario. However, this single value result of the NPV hides the variation that occurs between scenarios over the simulation years. Table 7.2 shows the NPV in the three scenarios for different simulation periods. The table shows that all three scenarios have negative NPV in the first few years and positive NPV after twenty years or longer. The minimum effort scenario is the first to move to a positive value while the TAC is the last. All scenarios remain positive over the remainder of the simulation. Further, the ranking in the table shows that during the first three years, the maximum effort scenario shows better results than the other two scenarios. After seven years, the NPV of the minimum effort scenario has the highest relative rank,

Table 7.2 Comparison of NPV in Million Euros in the Three Scenarios for the North Sea Case

Scenario	One year	Three years	Seven years	Ten years	Twenty years	Thirty years
Effort Min	–4.5	–16.4	1.1	63.8	368.4	670.7
TAC	–5.8	–25.0	–41.6	–4.7	228.2	483.9
Effort Max	–0.1	–9.6	–4.3	40.2	196.3	329.2

followed by the effort maximum and TAC scenarios. After twenty years, the effort maximum moves from second to third rank.

7.4.2 Faroese Case Results

Table 7.3 shows the total net present value in the Faroese case for each fleet segment in each scenario. The results are clearly surprising when considering the hypothesis, i.e. the effort-control system does not seem to improve the economic performance of the Faroese fleet. On the contrary, in an effort-control system where the fishermen overexploit two species (using the maximum number of sea days), the economic performance clearly decreases. When the fishermen on the other hand use the minimum number of fishing days the economic performance does not change much relative to the TAC scenario.

Table 7.3 Net Present Value Profit (Million Euros) for the Faroese Fleet Segments in Each Scenario

	Pair trawl	Long line
TAC	46.8	31.9
Effort Max 10%	–9.2	9.6
Effort Min 10%	43.0	32.6
Effort Max 20%	–5.6	10.6
Effort Min 20%	46.4	30.2

7.4.3 Western Shelf Case Results

Table 7.4 shows the result from the Western Shelf case study assessments. It is firstly observed that both fleets have higher economic performance when participatory governance is introduced in the TAC management case (scenarios TAC observed discard, and TAC observed Discard, reduced underreporting), when compared to the TAC base case (TAC base case). Furthermore it is seen that the scenario

Table 7.4 Net Present Values (Million Euros) Aggregated Over all Vessels for Basque Baka Fleet and Pair Trawlers in the Western Shelf Case Study

	Simulation period	TAC base case	TAC observed discard	TAC observed discard reduced underreporting	Effort base case	Effort observed discard reduced underreporting
BAKA	2008–2010	–31.9	–1.3	15	15.1	9.5
	2008–2014	–34.3	–0.8	17.7	18.2	11.7
	2008–2028	–36.7	–0.4	20.2	21.3	13.9
	2008–2040	–38.9	0.1	22.7	24.5	16.3
PAIR	2008–2010	53.5	103.6	98.4	90.8	86.4
	2008–2014	128	247.1	236.3	223.6	220.5
	2008–2028	195.5	384.3	390.9	374.1	366.9
	2008–2040	358.9	716.7	822.1	795.7	748.1

with reduced underreporting and observed discard (TAC observed discard, reduced underreporting) yields higher economic performance than the scenario only including observed discards in both the short and long run for the Baka fleet and in the long run for the pair trawlers.

In the case where effort management is used, the simulations indicate that it actually does not increase the economic performance of the Spanish fleet segments to include participatory governance. It is also interesting to observe that the effort base case yields significantly higher net present value than the corresponding TAC base case simulations, and that the effort base case actually has NPVs comparable with the TAC case including observed discards as well as reduced underreporting.

The results indicate that participatory governance seems to have a positive influence on fleet economic performance when a TAC management regime is used, but that participatory governance will actually reduce the economic performance in the case of effort management.

7.4.4 Baltic Case Results

In the Baltic case, the total net profit for the entire Danish fleet is 0.850 Million Euro in Scenario 0, and 0.464 and 0.495 in Scenarios 1 and 2, respectively, i.e. the hypothesis of decreasing economic performance with increasing MPA's is confirmed. Table 7.5 shows a more detailed picture of the economic effects of the closures for the Danish fleet. The table shows the total net present value per vessel for each segment of the Danish fleet. It can be seen that some of the smaller vessels

Table 7.5 Average Present Value Profit per Vessel (Million Euros) in Scenario 0, 1 and 2 in the Baltic Case

	Scenario 0	Scenario 1	Scenario 2
Netters and Liners < 12 m	1.38	0.67	0.87
Dingy < 12 m	2.79	**3.05**	**3.21**
Multi purpose vessels < 12 m	3.27	**3.89**	2.58
Trawl < 12 m	1.82	1.79	**2.05**
Netters and lingers 12–15 m	2.60	1.11	1.26
Multi purpose vessels 12–15 m	2.07	2.05	**2.56**
Trawl 12–15 m	0.46	0.34	0.36
Netters and liners 15–18 m	0.17	–0.35	–0.33
Multi purpose vessels 15–18 m	1.01	0.52	0.55
Danish Seine 15–18 m	0.15	**0.20**	–0.02
Trawl 15–18 m	1.15	0.24	0.11
Danish Seine 18–24 m	–0.44	–0.44	–0.44
Trawl 18–24 m	1.18	–0.13	–0.18
Industrial trawl 24–40 m	1.18	–0.36	–0.33
Other trawl 24–40 m	1.97	0.97	0.96

Note: Scenario 1 has extended closures around the Bornholm and Gotland deeps, while scenario 2 further extends the closure around the Bornholm deep.

(with NPV highlighted in bold) actually profit from the extended closures, especially in scenario 2. This may be due to changed catching possibilities for the small vessels when the big vessels shift their effort away from the Baltic Sea.

7.4.5 Conclusion

The North Sea and Faroese case studies both indicate that whether days-at-sea restrictions result in better economic performance than TAC management depends on how the effort restriction is set. For the Western Shelf case, however, the effort maximum management generally produces better economic performance than TAC management. In the North Sea case it is shown that while the effort maximum scenario shows relatively good results in the short run, it shows worse economic performance than the TAC scenario in the long run. However, in both the short run and in the long run, the more cautious effort minimum scenario shows better economic results than the TAC scenario in this study. In the Faroese case, the minimum effort scenarios show approximately the same economic performance as the TAC scenario, while the maximum-effort scenarios have a clearly negative effect on economic performance. The latter result corresponds with what has been observed in the North Sea case study.

Participatory governance (observed discards) improves economic performance for the Western Shelf fishery in the TAC scenario, an effect that is even stronger when the underreporting level is also reduced. Moreover, in the long run this improvement is higher the more profitable the fleet (pair trawlers). Contrary to this, participatory governance has a slightly negative effect on economic results in an effort scenario.

Finally, in the Baltic case the overall effect of increased MPA's on the economic performance of the Danish fleet is negative, even though some small fleet segments profit from the change.

7.5 Summary

The management regimes considered in this chapter give a mixed picture of the advantages of management innovations in terms of the economic performance of European fishing fleets. Given the above assessments, participatory governance and carefully planned days at sea management regimes may be profitable for the European fleets in the long run. However, days at sea regulations may also, when aimed at the target effort of the most abundant species, lead to decreasing economic performance in the long run because of a rapid decline in less abundant stocks. This scenario is, however, not a realistic option in many cases. Finally, it seems that MPAs may generally have a negative impact on the profitability of most fleets over a period of 10 years.

It is clear that management innovations are most often aimed at stock recovery for threatened stocks, i.e., they have biological objectives. However, the analysis performed in this chapter indicates that such innovations should also account for the economic implications for the affected fishing fleets, and should thus focus on maintaining biological as well as economic sustainability. In this respect, the work presented in this chapter shows what a great potential there is in combining biological and economic assessment models. The four models used in the evaluations all have the same basic structure, even though they are implemented differently to serve different purposes, and this structure has proven efficient in obtaining relevant and instructive results. This said, it is of course clear that all models used are simplifications of the real dynamics between biology, economy and management, and that there is still room for improvement in all models. Careful sensitivity analyses should be performed in real assessment situations to monitor how the results react to parameter values; and model results should, of course, always be seen as guidelines rather than direct qualitative results.

References

Baudron, A. (2007). *Evaluation of effort-based management in the Faroe Plateau cod fishery: Comparison with TAC management using FLR*. Dissertation, DIFRES/Agrocampus Rennes, September 2007.

EC. (2006). *Proposal for a Council Regulation Establishing a multi-annual plan for the cod stocks in the Baltic Sea and the fisheries exploiting the stocks*. Brussels 24.7.2006, CON (2006) 411 final, 2006/0134 (CNS), from http://eur-lex.europa.eu>/LexUriServ/site/en/com/2006/com2006_0411en01.pdf

EC. (2007, June). Council regulation (EC) No 676/2007 of 11 June 2007 establishing a multiannual plan for fisheries exploiting stocks of plaice and sole in the North Sea. *Official Journal L 157*, 19/06/2007, from http://eur-lex.europa.eu/LexUriServ/LexUriServ.do?uri=OJ:L:2007:157:0001:0006:EN:PDF

FOI. (2005). Accounts Statistics for Fishery 2005. *Institute of Food and Resource Economics, University of Copenhagen, Series F*, no. 11.

Frost, H., Buisman, E., Hoff, A., & Murillas, A. (2008). *Review on economic indicators*. Deliverable D7 and D9 from CEVIS., from http://www.ifm.dk/cevis/Downloads/WP4/Review%20on%20economic%20indicators_final.pdf

García, D., Santurtún M., Iriondo, A. and Quincoces I. (2008). *Northern Hake Long Term Management Plan Evaluation*. Working Document presented in ICES WGHMM, Copenhagen, May 2008.

Gordon, H. S. (1954). The economics of common property resource: The fishery. *Journal of Political Economy*, *62*, 124–142.

ECONOWS. (2008). *Final report from the economic workshops of EFIMAS*. 6'th framework programme project EFIMAS (Operational evaluation tools for fisheries management options), from http://wiki.difres.dk/efimas/doku.php?id=efimas1:wp3:3–6:econows:main&s=econows

Harrison, G. W., Lau, M. I., & Williams, M. B. (2002). Estimating individual discount rates in Denmark a field experiment. *American Economic Review*, *92*, 1606–1617.

Hillis, J. P. & Whelan, B. J. (1992). *Fishermen's Time Discounting Rates and Other Factors to be considered in Planning the Rehabilitation of Depleted Fisheries*. (Paper presented at the sixth conference of the International Institute of Fisheries Economics and Trade, IFREMER, Paris).

HM Treasury. (2003). *'Green book, appraisal and evaluation in central government'*, UK, from http://greenbook.treasury.gov.uk/

Kjærsgaard, J., & Frost, H. (2008). Effort allocation and marine protected areas: is the North Sea plaice box a management compromise? *ICES Journal of Marine Science*, *65*, 1203–1215.

Løkkegaard, J., Andersen, J., Boje, J., Frost, H., & Hovgård, H. (2007). *Report on the Faroese Fisheries Regulation – The Faroe Model*. Report no. 193 retrieved from the Institute of Food and Resource Economics, University of Copenhagen, official web site: http://www.foi.life.ku.dk/Publikationer/Rapporter/~/media/Foi/docs/Publikationer/Rapporter/Nummererede%20rapporter/2007/Report_193.ashx

OECD. (2004). *Tradeable permits: Policy evaluation, design and reform*. OECD Publishing.

Oostenbrugge, J. A. E. van Powell, J. P., Smit, J. G. P., Poos, J. J., Kraak, S. B. M., & Buisman, F. C. (2008). Linking Catchability and Fisher Behaviour under Effort Management. *Aquatic Living Resources*, *21*, 1–9.

Seijo J. C., Defeo O., & Salas S. (1998). Fisheries bioeconomics. Theory, modelling and management. *FAO Fisheries Technical Papers – T368*, from http://www.fao.org/docrep/003/w6914e/W6914E00.HTM

Varian H. (1999). *Intermediate microeconomics: A modern approach*. Norton: New York.

Chapter 8
Understanding Social Robustness in Selected European Fisheries Management Systems

Anne-Sofie Christensen, Martin Aranda, Bonnie McCay, H. Anne McLay, Carl Rova, Andrea Leme da Silva, and Franziska Wolff

Abstract Social robustness is a neglected but crucial component of fisheries management. We present a conceptual framework for evaluating social robustness and we apply it to the analysis of four case studies. We understand social robustness to be a combination of two factors that allow a management regime to adapt to a broad range of potential ecological, economic and political situations: acceptance by stakeholders, reflected in how they perceive and respond to management, and capacity for institutional learning, the process in which institutions change in reaction to internal or external socio-economic or ecological pressures.

We apply five hypotheses about social robustness to four European case studies of innovations in fisheries management in the Baltic Sea, the Faroe Islands, the North Sea and the Western Shelf. The innovations represent a range of systems that incorporate both rights-based management, including transferable effort allocations, and participatory governance. The overall conclusions are that the innovations of the Faroe Islands and the North Sea are socially robust with relatively high degrees of stakeholder acceptance and the ability, in many situations, to institutionally learn. In the Basque fisheries, innovations seem to be socially robust with high institutional learning, but low in stakeholders' acceptance. The Baltic innovations seem to be less socially robust compared to the other cases.

Keywords Fisheries management · Participatory governance · Rights-based management · Stakeholder acceptance · Institutional learning

8.1 Introduction

This chapter addresses the social robustness of four innovative fisheries management systems in Europe. What does it mean for a management regime to be socially robust? How can social robustness be measured? A conceptual framework

A.-S. Christensen (✉)
Institute for Innovative Management (IFM) – Innovative Fisheries Management – An Aalborg University Research Centre, Willemoesgade 2, 9850 Hirtshals, Denmark
e-mail: asc@ifm.dk

K.H. Hauge, D.C. Wilson (eds.), *Comparative Evaluations of Innovative Fisheries Management*, DOI 10.1007/978-90-481-2663-7_8,

is developed: In this chapter, social robustness combines two factors that allow a management regime to adapt to a broad range of potential ecological, economic and political situations: (1) the acceptance of the regime by stakeholders, and (2) the regime's capacity for institutional learning. Stakeholders' acceptance involves how stakeholders perceive and respond to management. Institutional learning is the process in which institutions change in reaction to internal pressures or external changes in the ecosystem or socio-economic contexts.

The conceptual framework on social robustness is then applied to four case studies of innovative fisheries management systems in Europe: (1) the closed areas and division of cod (*Gadus morhua*) stock in the Baltic Sea; (2) the fishing-days system on the Faroe Islands; (3) the Dutch 'Biesheuvel' system of co-managed individual transferable quotas and the role of UK Producer Organisations in quota co-management, both operating in the North Sea; and (4) the fishing rights of the Basque industrial fleet in the North East Atlantic Fisheries Commission (NEAFC) grounds, and the territorial rights of the Basque Cofradias on the Western Shelf. These innovations all combine different and often multiple aspects of rights-based management (RBM) including transferable effort allocations management and participatory governance. The case studies are introduced and later described in more detail with regard to their rights-based and participatory characteristics. Social robustness of the four management systems is analyzed in more depth by discussing the findings in relation to specific hypotheses. These hypotheses link the rights-based and participatory characteristics of the management systems to the dimensions of social robustness – i.e. stakeholder acceptance and institutional learning.

The conclusion synthesises findings from the case studies on social robustness, with regard to the hypotheses. Briefly, the findings are that RBM systems tend not to have broad stakeholder representation; systems with narrow stakeholder representation seem to be highly accepted among those stakeholders involved; and the perceived practicality of a management regime and its maintaining the status quo of fishing opportunities are more relevant for stakeholder acceptance than other effects, such as whether the management system facilitates new entrants and provides retirement options. With regard to institutional learning, significant steps of institutional learning did take place in management systems that involve only narrow circles of stakeholders, against hypothetical expectations. Moreover, RBM does not prevent management regimes from learning, although learning tends to take the specific path of making rights more easily transferable and/or more secure and exclusive.

8.2 Methodological and Conceptual Framework

8.2.1 Methodology

The four case studies were carried out using a common methodological framework. Firstly, the working group reviewed existing literature including scientific documents, grey literature and press reports relating to social robustness and fisheries management in the case study areas. This allowed placing the innovations in the context of the fisheries systems and to identify key informants, who would help us identify relevant people and institutions.

The second and most important source of information came from interviews with different stakeholders including fishermen, conservationists, scientists and managers. The aim of the interviews was to gather insights into the social robustness of the respective innovations. But in order to do so, it was necessary to develop a general understanding of how the systems work and of possible trade-offs in the systems. Moreover, the interviews sought to identify day-to-day issues in fisheries management, as well as contingency measures undertaken to counteract threats to the well-being of the resource such as non-compliant behaviour. The interviews covered two important aspects: the history and development of the innovations (institutional learning), and the views and opinions of fishermen, the wider industry, managers, and civil society stakeholders on the management system and compliance with it (stakeholder acceptance). Approaching the issue of stakeholder acceptance involved inquiry into changes in costs and benefits for fisheries management operations associated with the innovation: What indicators were used to monitor and improve outcomes? And what actors see as the best practices in implementing, monitoring and enforcing the innovations and the resulting management measures?

The interviews were in-depth and open-ended allowing interviewees to express their views and relate the story from their varied professional and academic perspectives. They were carried out during field trips that took place between the summer of 2007 and early 2008. Table 8.1 lists the number and professional profile of the interview partners.

8.2.2 What is Social Robustness?

For the purpose of this book, the *social robustness* of a fisheries management regime is defined by two dimensions: stakeholders' acceptance of the regime, and institutional learning of the regime.

Table 8.1 Professional Profile of Interviewees

	Representatives of					
Profession Case study	Fishing sector	Other industry or intermediate organisations*	Fisheries administration	Conservation organisations	Research organisations	Total
Baltic Sea	20	–	2	2	1	25
Faroe Islands	6	1	8	–	6	21
North Sea	19	6	10	2	0	37
Western Shelf	18	–	–	1	1	20
Total	63	7	20	5	8	103

* For example, industry chambers, industry marketing organisations, fish auctions, quota traders/vessel agents etc.

More concretely, *stakeholder acceptance* of a management regime describes the position that fisheries stakeholders take – either in support or opposition – vis-à-vis that management regime. Fisheries stakeholders are groups and individuals that have an interest in the decision-making process and that are potentially affected by the decisions (Pomeroy & Riviera-Guieb, 2006). Most notably, these are: commercial fisheries interests both in the fisheries sector and the processing sector; fisheries management actors, including scientists/advisors, government; and non-commercial interests, such as conservationists, recreational fishermen and communities (Borrini-Feyerabend, 1996). Stakeholder acceptance may be assessed through analysis of several factors, including: compliance/non-compliance with the management regime; the views expressed by the various stakeholders; stakeholder participation in management processes; and direct actions taken by the stakeholders in favour or against a management regime, e.g. protests or lawsuits. Thus, this approach to stakeholder acceptance assumes a link between compliance and acceptance of a management regime (Dietz, Ostrom, & Stern, 2003; Jentoft, 2000).

Institutional learning in this study is the process in which institutions change in reaction to internal pressures, such as pressures from rights holders or rights managers, or to external changes in the socio-economic context such as pressures from non rights-holding stakeholders or administrators or from the ecosystems themselves. Institutional learning is built on individual learning. It takes place when inferences from individual experiences are interpreted within networks and communities (Haas, 1992; Sabatier & Jenkins-Smith, 1993) and encoded into organisational routines (Levy, 1994). To evaluate institutional learning in the context of fisheries management regimes requires looking at changes in the management regimes over time, in particular at the processes and outcomes of these changes. First, on the process level, one should distinguish between simple learning and complex learning (Nye, 1987).[1] Simple learning describes changes in means in order to more effectively achieve given goals, while complex learning describes changes in goals. Complex learning includes fundamental questioning and redefinition of underlying values and ends, and the new specification of causal relationships, and it may even encompass 'reflexive learning' as a revision of the problem solving concept; i.e. the ability to learn how to learn. Secondly, at the outcome level, one should differentiate between learning processes that address the problem at hand successfully and learning processes that do not address the problem successfully, i.e. high and low problem-solving capacities.

The dimensions of stakeholder acceptance and institutional learning cover processes at the level of individual actors (stakeholder acceptance) and at the level of organisations and institutions (institutional learning). In the first case, the focus is on behaviour and attitudes of actors, in the second case on the permeation of pre-existing structures with such agency, referred to by Giddens (1984) as *structuration.*

---

[1] Or 'double-loop' learning (Argyris & Schön, 1978), 'meta-level' learning (Hedberg, 1981), or simply 'learning' as opposed to 'adaptation' (Haas, 1990).

Table 8.2 The Five Hypotheses Concerning Social Robustness

Management regime/ innovation	Social robustness	
	Stakeholder acceptance	Institutional learning
Rights-based management (including management of effort allocations)	1. RBM systems tend not to have broad stakeholder representation. 2. Commercial fisheries actors' acceptance of a RBM system is higher when (a) the management system is perceived by the fishermen to be practical [and necessary]; (b) the management system (in RBM: the initial allocation) reproduced the status quo of fishing opportunities when introduced; (c) new entrants are facilitated; (d) retirement options are provided for.	4. Rights-based management systems restrict capacity for institutional learning.
Participatory governance	3. The more diverse the stakeholder involvement in the development and/or operation of a management system, the lower its acceptability by the affected commercial fisheries actors.	5. The more diverse the stakeholders involved in the development and/or operation of a management system, the more institutional learning takes place.

Based on these definitions, a number of hypotheses on social robustness were formulated (Table 8.2). The propositions link the dimensions of stakeholder acceptance and institutional learning in the context of RBM systems and forms of participatory governance, including co-management.

8.3 Management Regimes in the Case Studies

Rights-based management (RBM) and participatory governance, or stakeholder representation in management, are the two broadly defined innovations that inform this study. It will be seen that almost all of the cases combine features of RBM with participatory governance, but there are important differences among them.

8.3.1 Rights-Based Management (RBM)

RBM is being applied more and more widely in fisheries (Christy, 1996; Arnason, 2000); the most studied and referred example of RBM systems is that of individual transferable quotas (ITQs) (Scott, 1988; Arnason, 2000). Rights are also applied, for example through territorial user rights (TURFs), to protect and keep community structures intact (Christy, 1982). Often, the purpose of implementing RBM is to enhance a fishery's economic efficiency (Scott, 2000; Arnason, 2000). RBM systems, especially those strongly market-based such as ITQs, can negatively impact the social context of a fishery by reducing the number of fishery participants, disrupting local fishing communities, or upsetting stakeholders that view the approach as privatisation of the commons (Le Gallic, 2003). On a social scale, RBM tends to lock in development of a management regime and immunise it against innovation because of the strong interests of rights holders in predictability, given their investments. In the study, three types of RBM systems have been empirically investigated: quota-based, territory-based and effort-based.

In the Basque fisheries, RBM is represented in two approaches: ITQs for the industrial fleets and territorial user rights for the coastal fisheries. The ITQ system in the industrial fleets has evolved from an effort quota system. The baseline came from a census performed in the early 1980s, which originated the '300 list', consisting of all the vessels with the right to fish in EU waters after Spain's entry into the European Union in 1986. Since then, the RBM system has evolved, and the introduction of transferability in the effort quota system reshaped the fishing fleet. Currently, the transferable effort system has transformed into an ITQ system, participation in which is still restricted to the original list of vessels. The other RBM system reviewed is the territorial user rights of the Basque Cofradias in the Bay of Biscay. These are rights with an old history and they limit entry to the fisheries under the jurisdiction of Cofradias. To fish in a given area, a fisherman must be a member of the Cofradia concerned. Cofradias also have the right, recognised by law, to propose technical measures to ensure sustainable exploitation of the resources. These measures are the basis of most of the technical regulations for anchovy (*Engraulis encrasicolus*) and other pelagic species. Rights of Basque Cofradias are widely recognised by the authorities and the civil society. These rights enable Cofradias to actively participate in the management of pelagic resources.

The RBM systems studied in the North Sea case, the Dutch Biesheuvel groups and the UK producers' organisations (POs), can be characterised according to the type of right, initial allocation, transferability, security and durability and further features (Scott, 1996). In both cases, the rights are related to catch quota. The initial allocation of Dutch sole (*Solea solea*) and plaice (*Pleuronectes platessa*) quota was 'grandfathered' on the basis of historical record; one year later it was adjusted to 50 percent engine power and 50 percent historical record as the basis for the initial allocation. The UK Fish Quota Allocations (FQAs) are based on catch records over a fixed three-year reference period.

Regarding transferability of quota rights, trade and ownership of both the Dutch ITQs and UK FQAs are restricted through several regulations. Dutch ITQs can only be traded among owners of EU registered and licensed vessels; are subject to ministry approval; must be traded within limited periods during the year; are traded jointly for related species, i.e. ITQs for sole are connected to ITQs for plaice; and are formally traded only as whole units. Selling part of a sole or plaice quota to vessels that do not have such ITQs is not allowed (Davidse, 2001). Fishermen exiting the fishery for good are obliged to sell their ITQ shares within three years, but this requirement can and often is circumvented. In the UK, for vessels over 10 m, the FQA unit is attached to a license entitlement. Only since 2002 is it possible, in certain circumstances, to transfer FQAs separately from licences, usually as part of a licensing transaction. FQA units may be transferred among others, to 'dummy licences' held by a PO. Quota rights are divisible and transfers need to be registered with the responsible Fisheries Department. In regard to the security and durability of the quota rights, the Dutch ITQs are based on a Ministry regulation and have duration of one year. Against the backdrop of the legal concept of 'legitimate expectations', the entitlements are evaluated as being relatively secure (Arnason, 2002:41). This is different in the UK, where the status of FQAs as property rights is weaker and the subject of an ongoing debate (Cabinet Office, 2004). Formally, FQAs are governed by rules of the UK Fisheries administrations. A noteworthy additional characteristic of the Dutch system is a, 'national reserve' of circa five percent of the national quota that is not turned into ITQs but used for compensating overshoots.

Most fisheries management systems are focusing on the output of the fisheries – namely on the fish. Hence, the unit of management is usually quantities of fish. The fisheries rights that are distributed in quotas – shares of total allowable catches (TACs) – designate how many fish can be landed through the system. In effort-based fisheries management, the focus is different: Effort-based management systems are so-called input controlled systems that focus on the effort the fishermen apply to fisheries. It is formulated in number of days the individual fishermen have the right to fish. Hence, the system is rights-based, and the fishing-days are tradable within certain restrictions. The fishing-days system has many of the same qualities of, for instance, an ITQ system approached from an economic and legal perspective.

8.3.2 Participatory Governance

Participatory governance in fisheries management means an institutional context in which fishermen take part in the making of various fisheries management decisions (Gray, 2005a; Kearney, Berkes, Charles, Pinkerton, & Wiber, 2007; Mikalsen & Jentoft, 2008; Ostrom, 1990; Symes, 2006; Wilson, Raakjær, Nielsen, & Degnbol, 2003). Participatory governance is held to internalise societal concerns and cope with uncertainty and change (Grote & Gbikpi, 2002; Heinelt, Getimis, Kafkalas, Smith, & Swyngedouw, 2002; Kooiman, 2002). The hypothesis is that it may foster innovation and institutional learning. Participatory governance can be institutionalised in a number of ways. Within the 'policy cycle' (May & Wildavsky, 1979; Sabatier, 2004), the scope of involvement may range from setting the agenda, to consultation and advice, decision-making, implementation and/or the evaluation of a management regime. According to the diversity of stakeholders involved, co-management (which involves the fishing sector and managers) or cooperative governance (which involves a more diverse range of stakeholders) may be appropriate (Gray, 2005b). There may be different levels of stakeholder involvement, including local, national, EU or international scales.

The two innovations analysed in the Baltic case have been implemented through a hierarchic top-down process with little involvement of stakeholders at the local levels. In Baltic fisheries, it is often high-level political negotiations that result in complex compromises. As often in legislation on the European level, and according to some of the respondents in this study, it has been a quite intensive lobby campaign from both green NGOs and fisheries interests in legislations concerning fisheries in the Baltic. The trouble is that local fishermen have little insight into the political negotiations that set the rules. This has created a situation where fishermen do not feel invested in the management process or feel that they have a say. It is the classical tendency of a purely centralised top-down management system.

The fishing-days system on the Faroe Islands has strong elements of participatory governance. Stakeholders have been consulted about both complex and trivial matters – e.g. the development and the daily operation of the system.

In the North Sea case study, participatory governance was studied to the extent that it relates to quota management within the Dutch and UK RBM systems. In the Netherlands, co-management was introduced to increase the stakeholders' acceptance of and compliance with the RBM system and ultimately with the EU quota regulations (van Ginkel, 2005; Dubbink & van Vliet, 1997). In the UK, the participatory function of POs existed prior to RBM, but initially related only to industry self-management of market supply and withdrawal schemes. The nine Dutch co-management groups are smaller and more homogenous in terms of regional basis, targeted species, and vessel and gear type than the 19 UK POs. Their prime functions are control and management of the groups' quota allocations and, in the Netherlands, of days-at-sea. This includes facilitation and monitoring of quota transfers within and between groups, and annual submission to the administration of a joint fishing plan. Development of internal rules, including sanctions for when members overshoot their quota, was coordinated and is hence harmonised among the Dutch co-management groups. This is not the case in the UK POs, each of which set its own rules. The main difference between the Dutch groups and the UK POs is that the Dutch system of quota management is ITQ-based only, while the UK POs operate with ITQs, quota-pools and mixed systems for quota management. Beyond their functions in quota management, the Biesheuvel groups have recently acquired some capacity control and technical responsibilities, whereas some of the UK POs continue their traditional engagement in marketing, including the operation of processing facilities.

Stakeholder participation is narrow in Basque fisheries, involving only fishermen. No conservationists have an official role in management of local fisheries. Industrial and coastal fishermen participate actively in management at the local level and through the Basque government in management at national level. Basque stakeholders are active in participation at the Community level through the Regional Advisory Councils (RACs), which are participatory platforms devised in the reform of the Common Fisheries Policy (CFP) in 2002. Basque industrial and coastal fishermen participate actively in RACs, where they play leading roles and work in developing of new management tools. Such is the case of industrial and coastal stakeholders' involvement in the development of the management plans for hake (*Merluccius merluccius*) and anchovy.

8.4 Fisheries Management and Their Innovations

This section provides information on specific innovations in each of the cases and how they relate to institutional learning and stakeholder acceptance in each of the cases.

8.4.1 Overview of the Baltic Sea Case Study

The two key management innovations' considered in the Baltic Sea are: (1) closed areas and seasons for cod fisheries; and (2) a new management regime based on the division of the cod TAC into east and west Baltic components. To analyse stakeholder acceptance and institutional learning regarding these innovations, two fishing communities, Nexø on the Danish island Bornholm and Simrishamn in Sweden, were selected. These communities are dependent on the cod fishery and there is – or at least used to be – a cultural preference for fisheries over other occupations (Delaney, 2007). In the Baltic case, stakeholders' acceptance was assessed through analysis of views expressed by various stakeholders (positive, neutral, negative) and through actions taken by the stakeholders as well as (non-) compliance with the management regime.

The main reasons behind the implementation of the two innovations in the Baltic Sea have been to cope with the severe situation and the high fishing pressure on Baltic cod. Although the stakeholders involved were not diverse, there is evidence of institutional learning mainly as a result of external changes and new knowledge about the Baltic Sea ecosystem. Accordingly, these innovations have been implemented primarily as a response to new scientific knowledge related to spawning areas, external factors affecting the cod resource, and genetic differences between the eastern and western stocks. A lot of this learning seems to take place within ICES and other ecological research institutes working at the national level. These scientists have an advisory role and are not involved in decision-making or in fishing activity. On the other hand, there is a low level of learning about how to generate trust between stakeholders and authorities, how to promote regime acceptance and regulatory compliance, as well as how to communicate the necessity of different management innovations at the local level.

The Baltic Sea case indicates that low stakeholder acceptance of management innovations affects support and compliance negatively. The division of the cod TAC in particular has had negative impacts on stakeholders' acceptance because of its practical implications for fishing. There are also indications that, even with a rule that is perceived as necessary among the stakeholders, acceptance may be low if the practical implications for day-to-day fishing activities are too severe. For example, the importance of protecting spawning and juveniles is unanimously recognised and is highly accepted, but the complex management regime with many temporal closures also reduces the stakeholders' acceptance of this innovation.

8.4.2 Overview of the Faroe Islands Case Study

The Faroe Islands fishery has an input-controlled system that focuses on fishing effort, the number of fishing-days. The fishing-days system segments the fleet into vessel groups, based on size of vessel and gear type. Each of these groups is annually allocated a number of fishing-days per year, which are in turn allocated to individual vessels. Fishing-days are tradable within the group and, at the end of the year, between the groups. The fishing-days system is supported by a number of technical measures, for instance minimum mesh sizes and gear restrictions.

The development of the system of fisheries management in the Faroes and elements of institutional learning identified during the study are summarised in Table 8.3. Fleet capacity was frequently discussed in the interviews. Most

Table 8.3 Key Changes and Institutional Learning in the Faroe Islands Case Study

Year	Management system	New situation	Institutional learning
Until 1994	Technical measures such as area closures and regulation of mesh sizes	Capacity was high, cod stock low. Over-investment in the fleet led to collapse of several banks.	Denmark interfered in Faroese fisheries policy, as DK demanded that the Faroe Islands set up a management system for their fisheries in return for loans.
1994–1996	ITQ system	Political demand from Denmark that the Faroe Islands should set up a management system. A Faroese group suggested an ITQ system, which was adopted politically.	The fisheries on the Faroe Islands are mixed and the TACs did not reflect the catch opportunities: the cod quota was low and catch rates were high. Hence, the ITQ system was abolished as it had no stakeholders' acceptance.
1996	Fishing-days system	Two very strong year classes of cod led to high catch rates and too small quotas. Both fishermen and politicians worked to change the system. Today there is a high degree of acceptance of the basic idea of fishing-days among interviewees, but they disagree on parts of the system, e.g. capacity policy and setting the number of fishing-days.	Since its introduction, the system has not changed much. The lack of measurement of capacity is often mentioned as the key flaw of the system. None of the interviewees wanted to take the first step, yet the debate is now going on regarding the issue.

informants agreed that capacity had increased and that the main flaw of the system was the lack of ability to measure and compare capacity over time. This is supported by Jákupsstovu, Cruz, Maguire, and Reinert (2007) and Løkkegaard, Andersen, Boje, Frost, and Hovgård (2004).

Given that the Faroese exclusive economic zone (EEZ) is under Faroese jurisdiction, the Minister and the Parliament are powerful when setting the number of fishing-days. But they do so in consultation with biologists and active fishermen. All the people interviewed assessed that the industry had greater authority when setting the number of fishing-days than the biologists. All participating stakeholders are commercial stakeholders and commercial fishing interests have particular influence in the decision-making processes. Stakeholder acceptance was high among this narrowly defined stakeholder group.

8.4.3 Overview of the North Sea Case Study

The North Sea study looked at the Dutch Biesheuvel groups and the UK Producer Organisations, which are two management systems that govern the allocation of national quota. Both combine RBM and participatory governance in what are in effect 'co-managed RBM systems'.

In the Netherlands, a system of individual transferable quotas (IQs), transferable only along with vessels, was introduced in 1976 for plaice and sole. With full transferability of IQs allowed in 1985, an ITQ system was created. However, TAC reductions and national overcapacity (Davidse, 2001) fostered non-compliance, continuous overfishing of the Dutch quota, and worsening state-industry relations in the 1980s (van Ginkel, 2005). In order to improve the situation, responsibility for quota management was devolved to industry groups in the late 1980s and, more systematically, in 1993 (Dubbink & van Vliet, 1997; Hoefnagel, 2005). The nine so-called 'Biesheuvel groups' ensure compliance with the group quota and manage quota transfers. Recently, their responsibilities widened somewhat.

Acceptance of the system is high both among the Dutch fishing industry and governmental stakeholders. Interviewed members and managers of various Biesheuvel groups praise efficient quota uptake, an end to the race for fish, stability of expectations, high levels of compliance and better fish prices. They criticised the practice of avoiding the final sale of ITQs when fishermen stop fishing and the resulting low level of new entrants to the sector. Membership within the Biesheuvel groups is about 97% of those eligible and is stable. Although individual incidents of non-compliance with quota rules have been reported over the years, the groups' self-policing is considered to function well. For fisheries managers, the system not only improved state-industry relations but also reduced the costs of public enforcement. Conservation groups accept that the Biesheuvel system functions well with regard to quota management, if less so with regard to engine power limitation.

In the UK, industry self-governance of market supply and market withdrawal schemes within Producer Organisations (POs) started in 1973 (Goodlad, 1998; Phillipson, 1999, 2002). The PO structure was used as one pillar when a system of sectoral quota management was introduced in 1984. Thereafter, allocations were based on vessels' most recent three years of catches, and as of 1999 catches over a fixed reference period, known as 'Fixed Quota Allocations' (FQAs). The allocations are assigned to three groups: 'the sector', i.e. fishing vessels over 10 m and belonging to POs; the 'non-sector', which includes over 10 m vessels not in PO membership; and vessels under 10 m. Co-management is limited to the 'sector', i.e. to POs. These manage their members' cumulated allocations (Hatcher, 1997), which amount to 96 percent of the UK quota. Different systems of quota management have evolved within different POs, with ITQ-style systems slowly replacing the 'traditional' pooling of IQs (Nautilus Consultants, 2006).

Among those interviewed, members and managers of four POs, the co-managed RBM system is generally accepted, although less unanimously than in the Dutch case. While some interviewees appreciate the opportunity to fish against their own quota share and buy and lease quota, others see benefits in the pool system. As in the Dutch case, criticism relates to FQA-holders not actively involved in fishing and difficulties experienced by new entrants. In addition, there seems to be unease among some about the leasing and buying of quota (Hatcher, Pascoe, Banks, & Arnason, 2002; Anderson, 2006); concerns about insecure ownership status of FQAs; and quota being bought off by foreign flag vessels. PO membership is high although less stable than in the Dutch groups. Self-policing and enforcement of PO quota management rules is said to have increased in recent years in relation to reduced TACs and the introduction of fish buyers and sellers registration in 2005. Among the other industry groups, in particular the under 10 m fleet, there are fears that PO members and the non-sector fleet could fish against their allocation. Fisheries managers confirmed that the POs' role reduced some of the administration's burden; however, changes in quota management were regarded as necessary and policy-makers have recommended a move to fully-fledged ITQs (Cabinet Office, 2004). At the time of this research, a consultation-based quota management change programme (UK Fisheries Department, 2005) had been suspended due to the Scottish Government's opposition to ITQs. Environmental organisations did not seem interested in UK quota management. Some groups, however, have more general positions on RBM, with World Wildlife Foundation stating a 'healthy scepticism', but not a general rejection of RBM (WWF, 2007).

8.4.4 Overview of Western Shelf Case Study

Two Basque fisheries have been examined for the purpose of this study: the Basque industrial fisheries in the NEAFC area (Gonzalez-Laxe, 2006) and the coastal fisheries in the Bay of Biscay (Astorkiza, Del Valle, & Astorkiza, 2000). The industrial fleet is managed through an ITQ system restricted to the census of vessels in the NEAFC area, and originally managed by effort quotas. Participation in this fishery is narrow and limited only to industry stakeholders – conservationists and others do not participate in management. POs involved in this fishery are active in participation at the local, national and European level through the Regional Advisory Councils (RACs).

The Basque coastal fleet in the Bay of Biscay harvests anchovy and other pelagic species. The management of the fishery receives meaningful input on suggested technical measures such as bans for certain technologies in certain areas (e.g. use of pelagic trawling in Spanish pelagic fishing in the Bay of Biscay) from the Cofradias. This is especially notable when setting regulatory parameters, such as technical measures. This fishery is considered to have a territorial user rights (TURF) approach in the sense that no one without membership in a Cofradia is allowed to fish in a given fishing area (Astorkiza et al. 2000). Cofradias and Associations of Cofradias actively participate in local and national management and in EU level management through RACs. Conservationists or other groups do not have any role in the management of this fishery.

Basque Cofradias and POs have adapted to EU requirements such as involvement in fleet reduction programmes. In the particular case of Cofradias, they exhibit a capacity to adapt to changes in the fisheries system and to react swiftly to challenges such as the Bay of Biscay anchovy fishery collapse in 2005. This event demanded the Cofradias to assume a unified position and request to the Commission the immediate closure of the Spanish and French fishery.

The Basque government seems to be active in supporting the local fisheries and presenting their needs to the central government. Management at the national level is much less palatable for stakeholders since the Spanish government continues executing the Treaty of Arcachon. This treaty was signed by Spain and France in 1992 and stipulates that 9,000 tonnes of the Spanish anchovy quota have to be exchanged every year for French quotas of other species such as the Northern hake (Aranda, Murillas, & Motos, 2006). Recent innovations within participatory governance, such as RACs, are widely supported and seen as improvements to management. Regarding fishing rights in the demersal fishery, they have evolved from transferable effort quotas to ITQs. This has naturally produced the predominance of the most efficient actors, a fact that is not palatable for all. Regarding territorial rights in the Bay of Biscay, they provide coastal stakeholders with a say and active participation in the management of the resource at local, national and community level.

8.5 Evaluating Social Robustness in the Case Studies

This section presents five hypotheses on the interrelations between management systems (RBM and participatory governance) and dimensions of social robustness (stakeholder acceptance and institutional learning). The hypotheses will be discussed against the empirical material from the four case studies, deepening understanding of their social robustness.

8.5.1 Hypothesis 1: RBM and Diversity of Stakeholder Involvement

The first hypothesis regarding social robustness is: *Rights-based management systems tend not to have broad stakeholder representation.* This hypothesis was tested in the cases of the Faroe Islands, the North Sea and the Western Shelf.

The thinking behind this hypothesis is that RBM systems create a sense of ownership and rights on the part of a narrowly defined group, e.g. vessel owners, that discourages the involvement of other stakeholders. RBM systems are mainly concerned with the allocation and management of fishing rights and are therefore often perceived to be of little interest to wider stakeholder groups, e.g. conservationists, processors, local communities without fishing rights. Hence, the perceived need for involvement of stakeholders with broader interests (such as fisheries management, marine conservation, securing the supply chain and community development during the development or implementation/operation phases) is modest.

All three case studies confirm the hypothesis: stakeholder representation is restricted to fishing industry members. Yet causality between the RBM and narrow stakeholder representation can be challenged as neither the Faroe Islands nor the Basque cases have strong traditions for a broad group of stakeholders participating in fisheries decision-making processes. In the North Sea cases, stakeholder representation was only investigated in relation to systems of quota management.

The case study of the Faroe Islands partly confirmed the hypothesis, as only a narrow group of commercial stakeholders are included in the decision-making processes. All participating stakeholders have commercial interests – each of them organised in their own association, e.g. for captains, fishermen on deck, ship owners, engine workers, people who work on shore etc. The breadth of stakeholder representation remains within these groups, and would, in an EU context, be considered narrow. So, on the one hand, the hypothesis is confirmed on the Faroe Islands; while on the other hand, stakeholder representation has historically always been narrow even before the introduction of the fishing-days system. It is difficult to assess whether it is the RBM system that created the narrow stakeholder representation or if it is due to other factors. These factors could have more to do with tradition, and reflect the tremendous importance of fisheries to the Faroe Islands and the economic crises that have occurred in the past whenever the fisheries were in a bad state.

The two quota management systems in the North Sea case supported the hypothesis. In the Netherlands only the fisheries administration and the concerned fleet were involved in the development and operation of the management system. A special role is that of the non-industry chairman that Dutch co-management groups must have. Often, the chair is a local dignitary. He or she might be seen as a community representative, who mediates between community and fishermen's interests. However, the chair can also be seen as a disinterested part of the fishing industry. The latter view is supported by the fact that there exists no formal feedback mechanism to the local communities and that the chair is mandated to act on behalf of the fishermen. In the UK, stakeholder involvement extends to the fisheries managers and to the PO-organised segment of the fishing industry only. Non-PO members, both from the so-called 'non-sector' and from the 'under 10 m fleet' are not involved in the system's operation. A certain role, however, is played by vessel agents, who act as non-fishing co-owners of vessels and of quota. They may exercise influence on quota management decisions through their business partners, who are PO members.

Finally, like in the Netherlands, environmental groups were not involved in the development or operation of the quota management systems – and were not specifically interested in being involved. The cause for the narrow stakeholder representation can be seen in the narrow definition of responsibilities within the management system (management of predefined quota shares only), which results in a narrow definition of 'stakeholder'. Examining the larger fisheries management system for the North Sea, one is more likely to encounter environmentalists and other stakeholders, as evident in the North Sea RAC.

In the Western Shelf case study, no representation by groups other than fishermen was found in the two cases studied: ITQs for hake and other demersal species, and TURFs for the Bay of Biscay anchovy. It is hard to assess whether the RBM system has created the narrow stakeholder representation or if it is due to other factors. A lack of participation of conservationists could likely be the result of the negligible room for non-traditional stakeholders in the management process at national or Community level. Basque green NGOs have views on fishery issues and convey their opinions to the civil society through the local media. However, they do not have any officially recognised consultative role in the management of Basque fisheries. Although stakeholder representation is narrow, RBM seems to have created a sense of ownership that has built strong and active participation on the part of industrial and coastal fishermen. They have an active consultancy role in fisheries management within the autonomous Basque jurisdiction. They are active in lobbying, and through their government channels, they express their needs to the central Spanish government and to the EU. Furthermore, Basque fishermen are active in four RACs and have leading roles in some of them.

8.5.2 Hypothesis 2: Stakeholder Acceptance and Characteristics of the RBM Systems

The second hypothesis regarding social robustness is: *Commercial fisheries actors' acceptance of a (RBM system is higher when: (a) the management system is perceived by the fishermen to be practical [and necessary]; (b) the management system (in RBM: the initial allocation) reproduced the status quo of fishing opportunities when introduced; (c) new entrants are facilitated; and (d) retirement options are provided for.)*

Our reasoning behind this hypothesis is that stakeholder acceptance is strongly related to fishermen's perception of the management system as being practical and necessary. Acceptance is also strongly correlated with the perceived preservation of economic opportunities by existing users and to the maintenance of economic opportunities by potential future users. The conclusions across case studies were that (a) and (b) are important determinants of commercial fisheries actors' acceptance of the management system, whereas (c) and (d) are less important.

In the Baltic case, the overall impression from the stakeholders interviewed is that it is a complex, hierarchic system with many different temporary and area closures, which make the daily fishing activities difficult and complicated. According to the respondents, the management innovations implemented are not perceived to be practical. The area closure is seen as necessary, but the division of the cod stocks is neither necessary nor practical according to a majority of the stakeholders interviewed. These innovations have reduced fishing opportunities, made fishing more complicated and have had a negative impact upon acceptance. Thus, these innovations have not reproduced the status quo of fishing opportunities when introduced. According to a majority of the respondents (particularly among the fishermen interviewed), they have also had an indirect impact on new entrants into cod fishing because the management system is more complicated than before – with unpredictability comes an unwillingness to invest. This is supported by the fact that the fisheries sector has had a very low recruitment in Nexø and Simrishamn during recent years. These two innovations have no impact on retirement options. Altogether this negatively affects the commercial fisheries actors' acceptance of the management system for Baltic cod, which is also verified in the interviews. Testing the hypothesis in the Baltic case strongly indicates that low stakeholders' acceptance of the innovations negatively affects support for and compliance with the management regime. The result also indicates that even when a rule is perceived as 'necessary' among the stakeholders, acceptance can be low if the practical implications for the daily fishing activities are too severe.

On the Faroe Islands there is an exceptionally high level of acceptance of the fishing-days system among the commercial actors. Fishermen found the system to be practical. Much of the information they needed appeared on the computer screen. Both fisheries inspectors and fishermen argued that it was not possible to cheat the system owing to the extensive satellite monitoring system. The system came in response to some difficult years for the fisheries with low cod stock and bankruptcies among vessel-owners, followed by the ITQ system with too low a TAC. Maybe this frustration created a willingness to engage in a new system as long as it addressed the problems of the old system. The economic situation for all fishermen improved when the fishing-days system was introduced; yet this is mainly owing to the fact that the cod stock increased at the same time. This could be one of the reasons why the allocation of fishing-days between the fishermen did not cause conflicts as it did in other regions, for instance, in New Zealand (see Chapter 2). Further, nobody was forced to leave fisheries after the introduction; even the small non-commercial vessels were included and given a common pool of fishing-days, which until now, has not been completely used. Facilitation of new entrants was not viewed as essential by the commercial actors. As in an ITQ system, new entrants have to buy/inherit fishing rights in order to enter the Faroese fisheries. None of the informants were concerned with retirement as an essential part of the system. However, people do have retirement options. If a vessel owner wishes to retire, he can sell his fishing-days and/or his vessel. Pension schemes are provided for the employed fishermen.

In the Dutch case, there is a high, and in the UK case, a moderately high level of acceptance among the fishing industry, which can be related to three of the above mentioned characteristics of the RBM system. In both countries, the systems are considered to be practical. The initial ITQ allocation in the Netherlands reproduced the status quo of fishing opportunities, although only after some early adjustment of the allocation basis. Similarly, in the UK, both the rolling share system and the fixing of the quota allocation came close to reproducing the status quo of fishing opportunities; although the time gap between the onset of FQAs and the qualifying period was substantial. Regarding retirement options in both countries, pension schemes for fishermen exist independently of the 'windfall' incomes generated by selling or leasing out quota rights at the end of a fisherman's professional life. Actually, many active fishermen were quite negative about retired quota-holders and other non-fishing quota-holders. In the UK a group of quota holders exists that is not part of the fishing industry. The fact that people viewed selling quota shares after retirement as more legitimate than leasing them out points to a broader moral issue about holding and speculating on quota. One condition, 'facilitation of new entrants' was not met. Fishermen in both countries expressed concerns about the difficulties and costs of entry, although possibly a bit more strongly in the UK. These concerns did not, however, undermine the systems' general acceptance, perhaps because they relate primarily to third parties rather than the fishermen themselves.

The RBM approach is partially accepted in the Basque cases. The system sets certain conditions to satisfy industrial fishermen: (a) there are clear rules; (b) initial allocation was based on historical criteria; (c) new entrants are facilitated although they are required to buy licenses or rights from vessels in the list of Spanish boats allowed to fish in NEAFC waters; and (d) the RBM system has facilitated retirement since fishermen have been allowed to sell their rights and receive a scrapping bonus. Nevertheless some stakeholders do not feel satisfied with the RBM system. Unconformity is, however, not a result of the RBM system, but of the unequal competition with other fleets. The evolution of the RBM resulted in the reduction of the Basque fleet due to transferability, which was introduced in 1997 and has allowed the Galician fleet to grow in terms of vessels and fishing rights. Stakeholders see that factors such as running costs may have determined the predominance of the Galician fleet. In this case, acceptance of the RBM system rests on external factors. In the coastal fisheries, factors that may produce the acceptability of the RBM system are: (a) the system is practical since Cofradias partially manage the fishery; (b) the rights allocation reproduced the status quo since Cofradias' historical rights to exploit and manage were respected; (c) new entries are allowed to join the Cofradia; and (d) retirement options, through incentives to decommissioning, are available. Acceptability of the system, however, depends on external factors. A source of complaint is the aforementioned Arcachon Agreement. Stakeholders blamed the national government for allowing foreigners to exploit resources considered to be Cofradias' historical rights.

8.5.3 Hypothesis 3: Stakeholder Acceptance of RBM and Diversity of Stakeholder Involvement

The third hypothesis regarding social robustness is: *The more diverse stakeholder involvement in the development and/or operation of a management system, the lower its acceptability by the concerned commercial fisheries actors.*

The reasoning behind this hypothesis is that broad stakeholder representation may lead to concerns from the most directly and historically involved group, the commercial fishers, that their interests are not being addressed, as well as to frustration with the complex process. This hypothesis has been difficult to test directly through the case studies. None of the case studies provided examples of diverse stakeholder involvement, and a high degree of commercial stakeholders' acceptance was evident in at least parts of each case study.

The Baltic Sea case study did not confirm this hypothesis. The division of the cod stocks has been implemented without diverse stakeholder involvement and is operated by a centralised top-down management regime. Even so and contrary to hypothesis 3, there is a low acceptance among commercial fishermen for these innovations – particularly for the negative practical implications of the division of the cod stock. Closed areas and seasons applied to the cod fisheries have a high level of acceptance. In the interviews that have been undertaken, 21 respondents are of the opinion that it is in some way necessary with these closures to protect spawning while only four respondents are entirely against a closure to protect spawning. Thus there is general agreement among commercial fishermen on the importance of protecting spawning and juveniles but the complex management process – with many temporal closures and stops – reduces the stakeholders' acceptance of this innovation. The conclusion from the Baltic case is that even in the absence of diverse stakeholder involvement in the development and operation of the management system, the frustration over complex processes and practical implications can be extensive. With or without stakeholder involvement, management must make sense for stakeholders affected by new management policies and rules, or at least not contradict fishing practices that may be highly valued within fishing communities.

The fishing-days system on the Faroe Islands shows both narrow stakeholder representation in the system and a high degree of acceptance and support from both the users and the managers. The fishing industry is strong when making decisions regarding fisheries management. No 'greens' or other non-commercial interest groups are represented in the decision-making processes. For example, the board that originally suggested the fishing-days system was composed of the administrative head of Fisheries Ministry, the chief biologist of the Faroese Fisheries Laboratory, and three fishermen representatives (one for the trawlers, one for the long liners, and one for the coastal fishermen). This board only functioned during the establishment phase of the system. Another example is the fishing-days board. This board is composed of a chairman and five active fishermen. The people interviewed expressed general acceptance of the fisheries management on the Faroe Islands. Yet many informants mentioned two flaws of the system: (1) that the biologists' advice was not taken into account properly when making decisions on the number of fishing-days; and (2) that the system had failed to set up a system for monitoring fishing effort. Although both flaws are potentially strong enough to undermine the system,[2] they were often considered of less importance in the overall picture by the commercial actors. Whether the decisions of these boards would have been less

---

[2]The cod stock is decreasing. ICES recommended no fishing for cod on the Faroe Bank in 2008 and 2009 (ICES, 2007).

accepted if they had included other non-commercial stakeholders is too speculative to assess, but the situation on the Faroe Islands does not contradict the hypothesis.

The situation in both North Sea countries is quite similar to that on the Faroe Islands: apart from fisheries managers, only specific industry segments have participated in the system's development and are involved in the operation of the co-managed RBM systems of the Netherlands and the UK. Stakeholder involvement can thus be classified as 'non-diverse.' Acceptance of the concerned commercial fishermen was high among the members of the Dutch Biesheuvel groups and moderately high among those of the UK POs. The hypothesis could be rewritten as follows: the less diverse the stakeholder involvement, the higher the level of acceptance on the part of the concerned commercial fisheries actors. In this form, the proposition is confirmed. However, causality is unclear. Is the acceptability of the system due to the lack of involvement of wider stakeholders or is it due to other factors?

Involvement of stakeholders is narrow in the case of the fisheries on the Western Shelf: conservationists and other groups are not involved in the management process. In the Western Shelf case stakeholders' acceptance of the RBM approach seems to depend on factors other than diversity of stakeholders' involvement. Acceptance of the Basque government intervention in the RBM management is high and due to commitment to support Basque stakeholders' interests. In contrast, low acceptance is found in relation to the national Spanish management for Basque fisheries. Basque stakeholders seem to disagree with central government management since they think it does not fully respect the rights of the Basque coastal fleet. This is especially notable in the case of the low acceptability of the aforementioned Arcachon Agreement. Hence, in this case, it is not possible to assess clearly whether or not involvement has an effect on acceptability of the RBM approach.

8.5.4 Hypothesis 4: RBM and Capacity for Institutional Learning

The fourth hypothesis regarding social robustness is: *Rights-based management systems restrict capacity for institutional learning.* The reasoning behind the hypothesis is that creation of property rights will create new expectations and demands for secure investments, and hence foster resistance to change which might affect the value of investment – e.g. through diluting or abolishing rights, creating a new pool of rights for other purposes, opening up the system to new entrants, or weakening the legal status of the rights. Such lock-in effects can be expected to be particularly strong when there are no sunset-provisions built into the allocation of rights. But even if sunset-provisions are given, rights holders' resistance may prevent changes to the system.

However, the case studies show that almost all the RBM systems studied have been able to institutionally learn following the introduction of rights. Specific paths of institutional learning indirectly confirm the hypothesis that RBM tends to narrow institutional development options. Learning was mostly geared towards making rights more easily transferable and/or more secure and exclusive, thereby locking the systems increasingly into place. The trend towards greater transferability can be explained by the fact that, once differences emerge between fishing capacity and available resources, pressures arise to trade rights, even when their trade was not originally intended. Rights become more secure and exclusive as rights holders strive to hedge the investments they have made over time. In some cases, institutional learning also included making the RBM systems more participatory. In such cases, the effect somewhat differs from that of strengthening rights: the management systems in principle become more open and responsive to the involved stakeholders' demands, rather than more locked-in. The reasons for institutional learning and the responses of the systems are different in the various case studies.

Complex institutional learning has taken place in the fisheries management system on the Faroe Islands during the last fifteen years. Prior to the national crisis in the early 1990s, the fisheries were managed by capacity restriction measures and spatial closures. In 1994, the ITQ system was introduced after a Faroese committee had recommended ITQs from the early experiences on Iceland and New Zealand. The fishermen and many of the politicians were not happy with the system because the scientists' assessment of cod abundance conflicted with fishermen's experience at sea. The ITQ system was abolished in 1996, and the fishing-days system was introduced. Hence, the Faroe Islands had an RBM system, which did not restrict the capacity for institutional learning. This could be the only example in the world of an ITQ system being abolished. To interpret this as a failure of the RBM system would be too hasty as a number of circumstances played a role in the change (1) The ITQ system was only in place for two years and the fishing-days system was developed within this period of time. Hence, the fishermen did not adjust well to the ITQ system. The lack of capacity adjustment also had to do with the general economic crisis on the Faroe Islands at the time. Nobody had made investments, which meant that nobody got caught too hard when switching systems. (2) The stock abundance of cod was exceptionally high at the time compared to how much cod the TAC (accompanying the ITQ system) allowed the fishermen to catch. Hence, everybody would gain from changing to the new system. So, on the one hand, the Faroe Islands fisheries management system has shown extraordinary capacity for institutional learning. On the other hand, since the introduction of the fishing-days system, not much has been changed in the system. Although most informants agreed that the plan failed to set up a system for monitoring the effort and that an increased focus on the fishing capacity development was required, nobody wanted to take the first step and many informants argued of the difficulties of measuring capacity. From this perspective, the fishing-days system has demonstrated a low capacity for institutional learning with regard to fishing efficiency.

In the North Sea cases, several steps of institutional learning can be identified to have occurred after establishment of the IQ system in the Netherlands (1976) and of the PO system in the UK (1973). The institutional learning that took place was mainly directed towards making the systems work more smoothly, through greater accountability and by making rights more easily transferable and secure. In the Netherlands, the first step of learning was the introduction of the transferability of quota in 1985. It was a reaction to the fact that fishermen had started to trade quota anyway although this had not been legally intended. The next step in initiating complex learning was the attempt to redesign the ITQ system into a participatory system by introducing quota management groups between 1988 and 1990. The idea was to counter both the issue of quota overshooting that the ITQ system had not been able to prevent, and the problem of crisis-ridden state-industry relations. The attempt failed at first. It succeeded the second time when, in 1993, an institutionally more sophisticated approach to quota management groups was developed and the Biesheuvel system emerged. Among others, the groups were obliged to publicly auction landed fish, which made for easier quota monitoring and enforcement. From 2000 onwards, building upon the systems' success, the system was further extended when the groups' responsibilities were widened from quota management to the control of engine power and technical measures. In the UK, the major steps of institutional learning included linking the existing industry self-governance structure of POs with the RBM system, and the subsequent change from a sectoral quota management system (based on a rolling track record quota allocation introduced in 1984) to a system based on fixed quota allocations in 1999. The political intention of this change, which was requested by at least parts of the industry, was to provide more stability. Emergence of quota trading and a quota market had not been intended by the fisheries administrations but intensified after successive rounds of decommissioning had 'freed' up quota (Hatcher & Read, 2001). These changes are unintended shifts in the system leading to new functions and practices rather than institutional *learning* as a conscious and intentional process. Interestingly, however, this unintended shift was geared, like in the Netherlands, towards making rights more transferable and secure, even though full-fledged ITQs are still resisted in the UK.

In the case of the Basque industrial fleet, it seems that the RBM system has not impeded the fishery in adapting to changes in the fishery system. Thus institutional learning has taken place regardless of the RBM mechanisms introduced. Rights have evolved since decommissioning of vessels produced an excess of effort rights expressed in days at sea, which needed to be transferred to other vessels within the census. The shift to transferability of rights has allowed a more efficient distribution of rights. The recent introduction of an ITQ system in the fishery has not been resisted, although it seems to be unpalatable for some of the stakeholders who have seen their effort rights absorbed by those more efficient – e.g. the Galician fleet. Thus, it is likely that rights will continue to be gathered by the most efficient actors to the detriment of the less efficient. It seems that the RBM approach has developed a sense of ownership that may propel the involvement of the industrial sector in RACs where they develop a multi-annual management plan for Northern hake together with scientist, managers and other stakeholders. In the case of the Basque coastal fleet, the sector has reacted swiftly to achieve a unified position on the closure of the anchovy fishery when it collapsed in 2005. They see a scientifically recommended closure as fundamental to the recovery of this stock and the sustainability of their fishing activity. In this context, they have questioned the opening of the fishery for the experimental surveys carried out by commercial vessels in 2007. It seems that in the coastal case, the TURF approach has created a sense of resource ownership that encourages stakeholders to participate actively in the RACs where they use alternative management measures such as the management plan for the anchovy stock.

8.5.5 Hypothesis 5: Stakeholder Diversity and Institutional Learning

The fifth hypothesis regarding social robustness is: *The more diverse the stakeholders involved in the development and/or operation of a management system, the more institutional learning takes place.*

The reasoning behind this hypothesis is that the involvement of more diverse stakeholders widens the range of alternative views in deliberations and negotiations. Alternatively, one could expect that the involvement of highly diverse stakeholders could lead to conflicts that forestall any significant change. There is a difference in whether the involvement pertains to advice or to decision-making. This hypothesis was hard to test empirically as all the case studies involve narrowly defined groups of stakeholders.

As with Hypothesis 3, the present hypothesis cannot be tested directly through the case studies: On the one hand, none of the case studies have found examples of diverse stakeholder involvement; on the other hand, some degree of institutional learning has taken place in all the case studies.

In the Baltic case, stakeholders do not feel involved in the management process or feel that they have a say. The main stakeholder group, the fishermen, feel distant from the political process and decision-making that governs their fishing operations, and from interviews it is evident that this could undermine compliance. In the Baltic case, there is a significant example of institutional learning – about spawning areas, external factors affecting the cod resource, genetic differences between the eastern and western stock, etc. – although the stakeholders involved were not particularly diverse. A lot of learning seems to be purely ecological and primarily performed within ICES and other ecological research institutes on national levels. These stakeholders have an advisory role and are not involved in decision-making. On the other hand, learning seems to be low with regard to creating trust between stakeholders and authorities, to stakeholders' acceptance and compliance with rules and regulations, and with regard to realising the necessity of communicating management innovations to the local level. The implementation of the Baltic Sea RAC can be seen as a way of trying to deal with these issues. If the RAC gets more influence in the future, there is a chance that stakeholders can have a better say in the decision-making process and that more institutional learning can take place. So far, the RAC has made little impression at the local level.

On the Faroe Islands, stakeholder diversity is lacking; nevertheless, significant complex institutional learning has taken place as described in relation to Hypothesis 4. The introduction of the fishing-days system was an example of complex institutional learning that was initiated in the absence of a diverse group of stakeholders to point out the weaknesses of the system. On the other hand, the lack of initiative to deal with flaws in the system (e.g. the blind spot with regard to fishing capacity) suggests that the system is slower to progress with simple institutional learning. One can only guess whether the system would have been more open to change if broader groups of stakeholders were active on the Faroe Islands.

In the North Sea case study, this hypothesis is not confirmed by this analysis. In both the Dutch and UK cases, stakeholder involvement was relatively narrow, basically encompassing only the relevant industry segments and government. However, significant steps of institutional learning did take place. These include widening the Dutch RBM system to co-management and combining the pre-existing UK industry self-governance structure of POs with RBM. Apart from these two approaches to participatory RBM systems, institutional changes within the RBM system in both cases are geared towards making rights more transferable and exclusive. Non-diverse stakeholder participation has favoured an even stronger movement toward market-based fisheries management.

Involvement is narrow in the case of the Basque fisheries. Conservationist groups, for example, do not participate. Even though participation is narrow in the management process of Basque fisheries, institutional learning has taken place in recent years in both fisheries. This fact does not confirm hypothesis 5. Simple learning has taken place in both fisheries, for example the swift adaptation to the new requirements by the EU on modernisation and withdrawal of fishing capacity and the involvement in RACs. In the particular case of coastal fisheries, institutional learning at a complex level has taken place when shifting objectives from rent maximisation to resource protection. Coastal stakeholders were active in requesting the closure of the anchovy fishery and supportive of scientific advice, even when it meant the indefinite closure of the fishery. Moreover, the swift reaction and effective opposition to what they perceive as threats (e.g. experimental surveys carried out by commercial vessels in spring 2007 has proved their high problem-solving capacity).

8.6 Conclusions

This chapter presents a framework for analysing the social robustness of fisheries management regimes – defined by the two dimensions, stakeholder acceptance and capacity for institutional learning. The framework was applied to four innovative management regimes in European fisheries, which all combined some form of RBM with participatory governance, using five hypotheses on the interrelations between these two management features and the two dimensions of social robustness.

Two of the management innovations – in the North Sea and Faroe Islands cases – seem to be socially robust with relatively high degrees of stakeholder acceptance and with the ability, in many situations, to institutionally learn. In the case of Basque fisheries, management seems to be socially robust with high institutional learning, but the stakeholders do not fully accept the system. The Baltic case seems to be less socially robust compared to the other cases: the innovations in the Baltic were implemented in a more traditional top-down fashion, and complex learning – that contains more fundamental questioning of redefining the underlying values and ends – has not taken place, affecting social robustness negatively. All the case studies only include narrow groups of stakeholders and it is easy to assume that a broader representation of stakeholders would have affected stakeholder acceptance and institutional learning and thus, social robustness.

Looking more closely at the factors influencing stakeholder acceptance, the North Sea, the Faroe Islands and the Western Shelf cases enjoy a generalised acceptance among, at least, industry stakeholders. The systems are all perceived to be practical and necessary by the people who have to work them, i.e. the commercial actors and, in some of the cases, the management. Conservation or green organisations do not play a central role in any of the cases studied although they are represented in some through the Commission's RACs. Yet, on the Faroe Islands critical voices that say that the fishing industry is too strong and that the biologists are ignored in decision-making processes can be found, even though no green organisations are represented in fisheries management. Stakeholder acceptance of the management in the Baltic case is much lower than in the other case studies. The management system is not perceived to be practical and necessary, and as a consequence issues of stakeholders' acceptance and compliance have arisen. These same issues do not seem to be as large in any of the other case studies.

Regarding institutional learning, the studied systems of the North Sea, the Faroe Islands and the Western Shelf have demonstrated capacities to institutionally learn and keep a fairly high stakeholders' acceptance among the commercial actors. This happened in spite of the involvement of narrowly defined groups of stakeholders. The finding was not consistent with the initial hypotheses. However, institutional learning within the RBM systems mostly took a very specific path: It was typically geared towards making rights more tradable and/or secure or exclusive. This actually creates a paradoxical situation where options for future learning in the system may be reduced since rights holders will want to maintain the value of their investment in the rights.

Relationships between rights-based management systems, participatory governance, stakeholder acceptance, and institutional learning in the four case studies are complex. Five hypotheses were developed to help disentangle the complexities (Table 8.2). The case studies turned out to be inappropriate for testing some of the hypotheses because of the lack of broad stakeholder participation in the governance systems studied. Therefore, it was not possible to come up with a conclusion about whether rights-based management precludes broad stakeholder representation (Hypothesis 1) or whether broad stakeholder participation in governance decreases the acceptability of the system for commercial fishers involved (Hypothesis 3). However, the research led to an appreciation of the importance of pre-existing traditions for or against broad stakeholder representation. In addition, support was found for an alternative to hypothesis 3, which is that RBM systems with narrow stakeholder representation seemed to have high degree of acceptance among those stakeholders involved.

The critical factors affecting commercial fishing actors' acceptance of a new management system were, as predicted in hypothesis 2, whether it is perceived by the fishermen to be practical and necessary and whether a new management system (in RBM: the initial allocation) reproduced the status quo of fishing opportunities when introduced. Somewhat less important were the facilitation of new entrants and provision of retirement options. The capacity for institutional learning was not, apparently, restricted by RBM systems, contrary to hypothesis 4, in so far as some kinds of institutional learning could be identified in the development of each of the RBM systems studied. Moreover, contrary to hypothesis 5, even though stakeholder involvement was narrow in all of the cases studied, they all showed some capacity for institutional learning.

Finally, the various case studies exhibit some factors that cannot be assigned to the management systems, and their characteristics have influenced the social robustness of fisheries management systems. On the Faroe Islands, cod was exceptionally abundant during the first years after the introduction of the fishing-days system – this took the pressure off the fisheries management system. In the case of Basque fisheries, the emergence of RACs is seen as a positive development that allows the Basque fishing groups to defend their interests and to participate in giving advice. The RAC could take the pressure off the regional fisheries management. In the North Sea cases, social robustness of the co-managed RBM systems was fostered by the fact that inequitable quota concentrations have, so far, been avoided. In addition, in both countries capacity reduction, days-at-sea schemes and strengthening of enforcement frameworks supported the systems' working over the years, maintaining economically viable fishing opportunities for those still involved. Looking at co-management, social robustness was promoted in the Netherlands in particular by the Dutch neo-corporatist and consensus-oriented culture, which pervades many aspects of social life.

Acknowledgments The authors are grateful to the European Commission and to all institutions and individuals in the Baltic Sea, the North Sea, the Western Shelf, and the Faroe Islands who have devoted their time and expertise to help us with our endeavours; and to Sarah Wise from Rutgers University for proof-reading.

References

Anderson, J. (2006). *The future of the market-based approach towards quota management in the UK – the case of the Shetland Isles*. Paper presented at the "Sharing the Fish" Conference, March 2006, Perth, Western Australia.

Aranda, M., Murillas, A., & Motos, L. (2006). Command and control quota-based regimes. In L. Motos & D. Wilson (Eds.), *The knowledge base for fisheries management* (pp. 143–161). The Netherlands: Elsevier.

Argyris, C., & Schön, D. (1978). *Organizational learning: A theory of action perspective*, Addison-Wesley: Reading MA.

Arnason, R. (2000). Property rights as a means of economic organization. In R. Shotton (Ed.), *Use of property rights in fisheries management*. FAO Fisheries Technical Paper 404/1. Rome: FAO.

Arnason, R. (2002). A Review of international experiences with ITQs. Annex to *Future Options for UK Fish Quota Management. Report to the Department for the Environment*, Food and Rural Affairs. CEMARE, University of Portsmouth, June 2002.

Astorkiza, K., Del Valle, I., & Astorkiza, I. (2000). *The regulatory capacity of the Cofradias in the Cantabric region. Management institutions and governance systems in European Fisheries* (pp. 196–210). Portsmouth, UK: CEMARE Miscellaneous Publications.
Borrini-Feyerabend, G. (1996). *Collaborative management of protected areas: Tailoring the approach to the context*. Gland, Switzerland: IUCN.
Cabinet Office. (2004). *Net benefits, a sustainable and profitable future for UK fishing*. Prime Minister's Strategy Unit. London, March 2004.
Christy, F. (1982). *Territorial use rights in marine fisheries: Definitions and conditions*. FAO Fisheries Technical Paper 227. Rome: FAO Fisheries Department.
Christy, F. (1996). *Paradigms lost: The death rattle of open access and the advent of property rights regimes in fisheries*. 8th Biannual Conference of IIFET, July 1996, Morocco.
Davidse, W. (2001). *The effects of transferable property rights on the fleet capacity and ownership of harvesting rights in the Dutch demersal North Sea fish*. In FAO Fisheries Technical Paper 412. Rome: FAO Fisheries Department.
Delaney, A. (2007). *Profiling of small-scale fishing communities in the Baltic Sea*. Denmark: Innovative Fisheries Management, IFM, Aalborg University.
Dietz, T., Ostrom, E., & Stern, P. C. (2003). The struggle to govern the commons. *Science, 302*(5652), 1907–1912.
Dubbink, W., & van Vliet, M. (1997). The Netherlands: from ITQ to co-management. Comparing the usefulness of markets and co-management, illustrated by the Dutch flatfish sector. In OCDE (Ed.), *Towards sustainable fisheries: Issue papers* (pp. 177–202). OCDE/GD(97)54, Paris.
Giddens, A. (1984). *The constitution of society*. Berkeley: University of California Press.
Gonzalez-Laxe, F. G. (2006). Transferability of fishing rights: the Spanish case. *Marine Policy, 30*, 379–388.
Goodlad, J. (1998). Sectoral quota management: Fisheries management by fish producer organizations. In T. S. Gray (Ed.), *The politics of fishing* (pp. 146–160). London: Macmillan.
Gray, T. (Ed.). (2005a). *Participation in fisheries governance*. Dordrecht: Springer.
Gray, T. (2005b). Theorising about Participatory Fisheries Governance. In T. Gray (Ed.), *Participation in fisheries governance* (pp. 1–25). Dordrecht: Springer.
Grote, J. R., & Gbikpi, B. (Eds.). (2002). *Participatory governance: Political and societal implications*. Opladen: Leske + Budrich.
Haas, E. B. (1990). *When knowledge is power: Three models of chance in international organisations*. Berkley: University of California Press.
Haas, P. (1992). Introduction: Epistemic communities and international policy coordination. *International Organization, 46*(1), 1–35.
Hatcher, A. C. (1997). Producers' organizations and devolved fisheries management in the United Kingdom: Collective and Individual Quota Systems. *Marine Policy, 21*(6), 519–533.
Hatcher, A., & Read, A. (2001). *The allocation of fishing rights in UK fisheries*. Portsmouth, UK: Centre for the Economics and Management of Aquatic Resources (CEMARE).
Hatcher, A., Pascoe, S., Banks, R., & Arnason, R. (2002). Future options for UK fish quota management. CEMARE Report 58. A Report to the Department for the Environment, Food and Rural Affairs. Portsmouth.
Hedberg, B. (1981). How organizations learn and unlearn. In P. C. Nystrom & W. H. Starbuck (Eds.), *Handbook of organizational design* (Vol. 1, pp. 1–27). Oxford: Oxford University Press.
Heinelt, H., Getimis, P., Kafkalas, G., Smith, R., & Swyngedouw, E. (Eds.). (2002). *Participatory governance in multi-level context: Concepts and experience*. Opladen: Leske + Budrich.
ICES. (2007). *Report of the ICES Advisory Committee on Fishery Management, Advisory Committee on the Marine Environment and Advisory Committee on Ecosystems, 2007*. Book 4. International Council for the Exploration of the Sea: Copenhagen, Denmark.
Hoefnagel, E. (2005). The Netherlands. In L. van Hoof et al. (Eds.), *Sharing responsibilities in fisheries management. Part 2 – Annex: Case studies* (pp. 161–211). The Hague.
Jákupsstovu, S. H. Í, Cruz, L. R., Maguire, J.-J., & Reinert, J. (2007). Effort regulation of the demersal fisheries at the Faroe Islands: a 10-year appraisal. *ICES Journal of Marine Science, 64*, 730–737.

Jentoft, S. (2000). Legitimacy and disappointment in fisheries management. *Marine Policy, 24*, 141–148.
Kearney, J., Berkes, F., Charles, A., Pinkerton, E., & Wiber, M. (2007). The role of participatory governance and community-based management in integrated coastal and ocean management in Canada. *Coastal Management, 35*(1), 79–104.
Kooiman, J. (2002). Governance: A social-political-perspective. In B. Gbikpi (Ed.), *Participatory governance: Political and societal implications* (pp. 71–96). Opladen: Leske + Budrich.
Le Gallic, B. (2003). *Why is it difficult for governments to move towards using market-based instruments in fisheries?* XVth Annual EAFE Conference, Brest, May 2003.
Levy, J. S. (1994). Learning and Foreign Policy: Sweeping a conceptual minefield. *International Organization, 48*(2), 279–312.
Løkkegaard, J., Andersen, J., Boje, J., Frost, H., & Hovgård, H. (2004). *Rapport om den færøske regulering af fiskeriet – Færømodellen*. Institute of Food and Resource Economics, Report no. 166 Copenhagen, Denmark.
May, J. V., & Wildavsky, A. B. (1979). *The policy cycle*. Beverly Hills, CA: Sage.
Mikalsen, K., & Jentoft, S. (2008). Participatory practices in fisheries across Europe: making stakeholders more responsible. *Marine Policy, 32*(2), 169–177.
Nautilus Consultants. (2006). A review of UK Producer Organisations: the effectiveness of service provision in Quota Management, Quota Trading and Market/Price Support. Authored by Carletin Crick, Southall Tristan and Cappell Rod.
Nye, J. S. (1987). Nuclear Learning and U.S.-Soviet Security Regimes. *International Organization, 41*(3), 371–402.
Ostrom, E. (1990). *Governing the Commons: The Evolution of institutions for collective action*. Cambridge, UK, Cambridge University Press.
Phillipson, J. (1999). The fish producers' organisations in the UK – A strategic analysis. In D. Symes (Ed.), *Alternative management systems for fisheries* (pp. 79–92). Oxford: Blackwell Science.
Phillipson, J. (2002). *Widening the net: Prospects for fisheries Co-management*. Newcastle: CRE Press.
Pomeroy, R. S., & Riviera-Guieb, R. (2006). *Fishery Co-management: A practical handbook*. Oxfordshire, UK: CABI Publishing.
Sabatier, P. (2004). *An advocacy coalition framework of policy change and the role of policy-oriented learning therein*. Netherlands: Springer.
Sabatier, P. A., & Jenkins-Smith, H. C. (Eds.). (1993). *Policy change and learning: An advocacy coalition approach* (pp. 41–56). Boulder, CO: Westview Press.
Scott, A. (1988). Development of property in the fishery. *Marine Resource Economics, 5*, 289–311.
Scott, A. (1996). The ITQ as a property right: Where it came from, how it works, and where it is going. In B. L. Crowley (Ed.), *Property rights and fishery management on the Atlantic coast* (pp. 31–98). Halifax, Nova Scotia: Atlantic Institute for Market Studies.
Scott, A. (2000). Introducing property in fisheries management. In R. Shotton (Ed.), *Use of property right in fisheries management*. Proceeding of the FishRight99 Conference, Western Australia, 11–19 November 1999, FAO.
Symes, D. (2006). Fisheries governance: A coming of age for fisheries social science? *Fisheries Research, 81*(2–3), 113–117.
UK Fisheries Department. (2005). *Securing the benefits*. The joint UK response to the Prime Minister's Strategy Unit Net Benefits report on the future of the fishing industry in the UK. London.
van Ginkel, R. (2005). Between top-down and bottom-up governance: Dutch beam trawl fishermen's engagement with fisheries management. In T. S. Gray (Ed.), *Participation in fisheries governance*. Netherlands: Springer.
Wilson, D. C., Raakjær Nielsen, J., & Degnbol, P. (Eds.). (2003). *The fisheries Co-management experience: Accomplishments, challenges, and prospects*. Dordrecht: Springer.
WWF. (2007). *The use of rights-based measures in fisheries management*. UK.

Chapter 9
Costs of Management in Selected Fisheries

Sarunas Zableckis, Tiit Raid, Ragnar Arnason, Arantza Murillas, Søren Eliasen, Sten Sverdrup-Jensen, and Emil Kuzebski

Abstract The research was carried out to investigate the management costs of selected European fisheries. First, the literature was reviewed for existing information, second, interviews were performed of experts from the countries of the case studies. The collected information from the case studies was used to assess the management regime innovations and changes in management costs after the new regimes had been introduced.

Findings show that data availability is the big issue in solving the question of how much fisheries management cost in the selected European cases. On the other hand, from the data acquired it became clear that the government spending on fisheries management usually does not depend on the change in fisheries management regime. This was observed in the Netherlands, Denmark, and Spain. Only the Polish case showed that a higher number of technical regulations applied to fisheries led to higher administration and enforcement costs and consequently research costs. This was observed when Poland joined the EU and had to adhere to CFP regulations. It is expected though that enforcement and research costs will decrease in the long run once the innovative regimes, for example participatory governance, are introduced.

Keywords Effort control · Participatory governance · Harvest control rules · Rights-based management · Fisheries management costs · Fisheries administration costs · Fisheries enforcement costs · Fisheries research costs

9.1 Introduction

This Chapter reports on research into the cost of fisheries management in selected European fisheries and attempts to relate these costs to the fisheries management innovations adopted. For this purpose, case studies were conducted in eight

S. Zableckis (✉)
European Commission, DG Joint Research Centre, Institute for the Protection and Security of the Citizen, Maritime Affairs, Via E. Fermi 1, I-21020 Ispra (VA), Italy
e-mail: szableckis@gmail.com

K.H. Hauge, D.C. Wilson (eds.), *Comparative Evaluations of Innovative Fisheries Management*, DOI 10.1007/978-90-481-2663-7_9,

countries where fisheries management innovations have recently been implemented. The aim of the exercise was to obtain measurements of actual changes in management costs following the implementation of four management innovations: participatory governance, rights-based approaches, effort control and harvest control rules. While this does not necessarily provide much information on the minimum costs of running the management systems in question, it offers insights into the costs experienced during the adjustment to the innovations and the early stages of the new fisheries management systems and may well be indicative of the real long term costs of running these systems.

The Chapter starts with a background discussion on fisheries management systems and the management innovations investigated. This is followed by a description of the main case studies conducted in Denmark, the Netherlands, Spain and Poland, and supplementary ones conducted in France, Sweden, UK and Faroe Islands. The first four cases were selected early in the research program as the most promising examples. The selection was also partly determined by the actual physical location of the partners involved. The supplementary cases, which are related to specific

Table 9.1 Personal Communication (Full Interviews Based on the Questionnaire, Phone Conversation and E-mails Based Communication)

Denmark:	Fisheries Directorate; Fishermens' association; Strandby Fiskeriforening and Strandby Quota Group
Netherlands:	Wageningen IMARES, Institute for Marine Resources & Ecosystem Studies Landbouw Economisch Instituut (LEI) Algemene Inspectie Dienst; Directie Visserij Ministerie Landbouw, Natuur en Voedselkwaliteit; Pelagic Freezer-trawler Association (PFA), Association of Dutch Pelagic Shipowners (RVZ), Dutch Fish Board (Pvis)
Faroe Islands:	Landsbankin; Ministry of Fisheries and Natural Resources; Faroese Fisheries Laboratory; Faroese Fisheries Inspection
France:	CEDEM-UBO; IUT – UMR Amure – Centre de droit et d'Economie de la mer; IFREMER – DEM – Département d'Economie Maritime; University of Brest
Scotland, UK:	Scottish Government, Marine Directorate; Fisheries Research Services; The Scottish Fisheries Protection Agency; Northern PO
Poland:	Department of Fisheries, Ministry of Agriculture and Rural Development, Warsaw Fisheries Monitoring Centre, Gdynia Regional Sea Fisheries Inspectorates, Gdynia
Sweden:	Swedish Board of Fisheries, Gothenburg

fisheries management innovations, were added to obtain additional empirical data for the final evaluation of the management costs associated with the fisheries management innovations. In all cases, data were collected from various data sources. They all made substantial use of face-to-face interviews with the people involved, including scientists, government and enforcement officials, and fisheries representatives. For the full list of institutions interviewed see Table 9.1. The interviews were conducted according to a uniform questionnaire prepared beforehand.

9.2 Background

The fundamental aim of fisheries management is to generate economic benefits by inducing fishers to behave more efficiently and sustainably in terms of fish stocks and the aquatic environment. In theory, most commercial fisheries should be capable of generating high salaries and good profits on a sustainable basis. This, however, does not happen automatically. The common property or common pool problem tends to get in the way of efficient utilization of most fish resources. Therefore, to realize the potential benefits of fish resources, it is necessary to implement a proper fisheries management regime.

Fisheries management must logically involve at least the following three main functions (Arnason, Hannesson, & Schrank, 2000; OECD, 2003).

1. Fisheries management administration (monitoring, designing, setting and modifying fisheries management rules and measures);
2. Research (biological, social and economic research to inform fisheries management decision-makers);
3. Enforcement (enforcing fisheries management rules).

The first function, the management function, comprises several sub-functions: system design; system implementation; adjusting management settings within an existing management system; recommending amendments or additions to the existing management system; administering the system; setting system measures such as the total allowable catch (TAC). Results from research (primarily biological, economic and social) provide the knowledge basis for carrying out this function. Common examples of these research activities include data collection, data analysis and stock assessment processes (Arnason et al., 2000; OECD, 2003). Thus, the research function is an inalienable part of any fisheries management regime. The enforcement part of the fisheries management involves surveillance of the fishing activity, on-site enforcement of the existing fisheries management rules and measures, and the processing and issuing of sanctions to alleged violators.

All these functions of fisheries management are costly. Typically, the enforcement function, monitoring fishing operations and enforcing rules, is most costly with research not far behind (Arnason et al., 2000; OECD, 2003). Compared to these two functions, the cost of the first function, setting fisheries management rules, is usually quite small.

Since fisheries management requires substantial economic resources, i.e. it incurs costs, socially sensible fisheries management cannot be rationally considered without including the cost of fisheries management. It should also be stressed that the different fisheries management systems require different research and enforcement efforts.

Limiting of fishing pressure to a level that stocks can sustain has been proposed as a key element in fisheries management within the context of the Common Fisheries Policy (CFP). With the 2002 reform, European fisheries management has shifted from a short-term approach to a longer-term approach based on multi-annual plans supplemented with emergency measures if the need arises to protect fish stocks in the short-term due to unforeseen circumstances. To promote the sustainability of fishing activities in EU waters and protect a specific stock or a group of stocks, the EU may, according to the CFP, apply a number of conservation measures as tools for multi-annual plans, regardless of the type of the plan. These measures include:

1. TACs to limit the maximum amount of fish that may be caught from a specific stock over a given period of time.
2. Technical measures, such as mesh sizes, selective fishing gear, closed areas, minimum landing sizes, and bycatch limits.
3. Limiting fishing effort by reducing the number of fishing days at sea.
4. Fixing the number and type of fishing vessels authorized to fish.

Certain types of TAC regulations (e.g. annual or multi-annual TACs) will most likely retain their position as a cornerstone in the CFP in the foreseeable future. Therefore, management innovations evaluated within the context of this research project should be understood as additional to those of a TAC regulation. In fact, the systems studied are hybrids of several management innovations.

9.3 Management Innovations

The study considered four management innovations which have been adopted within the EU in recent years and which are represented in the case studies. These are: Effort Control, participatory governance, harvest control rules (HCRs) and rights-based regimes. It should be noted that the innovations could represent different components of a management system and all could be combined. Effort control and HCRs restrict catches, participatory governance refers to the process of fisheries management decisions, and rights-based regimes refer to the rights of an individual fisherman or groups of fishermen.

9.3.1 Effort Control

Effort controls restrict fishing effort in a particular fishery and area. This measure is believed to reduce the fishing pressure on stocks or their components (e.g. spawners and immature fish). Effort control measures (usually limited number of days at sea or days of fishing) have been gaining attention in Europe as they are perceived, especially by the industry, to be working very well in neighbouring areas outside of the CFP, e.g. in the Faroe Islands. The implementation of effort control is also believed to reduce the cost of management compared to many other alternatives. Effort control-based management systems are sometimes also referred to as Total Allowable Effort (TAE). Effort restrictions can be regarded as 'input controls' vs. TAC limitation as 'output control'. Other measures such as limiting the number and size (engine power) of boats, size, type and number of fishing gear are related to effort restrictions, but are more properly characterized as capital or technical restrictions. The primary management measure such as TAC for demersal stocks in the Baltic Sea is accompanied by an extensive array of technical measures, including seasonally closed areas and days of closures to be allocated individually by the member states.

9.3.1.1 Research, Management and Enforcement Needs

A TAC management regime requires stock assessments. This is usually accomplished by monitoring of catches, collection of catch per unit effort (CPUE) data as well as fishery-independent data for tuning the stock assessment models. To run an effort control regime requires much of the same data with the addition of monitoring the actual fishing effort exerted by the fleets. Thus, provided a TAC regime or something similar is in place, an effort control regime does not require a great deal of additional research.

Note also that in order to impose the most appropriate effort regulation measures, costly collection of comprehensive information on technical performance of vessels and gears involved and the effect of fishing gears on fish stocks and their components might be needed. For instance in the Gulf of Riga in the Baltic Sea, a ban on herring (*Clupea harengus*) fishing was introduced following substantial additional research costs (Järvik, Raid, Shpilev, Järv, & Lankov, 2005).

In certain cases, particularly when environmental factors play a key role, for example on stock recruitment processes, comprehensive information on hydrological conditions is necessary in order to reallocate the fishing effort from areas of vital importance to stock reproduction. The case of Baltic cod (*Gadus morhua*), whose reproductive success is determined by the availability of suitable conditions for egg survival, also referred to as 'reproduction volume' (Köster et al., 2003) could serve as an example here.

This additional research effort would mostly be connected with costly work at sea. For instance, in the Netherlands, around half of a total research budget allocated each year for research is spent on research vessels, the other half on different activities ranging from the fisheries management to the phytoplankton surveys.

The information described above is usually not routinely collected during surveys, to obtain stock abundance indices or other stock assessment-related information. In order to obtain such information, special surveys are needed which would mean an increase in research costs. The cost implications certainly differ from case to case depending, *inter alia*, on the amount of existing knowledge but probably would be most substantial in the commencement phase of the implementation of effort management. In that respect, the incorporation of all available information from other stakeholders, from fishermen in particular, might possibly reduce the additional expenses.

The implementation of effort management might have implications for the administration costs. Indeed, according to Jennings, Kaiser, and Reynolds (2001), effort management licensing, establishing individual effort quotas and imposing vessel and gear restrictions would likely increase assessment, enforcement and administrative costs. Note, however, that effort restrictions, replacing or adding to quota restrictions, might increase the quality of stock assessments because misreporting of catches and discarding may decrease. The additional specific management needs linked to effort management are supposedly relatively low. However, the introduction of such measures would increase the administrative costs, e.g. coordination of implementation of the regulation.

The enforcement of effort restrictions, such as days at sea and engine power, imposed by governmental agencies, stakeholder groups, community, or combination of these under a co-management system, is relatively low cost and can be easily done using vessel monitoring system (VMS) data and available fleet databases. However, it should be realized that effort control without retaining continuous conventional control mechanisms on catches and technical measures is unlikely to produce the desired sustainable resource use (Jennings et al., 2001).

When fishermen are regulated in terms of effort and not in terms of landings, one would expect the incentive to discard to be comparatively low or determined only by the cost of retaining and landing the catch compared to the selling price. Therefore costly discard control at sea is not needed which means a decrease in enforcement costs. Moreover, one would expect a better agreement between catch and landings resulting in more reliable data for biological assessment. This, in theory and in the case of a pure effort management regime (without TAC regulation), would decrease the cost of biological research.

9.3.2 Participatory Governance

Participatory governance in fisheries can take many forms, for example co-management, where the fisheries managers collaborate with the fishing industry on decision-making. Co-management implies power sharing in the management process. It is sometimes referred to as a bottom-up management approach. The introduction of co-management is more than the introduction of a new management system, because it constitutes an institutional change. Thus, this innovation must be analysed and evaluated within the institutional framework around fisheries.

The most common participants in participatory governance are the stakeholders represented by fisheries authorities, fishermen and other aquatic resource users. Often scientists are involved in the process and sometimes even regarded as stakeholders. Participatory governance is becoming increasingly prevalent in European fisheries. Finally, the participation of environmental Non-governmental Organizations (NGOs) in fisheries management decision-making has proven important in the last decade to balance the short-term pressures on the industry. In Europe this has been recently developed through the Regional Advisory Councils (RACs).

9.3.2.1 Research, Administration and Enforcement Needs

Traditional research on biological and economic parameters aims to generate information for TAC management. Within the traditional scientific model, the TAC setting is based on the assessment of stocks and their regenerative power. Fishermen collaborate with scientists by providing data. However, it appears that data quality might be improved if fishermen participated in the decision making process. The participatory context implies a joint preparation of the scientific research agenda and the development of a collaborative framework, which could reduce the need for traditional formal research for setting TACs. The main research cost implication of such co-management is a possible decrease of the research costs. This gain should of course be set against the cost of operating participatory governance.

According to the new innovative context of participatory governance the process for designing fisheries management becomes more complex, although it is expected to enhance its effectiveness, adding new management needs. The reason is that, fishers' knowledge must be produced in response to government consultations and proposals and used as an additional input, to the traditional biological, ecological and even economic knowledge.

The traditional TAC regime requires the establishment of a robust control and enforcement system. Within the traditional process for setting and adopting TACs fishers sometimes misreport and discard fish. The participatory system adds legitimacy to the system from the perspective of the fishermen; they are more involved, which in turn increases compliance. The participatory system could make control and enforcement to both work better and to be less necessary, which can imply cost reductions in control and enforcement. It should be kept in mind, however, that to change fishermen's habits might be difficult in a short term, so the transition to less stringent enforcement might take longer.

9.3.3 Harvest Control Rules

Harvest control rules (HCRs) are defined as sets of laws that can be used for determining TACs (Cooke, 1999; Restrepo & Powers, 1999, Johnston, Parkinson, Tautz, & Ward, 2000). Thus, since HCRs define the how the TACs are set – both short term TACs and for a longer time period (multi-annual plans) – they are expected to reduce the administrative expenses to a certain extent. Often the HCR stipulate the TAC as a function of the stock size. Therefore, if a management policy can be expressed as an HCR, then the HCR provides means to determine the total allowable catch unambiguously (Housholder, 2004).

9.3.3.1 Research, Management and Enforcement Needs

Biological reference points of biomass and/or fishing mortality are often used in HCRs in order to evaluate the stock status and to trigger management actions. These reference levels can be used as limits or targets, depending on their intended usage (Caddy & Mahon, 1995).

According to Degnbol (2004), limit reference points relate to stock productivity and ecosystem dynamics and are thus considered as properties of nature. Therefore, the introductions of formalized management plans do not imply new approaches or methodologies for limit reference point estimation. However, due to apparent ecosystem dynamics, the revision of reference points/levels may be inevitable after certain time periods. Limit reference points should be updated when the understanding of the natural system improves.

Since the core elements of HCRs are generally similar to those applied in traditional TAC-based regulation processes, the implementation of HCR- regulation regime would not imply the extensive need for additional resources. The success of HCRs or TAC regulations depends on the effectiveness of the enforcement system, so that an introduction of TACs in a fishery would mean additional enforcement costs. Formal evaluation of cost efficiency of TAC regulations would imply costly data gathering, assessments, decisions and implementation against the benefits of objective achievements. This is not formally doable without extensive economic data (Degnbol, 2004).

9.3.4 Rights-Based Management

Public ownership of marine resources and freedom of access to the sea has contributed to the race to fish and eventually to biological overfishing. Management systems that restrict the right to harvest can help prevent overfishing. Individual transferable quotas, the ITQ system, usually restrict the right to harvest while ownership (of resources) remains with the government. Tenure systems also may allow fisheries to own the resources. ITQs and tenure arrangements seem to have encouraged sustainable fisheries for stocks that do not cross management or national boundaries (Costello, Gaines, & Lynham, 2008). Management systems that give property rights to stocks, or provide access rights to fishing grounds are often supported by fishers and they will take some responsibility for enforcement. One benefit of ITQs is that the fishery becomes market driven, fishermen that are not that efficient can sell their ITQs and thus receive compensation for leaving the fishery. Operating costs and fleet capacity generally fall. Other rights-based management examples (not examined in this work) are territorial use rights, community-based catch quotas, vessel catch limits, individual non-transferable quotas (IQ), limited non-transferable licenses, limited transferable licenses, individual non-transferable effort quotas, and individual transferable effort quotas.

9.3.4.1 Research, Management and Enforcement Needs

An important drawback of individual quotas is the enforcement effort that is needed to make them effective. Quotas, as all other restrictive management measures, generate incentives for avoidance and misreporting, as has been demonstrated wherever quotas, transferable or not, have been put in place; for example in Iceland, New Zealand and the European Community. To enforce individual catch quotas, catches or at least landings have to be monitored. Enforcement, which involves the inspection of actual catch or actual landings at numerous landing ports is often (but not always) quite expensive. The costs depend on the situation: if there are few landing places and the catch is homogeneous (not great variations in the value of a given volume of fish) or if the catch distribution chain is transparent, the cost of enforcing individual quotas is relatively small. In other cases, the cost of enforcing individual quotas is likely to be substantial to the point of being prohibitively high in places where fish are landed from numerous small vessels in a multitude of landing places with a minimum of technical devices and sold directly to consumers. In large-scale operations where elaborate landing facilities are required, and where most of the fish is processed on shore, it is not difficult to keep track of the fish and check different parts of the catching and processing system against one another. This method has been implemented in New Zealand, Iceland, Canada and Chile. Problems are also likely to arise at sea. Individual quotas sometimes give fishermen an incentive to 'high-grade' catches; that is, to throw away less valuable fish in order to accommodate more valuable fish within the limits of the quota. Besides being wasteful, such practices could have serious consequences for management, as the recorded catches would underestimate the quantity actually removed from the stock. This would weaken the factual base on which decisions about total allowable catches must be taken (Hannesson, 1993). Broadly speaking, it appears that individual harvest quota regimes usually require substantial enforcement activity and, therefore, may be more costly to enforce than many other management regimes. Therefore, it is interesting to note that in countries where individual quotas have been most extensively used, i.e. Iceland and New Zealand, overall fisheries management costs, as a percentage of landed values are amongst the lowest observed (Schrank, Arnason, & Hannesson, 2003; OECD, 2003). In any case, the substantial increases in economic gains, which are almost invariably generated by individual quota systems, should be set against the enforcement costs of the system.

The Biesheuvel group is a Dutch association of fishermen with individual fishing rights. The group provides the opportunity for individual fishers to rent and trade individual quotas under private law. The group is also responsible for adherence to the quota restrictions. From the perspective of the State, the costs of transfer and overfishing are internalized at the lowest possible level. Dutch fishers feel that the political value of the Biesheuvel groups is considerable. They feel that the improved coordination of micro-decisions in the collective interest has created opportunities for cooperation and interactive co-governance among partners to seek alternatives to command-and-control systems. It remains difficult to assess whether the 'sharing' of rent amongst the fishers operating within the Biesheuvel groups may have provided incentives for the sharing of information surrounding the natural resource (Venema, 2001).

9.4 Hypotheses

The following hypotheses (presented in Table 9.2) were developed in order to investigate fisheries management costs dynamics in selected case studies. The results of testing these hypotheses are presented in the country case studies.

The investigation of these hypothesis showed that it is possible to evaluate fisheries management systems in terms of costs, as well as by other indicators, for example staff time. The staff time is an indicator that can be used in almost all regimes investigated as a descriptive measure; it is universal in terms of what one person can do versus what one Euro can buy in different countries. Also the acquisition of data on staff employed in different fisheries management functions seems to be more readily available and more readily disclosed by stakeholders. While comparing

Table 9.2 Research Hypotheses. Each Hypothesis Assumes, Based on the *a priori* Information, that the Management Costs Would go Up, Down or Remain Constant with the Introduction of the Corresponding Management Innovation

Hypotheses				
Management measures	*Short/ long term*	*Administration costs*	*Enforcement costs*	*Research-costs*
Effort control	S	↗	↘	↗
1. Days at sea plus TACs (Baltic cod, Poland)	L	↘	↘	↗
2. Closed areas plus TACs (Baltic cod, Poland)	*S*	↗	↗	↗
	L	↘	↗	↗
3. Days at sea (Faroese 'clean' case)	*S*	↗	⇔	⇔
	L	⇔	⇔	⇔
Participatory governance	*S*	↗	⇔	↗
1. RACs combined with TACs (Northern hake case in Spain)	*L*	↗	⇔	↗
HCRs	*S*	⇔	⇔	⇔
1. TACs combined with closed areas (Baltic cod in Poland and Sweden)	*L*	↘	⇔	⇔
Rights-based regimes	*S*	↗	↗	↗
1. Combined with TAC (Danish pelagic fleet ITQs and UK FQAs)	*L*	⇔	↗	↗
2. Combined with cooperative management (Dutch Biesheuvel case)	*S*	↗	⇔	⇔
	L	⇔	⇔	⇔

Short Term (S): 2–3 years, Long Term (L): more than 3 years. ↗ – management costs increase, ↘ – management cost decrease, ⇔ – management cost remain stable or increase only due to inflation.

countries or fisheries innovations in term of costs, currency conversions are needed and money value adjusted over time and on country-to country basis.

9.5 Case Studies

The research focused on eight country case studies of which four countries (Denmark, the Netherlands, Spain and Poland) were investigated more thoroughly and the other four countries (Sweden, France, UK and Faroe Islands) in less detail. This was related to the availably of data, the complexity of the management system as well as the potential of useful findings.

The regulation of the **Danish** fisheries is primarily based on TACs, ITQs, days at sea and technical measures. The technical regulations of Danish fisheries include mesh sizes, minimum sizes, closed areas etc. and are based on EU regulations. The pelagic fishery has been under an ITQ management system since 2003. The demersal fishery came under a quasi ITQ system (quotas are transferable with fishing vessel only) in 2007. The costs of fisheries administration, enforcement and research have remained stable in the period of 2003–2005 (fixed prices).

In the **Netherlands** under the EU Common Fisheries Policy (CFP), the fisheries management is based on a system of TACs, complemented by technical conservation measures. Another important measure in place is an ITQ system. An IQ system was introduced in 1976, where quotas were officially only tradable with a vessel. The ITQs were allocated based on the average catch of the past six years according to a horsepower grouping and made individual quotas officially transferable by imposing certain restrictions on their transfer. Individual quotas became officially transferable in 1985. Fishers can only buy an ITQ from another ITQ holder if they are in possession of a fishing license.

The Biesheuvel Steering Committee introduced a co-management system in 1993 allowing fishers to rent and/or barter quotas within the Biesheuvel groups (10 in 2005), which operate under private law. The Biesheuvel Groups are groups of fishermen, making arrangements among themselves to divide the fish quota. The system is based on co-management and social control. This system enables fishermen to optimize the use of their ITQs by means of renting ITQs and days-at-sea within the co-management groups. While the introduction of the Biesheuvel system reportedly reduced uncertainty in the sector, increased profitability and transparency, the actual management costs were not reduced.

The specific case study for **Spain** was focused on the Northern hake (*Merluccius merluccius*) fishery in the Basque Country, which has been historically managed by a mix of TAC, total allowable effort (TAE) and technical measures. Participatory governance has been introduced in the hake fishery through the creation of North Western Waters Regional Advisory Council (NWW RAC) in 2005, and the South Western Waters Regional Advisory Council (SWW RAC) in 2007.

The main fishing areas of the **Swedish** fleet are the Baltic Sea and the Kattegat/Skagerrak. In terms of value, the fishery for cod has been the most important and accounted for about 20% of landings, followed by herring and sprat

– 20% and 10% respectively, *Nephrops* (10%), and northern prawn *(Pandalus borealis)* (10%) in 2007. TAC, fishing effort and licenses, technical measures and control and enforcement are the measures most commonly used in Swedish fisheries management.

The Polish demersal cod fishery alike the Swedish Baltic cod fishery is managed through a TAC system. From 2008, TAC for cod is set according to rules defined in a multi-annual plan for the cod stocks and closely linked with fishing mortality rate. There are various effort control regime restrictions in place (i.e. limitation of fishing days, vessels number, areas closed for fishery) that have gradually gained the importance in recent years.

The UK case focuses on introduction of the Fixed Quota Allocations (FQAs), which were introduced in January 1999 as a rights-based system. With the FQAs a fixed quota share (quota unit) was linked to a specific vessel, instead of the previous yearly allocation of quota based on historical data. The system has been criticized for not being clear on the precise ownership of the fish resource and that the FQAs can only be sold together with the vessel and the license to fish. But as vessels (with license and quota) can be sold to Producers Organizations (PO), the quota can be re-distributed among the members of the PO, and therefore in theory be redistributed to several vessels.

The fisheries management system of **Faroe Islands** is based mostly on effort control through the use of individual and transferable days at sea since 1997. The management system also comprises technical measures and area closures based on access rights to fishing areas. The area regulation is detailed by fleet segments and gives priority to smaller vessels and long-line fisheries. The fishing days are allocated to fishing vessels within five fleet segments on a long-term basis. In that sense it is also a rights-based management (RBM) system. The effort regulation is applied only to the fisheries within EEZ targeting demersal species.

9.6 Discussion, Hypothesis Testing and Conclusions

Most existing studies on fisheries management costs (such as Arnason et al., 2000; Schrank et al., 2003; OECD, 2003) limit themselves to presenting and comparing data on fisheries management costs and indicating a few theoretical implications. The CEVIS project focuses on exploring actual cost implications of certain management actions in selected fisheries and comparing them to hypothesized effects.

The management innovations are often linked to a certain fishery in a given country. The main methodological problems encountered stemmed from the lack of available data on the costs associated with these innovations. Also there were substantial national differences observed in how these costs were defined, measured and compiled. Furthermore, the innovations are very often just a supplementary element to existing fisheries management tools, while the data available were usually for all the fisheries management activities on a national level. Therefore it was difficult to link the costs of management to a specific innovative management system

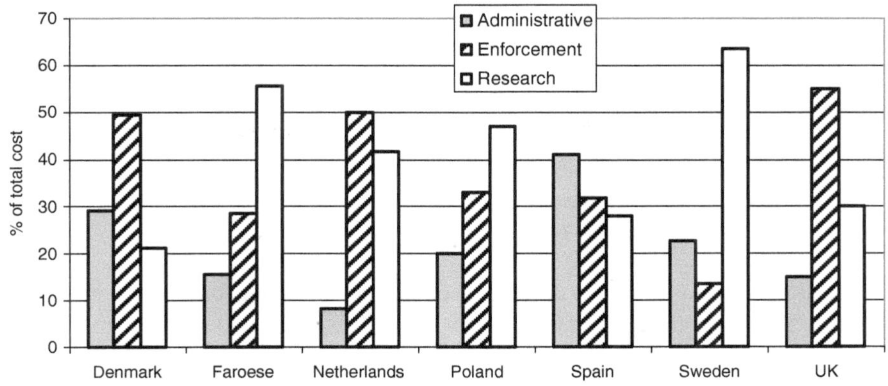

Fig. 9.1 Relative Average Spending on Administration, Enforcement and Research in Selected Case Studies

in a quantitative way. To counter this, more qualitative methods, such as interviews, have been used as supplementary evidence in many cases.

The general overview of fisheries management relative spending (broken down according to administration, enforcement and research) in country cases is presented in Fig. 9.1.

Spain and Denmark spend 40 and 30% of their fisheries management costs respectively on fisheries administration. The Netherlands spent less than 10% of total management spending on administration. The UK, Denmark and the Netherlands spent around 50% of the total management costs on enforcement, while the lowest relative enforcement effort was observed in Sweden. This might be due to the fact that most of the Swedish fisheries enforcement is done by the national coast guard. The research spending, as relative share of total spending, was the highest in Sweden – 60% followed by Faroe Islands – 55%. The lowest relative research spending was observed in Denmark.

9.6.1 Hypothesis Testing and Country Case Studies

9.6.1.1 Denmark

In the **Danish** case, based on the pelagic fisheries share of selected OECD cost drivers, the task was to test the hypothesis that rights-based regimes will increase all long-term and short-term management costs, except that long-term administration costs will stay the same. The calculation indicates lower costs of fisheries administration, enforcement and research under the ITQ management regime, than under the remaining fisheries, where the quotas were allocated to each vessel according to size and type at a weekly basis. The interviews with fisheries managers have indicated that management cost is not directly connected to the management system, for example the obligation to report catches before landing, which influences administration costs. In the initial implementation phase though, a link between management system and management cost (or effort) was felt, as the transferability of quotas increases the administrative burden. But generally it was not possible in the Danish case to reach a clear conclusion on the cost of management of an ITQ-system compared to the previous a ration-based catch allocation system.

Calculations related to OECD identified management 'cost drivers' showed lower costs within all three areas of management, thus supporting the hypothesis that implementation of rights-based management would decrease the management costs.

The qualitative assessment of the consequences for management costs of implementing an ITQ regime, however, is less supportive of the hypothesis. Though the public budget for management and enforcement has not changed, there seems to be more administrative work involved at least in the short run. The consequences on the costs of enforcement have been difficult to assess, as some of the control of effort has been changed from control at sea to land-based control. No change in research activity related to the regime change has been observed. This is in support of the hypothesis.

9.6.1.2 Faroe Islands

In the Faroe Islands case the development in the fisheries management costs from 2000 to 2006 shows the administrative costs increasing by 28%, research costs by 28% and the enforcement costs by 34% over the period. The fisheries management costs cannot be allocated to particular fisheries or fleet segments. Also, the available management cost data do not allow for a segregation of the costs associated with effort management and those associated with other management systems. Moreover, the time series data available only cover the period 2000–2006 when effort management was already well established in the Faroese fisheries. Therefore it was not possible to separate the effort management cost from rest of the management costs, and to check the cost effects of the introduction of the new system.

The (soft) evidence from the Faroese case study shows that the administrative burden associated with the implementation of the effort control system has been high because of the need to establish a system for comparison of effort between different vessel types and fleet segments (system complexity). This would indicate an increase in the administrative costs (in the shorter term) in support of the work hypothesis.

The same evidence indicated that research costs related to stock assessment and resource conservation have not increased. However, the data acquired show a 25% increase in research costs from 2000 to 2002 and stable costs thereafter. As the latter indicates a reduction of the research costs in real terms there is no indication of increase in the research costs related to stock assessment and advice. It should be mentioned that the research institute has not been able to investigate the possible increase in (total) effort that comes with technical development. 'Technology creep' is a major risk/concern associated with effort management systems.

9.6.1.3 Netherlands

In the Netherlands case it has been difficult to allocate costs of management to the specific management regime or specific fisheries. Also, no clear indications of the change of costs due to changes in management systems were observed, except the increase of the costs for the fishermen participating in the co-management groups.

From the country interviews with those responsible for administration, enforcement and research, it became evident that the management costs or the government allocations to specific management functions over the last several years changed only due to inflation. The cost breakdown showed that enforcement costs were 50%, research costs 41.6% and administration costs 8.4% of which Fisheries Department and Co-management groups both 4.2% of total costs of management system.

Objectives of the ITQs introduction were better economic performance of the fleet and greater flexibility of fishermen with respect to quota. One might also have assumed that the involvement of the state would be reduced. In spite of the successful introduction of the ITQ system, the government in the Netherlands is still deeply involved in management functions and the costs of management have not decreased in the past several years.

The results of the Netherlands case study indicated that, despite it being rather difficult to allocate management costs to specific management innovation or specific fisheries, the regulation did not decrease management costs. On the contrary, several programs implemented in the Netherlands by fishermen themselves, increased management costs and also the time used in the process, for example, fishermen coordination meetings. The same was concluded by Arnason (2007).

9.6.1.4 Poland

In the case of the Baltic cod fishery in Poland where area closures and days-at-sea regulation have been widely implemented (besides to TAC system), the overall costs have increased in all aspects of management.

The administrative costs increased from 0.54 to 1.57 million euros (290%); the enforcement costs went up from 1.0 to 3.52 million euros (352) %, and the research costs from 1.82 to 3.08 million euros (169%) in 2002–2007.

A particularly fast rise, by 25% annually, has been observed since 2004. It was, however, difficult to link the observed trend with changes in the system of fisheries management. Along with accession to the EU and acceptance of the principles of the Common Fisheries Policy, Poland was committed to adapting its standards of fisheries management to EU regulations. Thus, the strengthening of the fisheries administration and enforcement system as well as research, in order to meet the EU requirements, caused the major increase in costs. Therefore, the increase in management costs in Polish fisheries can more likely be seen as a result of new obligations connected with data collection and fisheries control after her accession to the EU.

Due to the lack of detailed statistical data it has been very difficult to allocate management costs to specific fisheries of research activities. It may be said, however, that the new responsibilities in fisheries enforcement, followed from the accession of Poland to EU in 2004, almost exclusively concerned cod fishing.

9.6.1.5 Spain

In the Spanish case, research costs relating to Northern hake fishery continue the historic trend. The following trends were observed: (i) The Basque Government has traditionally financed biological research activities in demersal fisheries, mainly Northern Hake, Anglerfish and Megrim. The budget for this research activity has been increasing annually by about 7% after the North Western Waters Regional Advisory Council (NWW RAC) implementation. This annual increase is higher than the one observed before the RAC creation; (ii) Additionally, research costs for data collection and data analyses increased annually on average by 9%; (iii) The European Union and the Basque Government have also continued co-financing other Northern Hake research activities.

Empirical data on administration costs showed increase although not significantly. The public budget for administration has been increased because an additional amount of money has been annually provided by Member States for supporting RAC activities. The NWW RAC receives from the Spanish government 5,000 euros a year, which is around 10% of the total public budget for the RAC. The remainder 90% is supplied by the EU. The South Western Waters Regional Advisory Council (SWW RAC) is funded by the Spanish government by 4,000 Euros a year. Besides, the Basque Government increased funding for the management of the created RACs. In 2007 the contribution was around 50,000 EUR, going up to 150,000 EUR in 2008. The costs related to RACs are not only due to the Northern hake management but also due to other fisheries and general issues related to the fishing activity in the sea areas covered by the RACs.

Empirical enforcement cost data related to the fishery of the Northern hake in the Basque Country and, in general, at the Spanish level seem not to have changed since the implementation of the RACs. However, some issues were noted. For example, the Northern hake enforcement costs are relatively less expensive and possibly not sufficient as compared to other regions of Spain. For example, in 2007 in Galicia, 73% of the total annual enforcement time went on the hake fishery control; while in the Basque country only 56%.

Thus, the empirical evidence in the Northern Hake fishery not always supported the hypotheses tested. One of the possible reasons might be that the recent creation of the RACs: the SWW RAC was created in April 2007 while the NWW RAC began operating in 2005.

9.6.1.6 Sweden

The management costs of Swedish fisheries have increased over the 2005–2007 period by 27% (from 22.5 to 28.6 million euros), of which research costs rose by 20%, administration costs by 21% and enforcement costs by 74%. TACs as well as fishing effort limitation (in days and areas closures) and other technical measures are most commonly used in Swedish fisheries management.

Allocation of national costs of management to the specific segment of demersal fisheries was found to be unfeasible. This is a result both of insufficient statistical data, and of the specifics of Swedish multi-species fisheries. According to OECD and SBF data steady increase of the fisheries management costs has been observed in Sweden in 2001–2007.

The Swedish case study indicated that the introduction of marine protected areas, according to the multi-annual management plan, had only a limited impact on increase of steady management costs. Gradual growth in expenditure on fisheries research, administration and enforcement may be explained by increased requirements and obligations imposed by the European Commission, especially those directed at better protection of cod stocks and to decreasing unreported and misreported landings. Also inflation might be to some extent another explanation for increasing of overall management costs.

9.6.1.7 United Kingdom

In the UK case the data on distribution of overall management cost between research, enforcement and administration are poor. OECD data indicate that of the total cost of management in 1997 was 83 million GB pounds. Of this sum, administration costs were approximately 15%, research 30%, and enforcement 55%.

In the years of introduction of the FQA system, the total management costs, indicated by the OECD data on general services in UK were stable. But as several factors could influence the cost level this is not a clear indication of cost-related consequences of the FQA introduction. The interviews with the officers revealed that the costs did not change after introduction to FQAs. None of the interviewees recall specific changes in cost of management in their area, which can be directly linked to the FQAs. The officers indicated, though in general terms, that such a managerial issue probably would not influence the cost level more than marginally, as there might be other factors, such as the cod recovery plans, that more seriously influence priorities and cost level in the institutions. There were no direct implications on research and enforcement/control related to the implementation of FQAs.

In relation to the hypothesis, the UK case cannot confirm the hypothesis, that the cost of management would increase in the private sector and decrease in the public sector, nor that the research costs would decrease. In regard to enforcement it was hypothesized that the cost would be neutral in the short run and reduced in the long run. It is not possible to say if this has been the case, as the long run effect of the FQAs cannot be filtered from other factors.

9.6.1.8 France

The only available data on the fisheries management costs in **France** cover the period of 1997–1999, and show a slight increasing trend in overall costs. The overall outcome from the case study was that the high number of regulations and increasing participation of fishermen in the management process would increase the total

Table 9.3 The Result of the Hypothesis Testing. Analysed Change in Management Costs when Implementing New Management Regimes

	Management	Enforcement	Research
Effort control	⇔	⇔	⇔
Participatory governance	⇗	⇔	⇔
Harvest rules	⇔	⇔	⇔
Rights-based regimes	⇗	⇔	⇔

management costs. However, the lack of information did not allow to conclude the effect of particular management measures on costs.

The summary of the hypothesis testing and the actual change in the management costs as observed in case studies is presented in the Table 9.3.

9.6.2 Management Innovations Conclusions

9.6.2.1 Effort Control

The case studies of the current project did not give a clear answer as to whether the implementation of the new effort regulation gave any clear advantage in terms of management costs. The case studies did not indicate decline nor increase in the management costs. In the case of the Baltic cod fishery in Poland and Sweden, where area closures and days-at-sea regulation have been widely implemented (besides to TAC system), the overall costs have increased in all aspects of management (administrative, research and enforcement). However, the additional commitments connected to the accession of Poland to the EU in 2004 seem to have been the main reason for such cost dynamics.

9.6.2.2 Harvest Control Rules

The experience obtained with HCRs, for example in the European cod management, has shown that uncertainties in the stock size and in catch estimates in particular may lead to substantial practical difficulties in implementing rules. This implies that full research process of conventional stock assessment, including comprehensive data collection, is an important prerogative for the implementation of management plans. Consequently, the implementation of HCRs as the major management tool would probably not reduce the research costs. Also a continuous need for monitoring how each fishery is following the TAC constraints imply that administrative and enforcement costs would not decrease.

The relevant CEVIS Case studies (e.g. on the Polish cod fishery), did not give clear picture on the possible effects of the implementation of HCRs.

9.6.2.3 Rights-Based Management

Transferable quotas appear promising as an incentive-compatible and cost-efficient system of management. Also, these systems are thought to reduce control and enforcement and overall administrative costs, since the governments can help to empower fishers to control their own fishery. If we take into account the number of administrative duties that fishermen's organizations (for example Biesheuvel Groups in the Netherlands) have, then we get a significant amount of money being spent on keeping the cooperatives functioning (office space, personnel), meetings and negotiations (meetings, hotels), etc. At the same time government agencies are still heavily involved in management functions and the consequent costs.

9.6.2.4 Participatory Governance

Findings from the Northern Hake Basque fishery case study indicate that administration costs increased after the introduction of the innovative regime, and that research and enforcement costs continue their increasing historic trend. However, it should be mentioned, that the effects of the participatory governance on management costs may not be evident yet given a very short time period after the implementation.

It is expected that in the medium to long-term horizon the administration costs will continue to increase, while the enforcement and research costs will have the tendency of decreasing or, at least remain stable.

9.6.3 General Conclusions

In general, the hypothesis testing has been hampered by the lack or shortage of data on specific management innovations, particularly in temporal aspects of cost developments. In most cases only national level fisheries management costs are available. Attempts to allocate costs to specific management innovation ended up in arbitrary numbers, coherence across countries was not possible so the end results could not be compared properly.

It was assumed, that there is a direct connection between the need for fisheries management, administration, research or enforcement and fisheries management spending. In other words the change in the need of services would result in the change in fisheries management costs. However, from the cases investigated it appears that there is no direct link between the need and the cost. At least in the short run the needs following management innovation implementation did not result in significant increase in costs. On the other hand, interviews with managers provided indications of how workload in various institutions changed following changes in the fisheries management system. This of course implies real cost changes, even if they are not accounted in the statistics.

The investigation of the fisheries management innovations revealed that the management cost effectiveness might not be a major driver for change in management systems. In several cases political reasons or degrading environmental conditions were the main drivers for management innovations. The response to the introduction of the innovations was not necessarily increase in funding but in several cases in the reallocation of efforts.

References

Arnason, R. (2007). Fisheries self-management under ITQs. *Marine Resource Economics, 22*, 373–390.

Arnason, R., Hannesson, R., & Schrank, W. E. (2000). Costs of fisheries management: The Cases of Iceland, Norway and Newfoundland. *Marine Policy, 24*, 233–243.

Caddy, J. F., & Mahon, R. (1995). *Reference points for fisheries management*. FAO Fisheries Technical Paper, 347, 1–83.

Cooke, J. G. (1999). Improvement of fishery-management advice through simulation testing of harvest algorithms. *ICES Journal of Marine Science, 56*, 797–810.

Costello, C., Gaines, S. D., & Lynham, J. (2008). Can catch shares prevent fisheries collapse? *Science, 321*(5896), 1678–1681.

Degnbol, P. (2004). *Management plan evaluations and the advisory framework – Issues to be addressed and the ICES approach for development*. Paper presented at Workshop on Harvest Control Rules for Sustainable Fisheries Management, Bergen, September 2004.

Hannesson, R. (1993). *Bioeconomic analysis of fisheries*. Oxford: Blackwell Publishing.

Housholder, D. J. (2004). *Evaluation of harvest control rules: Simple one-parameter versus complex multi-parameter strategies*. Dissertation, University of Bergen.

Jennings, S., Kaiser, M. J., & Reynolds, D. J. (2001). *Marine fisheries ecology*. Malden, Oxford, Carlton: Blackwell.

Johnston, N. T., Parkinson, E. A., Tautz, A. F., & Ward, B. D. (2000). Biological reference points for the conservation and management of steelhead, *Onchorhynchus mykis. Canadian Stock Assessment Secretariat, Research Document 2000/126.*

Järvik, A., Raid, T., Shpilev, H., Järv, L., & Lankov, A. (2005). Precautionary approach in Baltic herring trawl fishery: The effect of hauling techniques and engine power on unaccounted mortality estimates. In C. G. Soares, Y. Garbatov, & N. Fonseca (Eds.), *Proceedings of the 12th International Congress of the Association of the Mediterranean*(pp. 1223–1330). London, Leiden, New York, Philadelphia, Singapore: Taylor & Francis.

Köster, F., Möllmann, C., Neuenfeldt, S., Vinther, M., St. John, M. A., Tomkiewicz, J., et al. (2003). Fish stock development in the Central Baltic Sea (1976–2000) in relation to variability in the physical environment. *ICES Marine Science Symposia, 219*, 294–306.

OECD. (2003). The cost of managing fisheries. OECD. ISSN 9264099751.

Restrepo, V. R., & Powers, J. E. (1999). Precautionary control in US fisheries management: specification and performance. ICES *Journal of Marine Science, 56*, 846–852.

Schrank, W. E., Arnason, R., & Hannesson, R. (2003). *The cost of fisheries management.*Adlershot, Burlington: ASHGATE.

Venema, N. (2001). An overview of the Netherlands open sea and coastal fisheries: State of the art. In D. Zeller, R. Watson, & D. Pauly (Eds.), *Fisheries Impacts on North Atlantic Ecosystems: Catch, Effort and National/Regional Data Sets*. Fisheries Centre Research Reports 9(3). University of British Columbia, Canada.

Chapter 10
Legal Aspects of Individual Transferable Quotas

Miriam Dross and Hendrik Acker

Abstract Conformity of the innovations to the existing legal context represents an aspect of social robustness of fisheries management innovations. This chapter takes a look at the legal dimension of individual transferable quotas (ITQs). Since the chapter cannot analyse all legal issues that arise under the different legal systems that have introduced ITQs, the first subchapter tries to give a general overview of legal challenges ITQs can face under national law. The second analyses legal questions of European Community law in the specific context of the Common Fisheries Policy (CFP). Then, human rights and World Trade Organization (WTO) issues are discussed. Finally, considerations when legislating ITQs are elaborated.

Keywords CFP · European Community law · Individual transferable quotas · Legal constraints · Property rights · WTO

10.1 ITQs Under National Law

Although ITQ regimes worldwide vary widely, each ITQ system faces similar it is legal challenges when introduced. Stewart (2004) describes the gradual development of rights in fisheries management along typical lines. The development often moves from open access (i.e. anyone can fish) to a licensing system in which anyone can obtain a licence, to the conditional granting of licences, where licences can be limited to a particular group of holders, limitations on vessel or gear types and capacity, fishing seasons etc. Next, catch limitations or quotas may be applied to licences. Also, fishing rights may be made transferable although that is not necessarily the case, as the example of Canada shows, where in 1999 only about half of the IQs had permanent transferability (Burke & Brander, 2000). Finally, when quotas are decoupled from licences, they become something that could be called a form of tradable property. However, the unlinking is rarely comprehensive. This

M. Dross (✉)
German Advisory Council in the Environment, Reichpietschufer 60, 10785 Berlin, Germany
e-mail: miriam.dross@uba.de

K.H. Hauge, D.C. Wilson (eds.), *Comparative Evaluations of Innovative Fisheries Management*, DOI 10.1007/978-90-481-2663-7_10,

line of development is of course a theoretical model and in reality may not be so linear. Variations are often caused by geographical, political, social and economical conditions and considerations (Stewart, 2004).

10.1.1 Property Characteristics of ITQs

According to economic theory, ITQs have four characteristics, which are more or less distinct in the individual system: transferability, exclusivity, durability and security (Scott, 1988; Arnason, 2000). Transferability means that the right can be sold as a whole or in parts. It is seen as 'the key-defining feature of most fisheries rights' (Stewart, 2004). Depending on the system chosen in the respective countries, the right can be transferred freely or only to specific persons or entities. Exclusive fishing rights can also be designed not to be transferable, in which case they are referred to as 'individual quotas' (IQs).

Exclusivity refers to the ability to hold and manage the right without interference from others, e.g. regulatory bodies or other fishermen (Stewart, 2004). The meaning of the term exclusivity becomes obvious when looking at an important aspect: the possibility to enforce the right. This means that, for example, a fisherman could take a regulatory agency to court when he thinks that the agency denies him the possibility to fish against 'his/her' quota.

Duration alludes to the time span of the property right, during which the holder may exercise it. Finally, security of title refers to its strength as a constitutional or legislated right (Shotton, 2000). Generally, a right may be challenged by others, such as the state or other individuals. ITQs are not absolute, but they are present in different types of rights to varying degrees (Stewart, 2004).

Rights can be considered to be strong rights when they are durable, i.e. have long tenure; provide exclusivity of use, cannot be arbitrarily removed or diluted; and can be transferred (Shotton, 2000). Generally, the strongest property rights are those with the fewest constraints on the operation of markets.

When countries choose to introduce ITQs, different aspects play a role in the choices they make. The two countries with probably the strongest property rights in fisheries are New Zealand and Iceland. Both have similar general conditions: an isolated location and a unitary parliamentary system, which does not have to deal with the question of separation between federal and state legislatures.

Legal issues have a decisive influence on the design of the different systems. As regards the property characteristics of ITQs, a wide range of options exist. In the USA, the legal characterisation of quotas is as revocable privileges, in Australia, so-called statutory fishing rights were created[1] and in New Zealand, explicit property rights language is used for ITQs. The property characteristics of the rights-based systems can thus be more or less strong. The basic question whether ITQs can

---

[1]For a discussion of the legal nature of Australian Fishing Licences, see McFarlane (2000) and Fitzpatrick (2000).

represent property at all is hotly disputed in most countries. This question can be approached from an economic or from a legal point of view. The latter might diverge considerably between different jurisdictions and different models exist.[2] From a legal perspective there is thus not one answer to this question, but only an answer within the framework of a particular jurisdiction.

Whether the property characteristics of the rights-based systems are strong or not, in legal terms it is advisable to clearly define the attributes of the right. This is illustrated by the example of Iceland. Legal problems derived from the fact that on the one hand legislation explicitly stated that ITQs were not meant to constitute property rights, while on the other hand the ITQs had strong property characteristics. Interestingly, courts in one and the same country have come to different conclusions when asked whether the ITQ in question constituted property, depending on whether the adjustment of rights between private persons was in question or rights against the state. Furthermore, different national jurisdictions have also come to different conclusions despite similar conditions due to differing basic doctrines and constitutional provisions (Stewart, 2004).

Macinko and Bromley (2004) hold that ITQs in general are not property rights at all, because they are not enforceable vis-à-vis other fishers. Furthermore they maintain that fisheries do not need to be turned into rights-based systems, because fish resources are already governed by a rights-based regime, due to the fact that they are owned by the state. Instead they recommend discussing royalty leasing. This approach is supported by legislation in different countries, including the USA and Iceland, stating that ITQs are not to be considered as property. Whether or not one chooses to categorise ITQs as property (right), it remains important to clearly define the nature of the right that is bestowed to private individuals or groups in order to minimise legal challenges.

10.1.2 Lawfulness of ITQs

A fundamental legal aspect that is often discussed in countries that have introduced ITQs is the question whether the state has the right to introduce individual rights in fisheries at all (Stewart, 2004). This is sometimes negated because of legal traditions that grant the public the right to fish. Mostly, courts have not followed this line of argument but held that the state can abrogate this rule through statutes in order to conserve the fishery. 'The right of the states to legislate in respect of this fishing is based on sovereignty, not ownership, of the resource, and limited-access fisheries rights are appropriately established and regulated by statute. It is therefore for the statute[3] in question in each case to determine the legal nature of its creation' (Stewart, 2004).

---

[2]For an analysis of the different discourses in regard to ITQs as property, see Connor (2000).

[3]Statute is not necessarily the governing Act, but may be Regulations as in the US halibut and sablefish fishery, or a Management Plan under the Tasmanian Act.

Experience shows that ITQs are challenged in the courts mainly when the quotas are first allocated. Thus, the process and the conditions of allocation bring up legal concerns. Mostly, three legal issues come up in this context: national rules of non-discrimination (equal treatment), the right to a free choice of occupation, and the protection against deprivation of property. These are basic human rights which can be found in almost all democratic constitutions and human rights declarations.

The accusation that ITQ regimes are discriminatory stems from the fact that they always give rights to some and exclude others. The stronger the quota's property characteristics – such as exclusion and security – are, the more probable it is that the rights exclude others. Since one of the purposes of a quota system is to reduce the number of participants in the fishery in order to increase efficiency, the exclusion of some is an intentional effect (and probably the most contentious one). However, courts have generally not found that ITQ regimes were discriminatory, because the criteria upon which the quota were allocated were usually based on good reasons (e.g. in the case '*Alliance Against IFQs vs. Brown*'[4]).

The exception was the Icelandic Supreme Court, which in a first decision in 1999 came to the conclusion that the law allocating the quotas was in violation of the clause of equal treatment and the right to a free choice of occupation. Article 65 of the Icelandic Constitution states that everyone shall be equal before the law and enjoy human rights without regard to sex, religion, opinion, national origin, race, colour, property, family or other circumstances. The Supreme Court admitted that the legislature was allowed to limit access to fisheries to prevent the danger of over-fishing. Previously, however, fishing rights had been allocated for time periods of 2-3 years, which changed under the ITQ legislation. Regarding the latter, the Court did not see the necessity to transfer the rights indefinitely. This reasoning was not only applied to the equal treatment clause but also to the restriction of the right to a free choice of occupation. Interestingly, the Icelandic quota allocation system was not changed in response to the decision. Instead, the access to fishing licences was broadened.

In a second decision in 2001, the Supreme Court overruled its decision of 1999 and found the ITQ regime to be in conformity with the Icelandic Constitution. The Court ruled that the limitation of access rights was based on objective criteria. In particular, the Court said that 'considering the interest of employment and capital investment that are tied to the fishing industry, and to the experience and knowledge that goes with it, it has to be concluded that it was in conformity with the principle of equality to distribute the limited total catch among vessels which at that time were actively fishing for these stocks, even though the legislature had other options to choose from' (cited after Gudmundsdottir, 2001). This later decision is in line with the judicature in most other countries. Case law that found inequalities which resulted from an ITQ allocation processes were justified if the allocation was

[4] *Alliance Against IFQs v. Brown*, 84 F.3d 343 (9th Cir. 1996) (at: http://bulk.resource.org/courts.gov/c/F3/84/84.F3d.343.95-35077.html).

based on rational criteria that were applied uniformly and if the introduction of ITQs served the overall goal of sustaining the fish resources.[5]

The right against deprivation of property can be a legal issue if the country that introduces fishing quotas has a constitutional provision that no person shall be unjustly deprived of his/her property without fair compensation (Stewart, 2004). Here, two aspects have to be distinguished. The first question is whether the introduction of quotas can constitute a deprivation of property. The second question is whether quotas once allocated become property so that their repeal or diminishment could represent a deprivation of property.

With regard to the first aspect, it is clear that the access to a fishery as such, whether or not linked to a licence scheme, generally does not constitute property but only creates the rightful expectation to earn profits. The introduction of quotas will therefore not be considered a deprivation of property. Whether the allocation of fishing quotas creates property that can be detracted unconstitutionally under certain conditions depends on the exact nature of the right under national legislation. If fishing rights are qualified as private property, a diminishment or repeal can theoretically become an acquisition of property by the state, which can then be liable to pay compensation. This is the reason why the Australian Commonwealth legislation has explicitly stated that what is created is a right to fish but not property (Stewart, 2004). Similarly, the legislation of the USA and Canada either affirms that the fishing rights are privileges or that fisheries resources are public resources.

10.2 ITQs and European Community Law

The European Union decides on the total allowable catch (TAC) under its Common Fisheries Policy (CFP) and applies the principle of relative stability to these TACs. The principle of relative stability is based on historical catch levels and implies the maintenance of a fixed percentage of the total available and a traditional share of the catch for the main commercial species for each Member State. The rules that Member States apply to the domestic allocation of their national quota remain the basic responsibility and competence of Member States, in conformity with Community law and the CFP rules (Nordmann, 2000). These national allocation rules differ greatly between Member States. As Nordmann (2000) points out: This is 'not only because of the variety of fishing traditions and patterns but also because of the different political and socio-economic options which are not subject to common rules'.

Under European Community law, any ITQ regime has to fit into the existing CFP and must respect all existing EC law. Measures should be assessed in the light of their contribution to the objectives of the CFP, especially the 'exploitation

[5] For another more recent exception also related to Iceland see Section 3.1.

of living aquatic resources that provides sustainable economic, environmental and social conditions (Art. 2, Council Regulation No 2371/2002/EC)', as noted by the European Commission (European Commission, 2007a). Furthermore, a national ITQ system has to respect the fundamental freedoms of the EC Treaty and non-discrimination rules. In addition, the design of an ITQ system shall not contravene EC state aid rules. We will elaborate on these issues in the following.

10.2.1 Compatibility of ITQs and the CFP

One of the basic principles of the European fisheries policy is still the principle of relative stability. As the introduction of ITQs may fundamentally change the institutional structure with regard to fisheries, national ITQs could contravene this EC principle of relative stability (Frost & Lindebo, 2003). According to this reasoning, freely tradable quotas could lead to the situation where the resources are no longer captured by the nationals of the Member State that introduced the ITQs but by the fishing fleets of other Member States or even those from outside the borders of the EU. Accordingly, there are concerns that the EU fishery policy would no longer function properly. However, these fears might be exaggerated. Currently, introducing ITQs lies in the competence of the Member States and this will continue to be the case. If ITQs are introduced within and by a Member State, it has the right to limit the transferability of quotas, without infringing on the EC Treaty. However, the Member State might have to deal with cases of 'quota-hopping': the freedom of establishment and the free movement of capital have enabled EU ship-owners to purchase vessels and thus to fish against national quotas in other EU countries. These phenomena may only be reduced by individual Member States to a certain degree. Therefore, ITQ systems may endanger the principle of relative stability if vessels are registered in one Member State by nationals from another Member State. As long as the decision whether or not to introduce ITQs rests with the Member States, however, the choice to favour ITQs over relative stability also rests with the Member State.

In regard to the basic features of the CFP, the European Court of Justice (ECJ) ruled in 1976 that a Member State does not jeopardise the Common Fisheries Policy (as far as it already existed then) if it adopts measures involving a limitation of fishing activities with a view to conserving the resources of the sea.[6] In this case, Dutch courts had asked the ECJ for a preliminary ruling about the question whether the Netherlands had the right to introduce a quota system for their sole and plaice fisheries by fixing catch quotas for a given year.[7]

---

[6]Judgment of the Court of 14 July 1976, Joined Cases 3, 4 and 6-76, Cornelis Kramer and others.

[7]At the time in question the European Community had not exercised its competence to take measures for the conservation of the resources of the sea.

The ECJ found that the measures adopted by the Netherlands neither ran counter to the objectives established by the Regulations EEC/2141/70[8] and EEC/2142/70,[9] nor infringed on the freedom of goods/services of the EC Treaty. Especially in Council Regulation 811/76/EEC,[10] which was adopted after the questions had been referred to the Court and repealed by the Council at the end of 1976, the Council expressly authorised Member States to limit catches of their fishing fleets. The Court pointed out that the Member States have an obligation under the Common Organisation of the Market to keep effects on the functioning of the market through catch limitation to a minimum. This does not preclude them, however, from adopting a quota system.

An ITQ system should not contravene other CFP rules. A case, which also concerned the Netherlands, pertained to an aid scheme, which provided for the buying out of reserved licences. The Commission found this to prevent the Dutch authorities from complying with fleet capacity targets and therefore hindering the implementation of the CFP (European Commission, 2007b).

The ECJ also found that quota systems do not infringe upon the prohibition of quantitative restrictions on trade contained in the Treaty. The Court found the prohibition of quantitative restrictions in the Treaty not to be applicable to restrictions on catches, because the restrictions relate to different stages of the economic process, i.e. production and marketing respectively. Early on, the ECJ decided that the allocation of national fishing quotas is not an infringement on the basic freedoms of the internal market. This means that, in general, the introduction of ITQs is allowed under EC law, of course under the condition that the specific characteristics of the ITQs do not infringe on Community law.

This can also be derived from the basic Council Regulation (EC) No 2371/2002 of 20 December 2002,[11] which contains a rule in Art. 10 equivalent to the one that was contained in Council Regulation 811/76. It states that Member States can take measures for the conservation and management of stocks in waters under their sovereignty or jurisdiction if these apply only to fishing vessels flying the flag of the Member State, are compatible with the objectives of the CFP and are no less stringent than existing Community legislation. It can thus be concluded that it is generally admissible to introduce fishing quotas on a national level under the rules of the CFP, provided that the concrete design of the quota system does not infringe on EC law.

---

[8]Regulation (EEC) No 2141/70 of the Council of 20 October 1970 laying down a common structural policy for the fishing industry, OJ L 236, 27.10.1970, p. 1.

[9]Regulation (EEC) No 2142/70 of the Council of 20 October 1970 on the common organisation of the market in fishery products, OJ L 236, 27.10.1970, p. 5.

[10]Council Regulation (EEC) No 811/76 of 6 April 1976 temporarily authorising certain systems of catch quotas in the fisheries sector, Official Journal L 094, 09/04/1976, p. 1.

[11]Council Regulation (EC) No 2371/2002 of 20 December 2002 on the conservation and sustainable exploitation of fisheries resources under the Common Fisheries Policy, OJ L 358, 31.12.2002, p. 59.

10.2.2 Quota Systems and the Basic Freedoms of the EC Treaty

The national ITQ system should be designed in a way that does not infringe on European Community law due to its specific characteristics. Generally, the Member States are free to organise the distribution of their national quotas according to their own concepts. 'The Community fisheries sector is characterised by a variety of management instruments and mechanisms. Fairly comparable situations are dealt with in sometimes very different ways depending on the Member State, the region or the fisheries concerned' (European Commission, 2007a). In particular, national authorities distribute and manage licences, quotas and effort at the national and regional level (European Commission, 2007a). According to the European Commission, it is perfectly legitimate for each Member State to opt for a system that is sub-optimal in economic terms, causing barriers to the trade of rights. The European Commission explicitly refers to the protection of small-scale fisheries, which might be protected against more capital-intensive competitors as a political objective. However, the European Commission points out that any mechanism designed to this end has to be compatible with Community single market and competition rules (European Commission, 2007a).

For example, some of the rules that the UK had adopted in order to prevent foreign fishermen from fishing against the UK's national quotas were incompatible with the principles of the single market. The ECJ held in a judgment of 1989 that Community law does not preclude a Member State to adopt conditions for the issuing of licences to vessels authorising it to fish against national quotas.[12] These conditions can be designed to ensure that the vessel has a real economic link with that State. However, the link has to either concern the relations between the vessels' fishing operations and the fisheries-dependent populations and related industries or the fact that the vessel has to operate from national ports. The ECJ went on to detail that the Member State concerned is entitled to consider a vessel as operating from a national port, if it lands a proportion of its catches there or is periodically present in the national port, provided that the frequency with which the vessel is required to be present in those ports neither imposes directly or indirectly, an obligation to land the vessels' catches in national ports nor hinders normal fishing operations.

The same was held for the requirement that if a company was the owner it had to be incorporated in that State and that at least 75% of the capital in any such company be held by nationals of that Member State or by companies fulfilling the same conditions and that 75% of the directors of every such company be nationals of that Member State.[13]

In a case decided one year later, the ECJ also prohibited Member States from connecting a licence to the nationality of the crew members, irrespective of their

---

[12]Judgment of the Court of 14 December 1989, Case 216/87, The Queen v Ministry of Agriculture, Fisheries and Food.

[13]Judgment of the Court of 4 October 1991, Case 246/89, Commission of the European Communities v United Kingdom of Great Britain.

status as self-employed or employed.[14] The ECJ found that this infringed upon Art. 48, 52 and 59 of the then EEC-Treaty (now Art. 39, 43, 49 EC Treaty). The same was held for the requirement that at least 75% of the crew of a fishing vessel flying its flag had to reside ashore on its territory, because it constituted discrimination on grounds of nationality against nationals of other Member States.

10.2.3 State Aid

According to Art. 36 of the EC Treaty and Art. 19(1) of Council Regulation No 2792/1999/EC, laying down detailed rules and arrangements for Community structural assistance in the fisheries sector, state aid rules apply to the production of and trade in fisheries products. A block exemption regulation exists for certain types of aid that no longer have to be notified to and approved by the Commission.[15] In addition, under the 'de minimis' regulation for fisheries and agriculture, aid of up to 3000 € per enterprise over three years does not have to be notified to the Commission, provided that the sum of the aid remains under 0.3% of the turnover of the fisheries sector of the Member State concerned.[16] Aid not covered by the block exemption regulation or the 'de minimis' regulation is assessed by the Commission in line with the new Guidelines for the examination of State aid to fisheries and aquaculture.[17] The EC state aid rules do not explicitly state whether or not prohibition of state aid pertains to national quota systems.

With regard to state aid, two questions are relevant: firstly, the allocation of individual quotas for free (e.g. based on historical catch) could constitute state aid. Secondly, when introducing a quota system, the government may decide to support those not having been allocated a quota share in the first place and entering the fishery by allowing for buying or leasing.

10.2.3.1 Do Fishing Quotas Constitute State Aid?

The first question is thus whether introducing fishing quotas as such represents state aid. In order for a state aid to be considered incompatible with the Common Market, a number of criteria have to be fulfilled:

---

[14]Judgment of the Court of 17 November 1992, Commission of the European Communities v United Kingdom, Case C-279/89.

[15]Community Regulation on the application of Art. 87 and 88 of the EC Treaty to State aid to small and medium-sized enterprises active in the production, processing and marketing of fisheries products, OJ L 291, 14.9.2004, p. 3.

[16]Commission Regulation (EC) No 875/2007 of 24 July 2007 on the application of Art. 87 and 88 of the EC Treaty to de minimis aid in the fisheries sector and amending Regulation (EC) No 1860/2004, OJ L 193 of 25.7.2007, p. 6.

[17]Guideline for the examination of state aid to fisheries and aquaculture, 2004/C229/03, OJ C 229, 14.9.2004, p. 5.

(i) The measure has to confer an advantage to the beneficiaries;
(ii) It has to represent a state resource and to distort or threaten to distort competition, and
(iii) Has to affect trade between Member States (Art. 87 EC Treaty).

(i) Does the measure confer an advantage to the beneficiaries? In the case of the transfer of quotas to fishing enterprises, it could be argued that these quotas are generally allocated based on catch history and that thus the recipients of the quotas do not receive an advantage. However, that would ignore the fact that before the introduction of the quotas system, fishing enterprises had to compete with all other enterprises that wished to enter the market (provided the latter did manage to get a fishing licence) in the 'race to fish'. After the introduction of the quota system in contrast, the fishing enterprise holding a quota has a guaranteed share of the catch. Thus, the allocation of quotas to individual quota holders has to be considered as an advantage to the beneficiaries.
(ii) Furthermore, to be unacceptable under Art. 87, the fishing quotas allocated without payment have to represent a state resource. It is questionable whether the resource 'fish' can be deemed to represent a state resource, a question that might depend on the national definition of aquatic resources. However, it is not the fish as such that is allocated but only the fishing rights. These rights can only be conferred by the state, since only the state as legislator is able to create a quota system. This system would have to favour a certain industry or enterprise. Fishing quota systems favour the recipients of the quotas over those that do not receive any quotas. The allocation of quotas would have to distort or threaten to distort competition. The consequence of allocating fishing quotas is that only those enterprises receiving them can continue to fish, whereas before all enterprises (that had a fishing licence) could compete. It can thus be inferred that the introduction of fishing quotas distorts competition.
(iii) However, it is questionable whether a system of fishing quotas has an impact on the trade between Member States. This seems not to be the case. Under the CFP a Member State can only allocate its national quotas. These quotas will be fished by its nationals. The way in which Member States' quotas are allocated internally will consequently not affect the trade between the Member States.

It can therefore be concluded that the adoption of a system of national fishing quotas and the domestic allocation of the quotas does not represent state aid under Art. 87 EC Treaty.

10.2.3.2 Government Support Regarding Acquisition of Quotas

On the second issue – government support regarding the buying or leasing of quotas – governments within the EU might decide to protect vulnerable fishing communities by designing a quota system that can infringe upon EC state aid rules.

An example of a system that was designed to protect certain fishers outside the EU was the Alaskan low-interest loan program for new entrants and fishermen fishing from small boats in the halibut and sablefish fisheries.

The European Commission pointed out that it appears that Member States are encountering problems and are 'envisaging state aid measures or other types of public intervention to rectify unwanted effects of transferable rights' (European Commission, 2007b). This was the case on the Shetland Islands, where the UK government aimed to protect dependent fishing communities. In addition, it was difficult for new and young fishermen without prior track record to enter the fishing industry. The UK government introduced community quota schemes in certain 'fishery dependent' areas to try and sustain local fishing fleets and safeguard fishing opportunities for future generations. A community quota scheme is essentially a scheme implemented by local communities to purchase and distribute fish quota in a way that benefits local fishermen. Among several community quota schemes in the UK, one of the most established and largest schemes operated in the Shetland Isles (Anderson, 2006). Significant quota holdings were purchased from other areas of the UK and Shetland fishermen were given preferential access to these shares at a nominal price. The objective of this scheme was to give entitlements to annual fish quotas for the benefit of the local fleets in the community (to ringfence). They were set up because local fishermen were finding it difficult to obtain financial backing to purchase quotas (European Commission, 2001).

In 2001, the European Commission received complaints about the Shetland Community Quota Scheme from competitors within the UK fishing industry. Subsequent investigations into the Shetland Community Quota Scheme by the EC confirmed that the scheme did indeed constitute unlawful state aid under Art.87 (1) of the EC Treaty. The Commission considered that the letting of quotas to vessel-owners that are in the membership of a particular body is not compatible with the common market.[18] Fishermen who were not able to borrow money to buy track records benefited from the scheme. It enabled them to fish against quotas to which they would not otherwise have had access. The resources transferred came from state funds. The quotas from which the fishing enterprises benefited under the scheme reinforced their position vis-à-vis to other fishing enterprises, whether registered in the UK or in the other Member States, because it enabled them to land and sell more fish than they would have otherwise been able to. The implementation of the scheme therefore affected competition. It had given rights to fish for products that are sold on the Community market. The scheme also had an effect on trade in the products concerned. The scheme allowed the beneficiary fishing enterprises to maintain a share of the market that would otherwise have seen seized by competitors. Thus, trade between the Shetland fleet and producers from other Member States was affected.

[18]Commission Decision of 3 June 2003 on loans for the purchase of fishing quotas in the Shetland Islands (United Kingdom) 2003/612/EC, OJ L 211, 21.8.2003, p. 63.

Finally, the scheme did not fall under the block exemption for certain types of aid (linked to restructuring plan as defined in the Block Exemption Regulation).[19]

With regard to public intervention, the European Commission concludes that although a system of transferable rights in fisheries may be advantageous from a certain point of view, 'such a system can over time also be the cause of serious difficulties and put community law at risk' as Member States are set to try rectifying unwanted effects of such rights (European Commission, 2007b).

10.3 Public International Law

10.3.1 Human Rights

In October 2007, the Human Rights Committee under the International Covenant on Civil and Political Rights (ICCPR)[20] found the Icelandic ITQ system to be in violation of Art.26 of the Covenant.[21] The Committee was established in 1976 under Art. 28 of the International Covenant. Iceland is a signatory state to this, and has obligated itself to respect the Committee's opinions.

Art. 26 reads:

> *All persons are equal before the law and are entitled without any discrimination to the equal protection of the law. In this respect, the law shall prohibit any discrimination and guarantee to all persons equal and effective protection against discrimination on any ground such as race, colour, sex, language, religion, political or other opinion, national or social origin, property, birth or other status.*

Two Icelandic fishermen were not able to acquire fishing quotas after they had purchased a boat in 1989 because permanent quota allocation in 1990 was linked to vessel ownership during the reference years between 1 November 1980 and 31 October 1983. In these years they had worked on a fishing boat without being the owners. In 2001, they fished without the required quotas and after being convicted, they brought the case before the Human Rights Committee without first exhausting national remedies.

The Committee argued that the Icelandic legislation differentiated between groups of fishers that were allocated quotas and those that had to purchase them on 'grounds equivalent to those of property' (UN Human Rights Committee, 2007). While the Committee found that the 'aim of this distinction adopted by the State

[19] Community Regulation on the application of Art. 87 and 88 of the EC Treaty to State aid to small and medium-sized enterprises active in the production, processing and marketing of fisheries products, OJ L 291, 14.9.2004, p. 3.

[20] International Covenant on Civil and Political Rights, Adopted and opened for signature, ratification and accession by General Assembly resolution 2200A (XXI) of 16 December 1966, entry into force 23 March 1976.

[21] United Nations International Covenant on Civil and Political Rights, Human Rights Committee, Ninety-first Session, CCPR/C/91/D/1306/2004, 14 December 2007.

party, namely the protection of its fish stocks which constitute a limited resource, is a legitimate one', it did not deem the distinction to be based on reasonable and objective criteria.

In brief, the Committee ruled that a temporary allocation of the quotas may have been reasonable, the 'permanent' allocation, which 'transformed original rights to use and exploit a public property into individual property', was not. In the Committee's view the permanent allocation would have been acceptable if the allocated quotas that were no longer used by their original holders were reverted to the State for allocation to new quota holders in accordance with fair and equitable criteria instead of being sold or leased at market prices. According to the Committee, this particular design and the modalities of implementation of the quota system did not meet the requirement of reasonableness.

A number of dissenting opinions found that the Icelandic state had allocated the fishing quotas based on reasonable and objective criteria. Among other arguments, the dissenting Committee members found that 'economic benefits leading from the permanent nature of catch entitlements' represented a reasonable criterion for the differentiation between fishermen that owned vessels and others. Also, they argued that 'possibilities for assignment of catch entitlements and quotas will lead to gainful utilization of the fish stocks for the benefit of the national economy'. In sum, the dissenting members found that the state had carried out a careful balance between the 'advantages which the current system offers for the fishing management in Iceland, notably the need to have a stable and robust system, as well as the disadvantages of the system for the authors i.e. the restrictions on the author's (i.e. the plaintiffs') 'freedom to engage in commercial fishing', 'between the general interest and the interest of the individual fishers'. Moreover, the dissenting members held that the distinction between the two groups of fishers was based on objective grounds and was proportionate to the legitimate aim pursued.

It is difficult to foresee the practical relevance of this quasi-judgement against the Icelandic ITQ system. On the one hand, Iceland has ratified the Covenant and is bound to accept the Committee's opinions. On the other hand, the Committee has no possibility to enforce the opinion and thereby force Iceland to change its ITQ system. It remains to be seen if and how Iceland will implement the decision.

From a legal point of view, the Committee's opinion seems not to be well reasoned. From the fact that Iceland's constitution declares the fishing banks to be common property of the Icelandic nation it cannot be inferred that all Icelanders had an 'original right to use and exploit a public property'. The sovereign right of a state over its natural resources does not automatically imply that the use cannot be restricted. And this was the case in Iceland, where different measures of effort control and vessel catch quotas existed before the ITQs were made permanent in 1990. Furthermore, the allocation of ITQs does not turn the fish into individual property. Quite the opposite, the Fisheries Management Act of 1990 states once more that the fish stocks of the Icelandic Waters are the common property of the Icelandic people and that the allocation of ITQs to individual firms and vessels does not give them irrevocable property rights in the TAC shares allocated. The opinion also did not take into account that exceptions for special circumstances existed when quotas were allocated. In assessing the quota system, the Committee did not mention the reasons Iceland had for introducing the quotas and allocating them to vessel owners, such as the economic viability of the fishing industry, employment, the sustainable management of fish stocks and environmental concerns. The Committee did not evaluate whether the Icelandic state would have been able to adopt other measures to reach these goals. Moreover, it is contestable that the Committee is taking an extensive approach to reviewing economic policies on the grounds of 'human rights'. The focus of the Covenant is clearly on the protection of civil and political rights. As was also pointed out in a dissenting opinion, in order '[t]o effectively protect the important rights that fall within the aegis of the Covenant, the Committee also must remain true to the limits of its competence, both legal and practical'. In sum, the opinion does not regard the limited scope for review of economic regulatory matters under Art. 26 of the Covenant, does not discuss the legal issues involved in a comprehensive manner and thus cannot convince.

10.3.2 WTO Law

When an ITQ system is introduced, the free allocation of quotas could represent a subsidy, which could face constraints under public international law. Currently, there are no special WTO provisions relating to fisheries subsidies. These subsidies are disciplined only by the general subsidies rules found in the current WTO Agreement on Subsidies and Countervailing Measures (SCM Agreement) (Benitah, 2004). In the Doha round WTO Members agreed in November 2000 to launch negotiations with the aim to 'clarify and improve WTO disciplines on fisheries subsidies, taking into account the importance of this sector to developing countries'.[22] As Benitah (2004) remarks, the issue of fisheries subsidies has been pursued in the WTO Committee on Trade and Environment (CTE) for several years without results.

Negotiations on fisheries subsidies are currently taking place within the Negotiation Group on Rules with the aim to clarify and improve the existing disciplines as laid out in the WTO Agreement on Subsidies and Countervailing Measures of 1995. So far, the negotiation process has been marked by substantial disagreement, particularly on whether fisheries subsidies are responsible for over-capacity and overfishing and on the degree of Special and Differential Treatment (S&DT) that should be accorded to developing countries (European Commission, 2006). However, slow progress has been made as the world's trading nations negotiate a new approach to setting up and implementing new disciplines on subsidies that will go beyond the current agreement (WTO, 2005). At the end of 2007, a draft text on fisheries subsidies was circulated to WTO Members. This draft text took the form of an Annex (Annex VIII), which is to be inserted in SCM Agreement. This Annex is consid-

---

[22]Paragraph 28 of the Doha Declaration, 'Ministerial Declaration at Ministerial Conference', Fourth Session, Doha, 9–14 November 2001, WTI MIN(OI)/DECIW/I, adopted on 14 November 2001.

ered as an integral part of the SCM. The draft reflects a mix between the so-called 'traffic lights' approach and the 'special and differential treatment' approach. The most recent working document was issued by Chairman Valles Galmes on 2 June 2008 (Oceana, 2008). The 'traffic lights' approach places most fisheries subsidies in a 'red box', which means they are prohibited. ITQs are placed in the 'green box' of subsidies that are allowed. The draft text lists 'limited access privileges to individual and groups and other exclusive quota programmes' in the green box. Given the unclear state of the overall Doha-round negotiations, it is impossible to predict the outcome. However, taken the current state of discussions on fisheries subsidies, it can be expected that ITQs will stay exempt in the near future.

10.4 Considerations when Legislating ITQs

The literature discusses number of points are being discussed that should be addressed when designing a rights-based management program in order to ensure legality. These include, among others, the nature of the rights to be conferred to fishermen or vessel owners, management units, determination of total allowable catch, monitoring and enforcement, need for other regulations concerning particular conservation issues (e.g. fishing in nursery grounds or during spawning periods), rent extraction and cost recovery, and initial allocation (Anderson, 2000). At the outset of designing rights-based fishing regimes, a legislative approach has to be selected.

10.4.1 Legislative Approach

Different countries have opted for different approaches. One option is to address all issues in an act of Parliament. The drawback of this approach is its rigidity, because legislation is more difficult to change than regulations or guidelines. Also, such an approach might not be possible where a number of different systems will be introduced based on the legislation, which have different characteristics, as for example in Canada. The other extreme is to leave the decision about the nature and characteristics of the system completely to subordinate legislation or management plans and simply to ensure that the governing statute enables or at least does not prevent the implementation of fisheries rights as it was done in Canada. An intermediate approach would be to adjucate the essential matters in the governing statute and leave details to regulations or management plans. Also, account must be taken of the existing legislation, such as fisheries law, jurisdictional issues and constitutional matters. The advantage of the intermediate approach is that it ensures a stable legislative framework that can address basic issues such as the nature of the right, the transferability etc. without overloading the legislation with details (Stewart, 2004).

It might not always be necessary to draft new legislation to introduce rights-based regimes. Instead, it might be sufficient to amend existing law, especially when the law is not supposed to contain details of the system. One basic question is the issue of territorial jurisdiction in federal legal systems, i.e. whether the federal or the state legislature has the competence to regulate fisheries management. This certainly is an issue in many European countries.[23] If new legislation is drawn up, it should also refer to conservation and management principles, which can be used to interpret the statutes (Stewart, 2004).[24]

10.4.2 Nature of Property Rights

It can be questioned whether it is necessary to clearly define the nature of the property right through legislation. This might prove difficult, since the quota might have some aspects that closely resemble a property right whereas others are too weak to qualify the right as such. Thus, it seems advisable to define the characteristics of the right, while it might not be necessary or even prudent to define it as property. On the other hand, statutes that declare the fishing rights not to be property gave rise to problems as well, especially when in fact the quotas have characteristics that are very close to property.

Different views exist on the question whether or not regulation should define ITQs as property. The right can be recognised as a permanent property (e.g. New Zealand). The right can also be designed as a special property right that has clearly defined and limited characteristics, while the right to the resource remains explicitly with the state. For example it can be specified that no compensation will be paid should the right be revoked. Finally, ITQs may be classified as revocable privileges that are not property rights as in the US. In many instances, the legislation does not explain the exact nature of the rights and leaves it to the courts to define it as it was the case in Australia. As Stewart (2004) points out, if the purpose is to ensure that ITQs will be considered as property, it is important to strengthen their property characteristics, i.e. transferability, durability, security, and exclusivity. In sum, it is more important to have a clear concept regarding these four features of ITQs than to declare a right to be property or not, while one of the characteristics is very weak or very strong.

10.4.3 Management Units

Basic legislation or rules have to contain provisions on fisheries management planning. The first crucial question is which unit takes responsibility for planning. That can be regional fishery management councils (as in the USA), the Regional Advisory Councils in the European Union which however at present have an advisory function only, a central management authority (like the Australian Fisheries Management Authority), a supranational authority (like the Directorate General for

---

[23] For a discussion of potential conflicts with different concepts in non-Western legal systems, see Stewart (2004), p. 103.

[24] Stewart also cites examples of legislation in different countries regarding this and other characteristics of fisheries legislation on property rights discussed.

Maritime Affairs and Fisheries in the European Union) or the responsibility can remain with the government (as in New Zealand, where no provisions for management plans exist and where the Minister issues quota management rules). Mostly, planning is based on the distinction between different areas, fish stocks or vessel types. Lawyers recommend giving management plans legal effect in the governing statute or through regulation and that they include the process of determining TAC, which should not be left to subordinate legislation. However, this aspect is not relevant for EC fisheries as in the framework of the CFP, TACs for most commercially exploited species are determined by the Council. Public consultation of management plans is also considered important (Stewart, 2004).

10.4.4 Allocation of ITQs

A fundamental decision when drafting a fisheries rights system is the definition of the holders of the rights. Fishing licences are usually issued with respect to vessels. If the ITQ-system is to be based on the issuing of fishing rights to legal or natural persons, these two approaches have to be reviewed for possible conflicts. Besides individuals and vessels, communities or traditional right-holders are potential recipients of quotas. As Stewart (2004) explains community-based fishery management is taking place worldwide in many different forms. They are usually characterised by a 'relative weakness of the legal basis'. In each case, internal management procedures have to be designed, or existing management procedures have to be taken into consideration.[25]

Important aspects in the design of fisheries legislation are exclusionary factors because they are common subjects of lawsuits. The most important exclusionary factor is citizenship. This criterion has to be used carefully, taking into account bilateral, multilateral, regional and international fishing agreements. As discussed at length supra, in the European Community it is not possible to exclude fishing companies from other EU countries. Limiting access to vessel ownership to prevent 'quota-hopping' is restricted through the EC Treaty. Often the total holding of quotas is limited, e.g. through a cap on the amount of quotas one company or individual can hold. Sometimes fishing rights are separated from processing rights. It is advised to include any restrictions or limitations on the acquisition of quotas in the governing statute (Stewart, 2004). Quotas cannot only be attached to natural or legal persons but also to vessels, which should be reflected in the act as well.

Allocation methods are highly contentious as proven by the many court proceedings and appeal processes that followed the allocation of quotas in almost all countries. In most cases the 'fundamental legal principles underlying the property nature of fisheries rights' were at the centre of court proceedings (Stewart, 2004). The stronger the property characteristics (transferability etc.), the more contentious are

---

[25]For further aspects of creating legislation on community-based fisheries management, see Stewart (2004), p. 116–118.

the proceedings. There are many different ways to allocate fisheries rights (catch history, vessel/gear specifications, distribution of equal shares, lottery, tenders, investments), but most allocations are based on catch history. Catch history as well as vessel/gear specifications privilege those who are already in the industry. Generally, the allocation method will be selected according to the general set-up of the fisheries. If fleet reduction is necessary and there are already too many operators in the sector, lottery or tenders can be used. Where the sector is rather small, equal shares can be distributed. If allocation is based on catch history, attention has to be paid to a possible build-up of fleet effort in order to qualify for allocation. Case must be taken, when drafting exceptions in cases where an operator is prevented from fishing, because of exceptional circumstances. In order to facilitate allocation, special appeal mechanisms such as the appeal authority in New Zealand can be established. Since these appeals take many years in some countries, it is important to place time limits on appeal processes (Nielander & Sullivan, 2000). Nielander and Sullivan state that in general ‘there is no doubt that every effort should be made during the legislative and regulatory drafting stages to minimise possible litigation exposure. Unfortunately, ITQs are, at times, so valuable that individuals risk litigation costs for the possibility of obtaining initial or additional quota’. Therefore they recommend minimising the scope for successful legal challenges, by allowing for some discretion in the allocation formula and the allocation process to avoid gross examples of unfairness resulting from a strict and inflexible application of general rules. Moreover, the allocation decisions should be meticulously documented.

10.4.5 ITQ Characteristics

Quotas gain value through the possibility to be transferred and accumulated. The fact that quotas are generally transferred; whether or not that is foreseen in the legislation also shows that transferability is an important aspect that should be clarified in legislation. There are many variations of transferability in ITQs, from ful and partial transferability to leasing only. Usually, some limitations are included in legislation, such as the very common one that the transferee must already hold a fishing permit or vessel licence. All restrictions should be made clear in legislation, regulations or management plans. Since some limitation generally exists, a mechanism has to be foreseen to ensure compliance. The same is true if limits or caps on the number of quotas held by one person or entity are defined. These kinds of restrictions therefore require a management authority.

Regarding durability, quotas can be designed to be issued for a year or season only or for longer periods of time, or even in perpetuity. In economic theory, quotas that are issued infinitely better serve their purpose, because the rights holder will be more interested in conserving the resource. Any legislation should be based on a clear concept of the durability of the quotas. That also concerns the question whether compensation will be paid in case of withdrawal of quotas. Respective provisions exist in some jurisdictions, such as Australia. Generally, it is advisable to foresee reduction or suspension of the quota in case that the TAC is reduced, as it is the case in New Zealand, where quotas were allocated in perpetuity.

Security can be achieved through a register of quotas and their holders. Some quota registers have special features. Registers can be combined with already existing registers of licences and permits. In order to increase the value of the right, registration of mortgages, liens etc. in the register could be made possible.

Exclusivity is defined as the factor that assures 'the fisher much the same control over their resource as a farmer has over his land and its produce' (Stewart, 2004). Whether that is in fact possible is questionable. On the one hand, a quota holder has more possibilities to fish without outside interference than a mere licence holder. On the other hand, other quota holders can also fish and the fish will only become the property of the fishermen when they are on board their vessel. In addition, an increase in quota busting, misreporting etc. was described after the introduction of quotas (e.g. in the Netherlands and in New Zealand). That shows that the exclusivity of the right might also depend on effective enforcement measures. Also, it has to be pointed out that the exclusive nature of fisheries rights does not mean that the state cannot interfere, since it may suspend or cancel the right, e.g. for fisheries management reasons (Stewart, 2004). Also, the quotas can be linked to the holding of fishing licences, with the consequence that they will end if the licence is terminated.

10.4.6 Monitoring and Enforcement

Legislation on property rights in fisheries should also take into account administrative questions, such as the preparation and management of plans, quota registers, transfers of quotas, data collection, enforcement, and monitoring. As mentioned before, different approaches were chosen in different regions of the world, ranging from regional fisheries councils to centralised government management authorities. Finally, an issue to be tackled is whether fees and charges will be levied.

It is important to ensure sufficient flexibility to be able to amend and adjust existing rules, since ITQ systems often evolve over time. To some extent this may be achieved by allowing for some discretion in the application of the rules. In other cases the legislation and basic rules of the system may have to be changed. Another possibility could be to introduce review and reporting clauses as they are frequently used in Community law. However, as Stewart (2004) points out, 'it must be recalled that any major overhaul to a fisheries rights system carries with it the possibility of undermining the security and predictability associated with the system. The more the fisheries right is being capable of being viewed as property, the more guarantee of security will be required'.

10.5 Conclusion

In conclusion, legal concerns do not hinder the introduction of individual transferable quotas in fisheries management, when the respective legislation is drafted carefully. However, especially the initial allocation of quotas has proven to be difficult and is often challenged in the courts. Here, attention has to be given to national rules of non-discrimination (equal treatment), the right to free choice of occupation and the protection against deprivation of property. In most countries, quota allocation can be based on grandfathering, when the adoption of legal remedies and hardship regulations ensure that these basic rights are not violated. However, the recent opinion of the Human Rights Committee under the International Covenant on Civil and Political Rights shows that especially non-discrimination clauses can represent a hurdle for ITQ systems. The practical consequences of this ruling remain unclear.

European Community law does not stand against the introduction of ITQs in Member States. ITQs do not contravene the EC principle of relative stability under the CFP. ITQs are also compatible with the basic freedoms of the EC Treaty. However, Member States must be careful when they try to shield their quotas from being bought by other nationals. Conditions such as linking the registration of vessel to the nationality of the crew or the ownership of the vessel infringe on European Community law. The European Commission pointed out that the Member States have to be prudent when envisaging state aid measures or other types of public intervention to rectify unwanted effects of transferable rights. While the allocation of quotas as such does not represent state aid, supporting fishermen in order to enable them to buy quotas can be problematic.

In the WTO, the distribution of ITQs may also constitute a subsidy. However, given the current state of negotiations of fisheries subsidies under the Doha round, ITQs – described as 'limited access privileges to individual and groups and other exclusive quota programmes' -will most likely be exempt from any disciplines.

A number of points should be addressed when designing a rights-based management program. These include among others the nature of the property right, management units, determination of total allowable catch, monitoring and enforcement, need for other regulations, rent extraction and cost recovery and initial allocation. It is important to ensure sufficient flexibility to be able to amend and adjust existing rules, since often ITQ systems evolve over time.

At the European level, the debate on rights-based management tools in fisheries is still ongoing. Set off by the European Commission in February 2007, it is meant to improve knowledge of national regimes, identify best practices and eventually evaluate the scope for new initiatives (European Commission, 2007a, 2007b). Legislative competence regarding implementation measures and management regimes (i.e. fishing rights) lies with the Member States and currently it is therefore up to them to address the legal questions associated with designing rights-based management systems. However, there is a tendency in favour of the introduction of a management system at Community level.[26]

[26]See European Parliament resolution of 10 April 2008 on rights-based management tools in fisheries (2007/2111(INI)) and the Opinion of the European Economic and Social Committee on Rights-based management tools in fisheries of 13 February 2008.

References

Anderson, L. G. (2000). Selection of a property rights management system. In R. Shotton (Ed.), *Use of property rights in fisheries management*. FAO Fisheries Technical Paper 404/1 (pp. 26–38). Rome: FAO.

Anderson, J. (2006). *The future of the market based approach towards quota management in the UK – The case of the Shetland Isles*. Paper presented at the Sharing the Fish Conference, Perth Western Australia, March 2006. Retrieved October 15, 2008, from http://www.fish.wa.gov.au/docs/events/ShareFish/papers/index.html

Arnason, R. (2000). Property rights as a means of economic organization. In R. Shotton (Ed.), *Use of property rights in fisheries management*. FAO Fisheries Technical Paper 404/1 (p. 320). Rome: FAO.

Benitah, M. (2004). Ongoing WTO negotiations on fisheries subsidies, *The American Society of International Law*, June 2004. Retrieved October 1, 2007, from: http://www.asil.org/insights/insigh136.htm

Burke, D. L., & Brander, G. L. (2000). Canadian experience with individual transferable quotas. In R. Shotton (Ed.), *Use of property rights in fisheries management*. FAO Fisheries Technical Paper No. 404/1 (pp. 151–160). Rome: FAO.

Connor, R. (2000). Are ITQs property rights? Definition, discipline and discourse. In R. Shotton (Ed.), *Use of property rights in fisheries management*. FAO Fisheries Technical Paper No. 404/1. Rome: FAO.

European Commission. (2001). Commission finds UK aid schemes involving purchase and leasing of fish quotas not compatible with common market rules, Press release of 3.6.2003. Retrieved September 18, 2007 from http://ec.europa.eu/fisheries/press_corner/press_releases/archives/com03/com03_19_en.htm

European Commission. (2006). *Trade in agricultural goods and fishery products: fisheries subsidies*, Trade Issues. Retrieved June 9, 2008 from: http://ec.europa.eu/trade/issues/sectoral/agri_fish/fish/fs_en.htm

European Commission. (2007a). Communication from the Commission on Rights-based management tools in fisheries, COM(2007) 73 final. Brussels.

European Commission. (2007b). Commission Staff Working Document, accompanying the Communication from the Commission to the Council and the European Parliament on Rights-based management tools in fisheries, SEC(2007)247, Brussels.

Fitzpatrick, D. (2000). Property rights in relation to fishing licences in Australia from a legal perspective. In R. Shotton (Ed.), *Use of property rights in fisheries management*. FAO Fisheries Technical Paper No. 404/1.Rome: FAO.

Frost, H., & Lindebo, E. (2003). *Alternative management systems in EU fisheries*. Copenhagen, Denmark: Fødevareøkonomisk Institut.

Gudmundsdottir, E. (2001). The Supreme Court Rules the Allocation of Individual Transferable Quotas to be Constitutional. International Journal of Marine and Coastal Law, Vol. 16, No. 1 (p. 125).

Macinko, S., & Bromley, D. (2004). Property and fisheries for the twenty-first century: Seeking coherence from legal and economic doctrine. *Vermont Law Review, 28*, 623–661.

McFarlane, B. (2000). The legal nature of Australian fishing licences – Are they property rights? In R. Shotton (Ed.), *Use of property rights in fisheries management*. FAO Fisheries Technical Paper No. 404/1. Rome: FAO.

Nielander, W. J., & Sullivan, M. S. (2000). ITQs – New Zealand and United States: Allocation Formula and Legal Challenges. In R. Shotton (Ed.), *The use of property rights in fisheries management*. FAO Fisheries Technical Paper 404/02, p. 59. Rome: FAO.

Nordmann, C. (2000). The common fisheries policy of the European Union and fishing rights. In R. Shotton (Ed.), *The use of property rights in fisheries management*. FAO Fisheries Technical Paper 404/02. Rome: FAO.

Oceana. (2008). New WTO Fisheries Subsidies Document Creates Path for Negotiations, Environmental News Network. Retrieved June 2, 2008 from: http://www.enn.com/wildlife/article/37214

Scott, A. (1988). Conceptual origins of rights based fishing. In P. A. Neher & N. Mollet (Eds.), *Rights based fishing*. NATO ASI Series, Series E: *Applied Sciences, 169*.

Shotton, R. (Ed.). (2000). *Use of property rights in fisheries management*. Proceedings of the FishRights99 Conference. Fremantle, Western Australia 11–19 November 1999. Mini-course and Core Conference Presentations. FAO Fisheries Technical Paper No. 404/1. 355 pp.

Stewart, C. (2004). *Legislating for property rights in fisheries*, FAO Legislative Study (FAO), No. 83/FAO, Rome, Italy. Retrieved September 2007, from ftp://ftp.fao.org/docrep/fao/007/y5672e/y5672e00.pdf

United Nations Human Rights Commission. (2007). Views adopted on 24.10.2007, Doc. CCPR/C/91/D/1306/2004. Retrieved October 15, 2008, from http://daccessdds.un.org/doc/UNDOC/DER/G07/458/05/PDF/G0745805.pdf?OpenElement

WTO. (2005). Negotiations to clarify and improve disciplines, Hong-Kong WTO Ministerial 2005, Briefing Notes. Retrieved June 9, 2008, from: http://www.wto.org/english/thewto_e/minist_e/min05_e/brief_e/brief08_e.htm

Chapter 11
How to Compare (the Efficiency of) Fisheries Management Systems?

Ragnar Arnason

Abstract To realize the very substantial economic benefits offered by many fish resources an appropriate fisheries management system must be installed. The number of possible fisheries management systems is very large, as demonstrated in the paper. It is obviously not practical to compare the empirical outcomes of all these systems. It is therefore potentially very useful to develop a reasonably simple measure to assess the economic efficiency of the various fisheries management systems that may be proposed. Luckily, it turns out that by appealing to the basic theory of property rights and economic efficiency such measures are available. In the paper, two such measures are proposed; a simple graphical measure based on radar diagrams and a more flexible numerical measure, referred to as the Q-measure. The properties of these measures are discussed in the paper and their use illustrated by means of examples.

Keywords Comparison of fisheries management systems · Fisheries management systems · Property rights · Property rights quality · Property rights and economic efficiency

11.1 Introduction

Anyone who wants to assess and compare fisheries management systems quickly finds that these are complicated constructs and there is a huge number of them. This raises two problems. The first is what fisheries management systems to compare. The second is what features of the different systems to compare. Let us clarify:

A fisheries management system (FMS) is a set of formal and/or informal rules stipulating how fisheries may be conducted. These rules pertain among other things

R. Arnason (✉)
Department of Economics, Oddi v. Sudurgotu, University of Iceland, IS-101 Reykjavík, Iceland
e-mail: ragnara@hi.is

K.H. Hauge, D.C. Wilson (eds.), *Comparative Evaluations of Innovative Fisheries Management*, DOI 10.1007/978-90-481-2663-7_11,

to permitted fishing times, fishing areas, fishing equipment, fishing vessels, species, harvesting volumes, discards and so on. Each different combination of rules defines a fisheries management system. Since there can be a large number of such rules, the number of possible fisheries management systems is very great. Thus, assuming the very modest number of 10 possible fisheries management rules (anyone familiar with fisheries can think of at least double that number), the total number of combinations of these rules is over a thousand.[1]

A couple of pertinent conclusions immediately follow. First, it is easy to think of new FMSs. Each such FMS may be regarded as a fisheries management innovation. Therefore the total number of fisheries management innovations to assess and compare is very great. Second, it is rare to encounter identical FMSs in the real world. Although they may be broadly of the same type, they almost always differ in more particular respects. These observations suggest that an empirical comparison of FMSs is unlikely to be successful. Something more is needed.

Any given FMS has a number of attributes and implications. Presumably, however, in the context of this study, the interest in fisheries management systems stems primarily from their social impacts. The social impacts of fisheries management systems occur along many different dimensions. Often mentioned of these are economic efficiency, social stability, social equity and regional and community economic activity. However, as argued by John Hicks (1939) and Nicolas Kaldor (1939) many decades ago, economic efficiency is in a certain sense the most fundamental of these social dimensions. The essential reasons can be expressed quite simply. Society's ability to attain what it wants in terms of income, welfare, stability, equity and so on is always constrained by the net economic output. The greater the economic efficiency, the larger is the net economic output and, consequently, the ability (opportunity set) to achieve the aims of society and its members. Given this maximum possible output to play with, society has a great deal of room, albeit not an infinite one, to attain the distribution of income and social stability it desires. This argument suggests the fundamental social policy, which interestingly seems to have been broadly adopted by western societies, that industries should aim for maximum efficiency and other socially important objectives should be dealt with by other means. No rational argument seems to exempt the fishery from his social principle. Therefore, it seems reasonable to compare fisheries management systems with respect to their economic efficiency only.

So, from the above it should be clear that it is not really feasible, and certainly not practical, to assess the relative performance of different fisheries management systems solely on the basis of empirical investigation. What is needed is a reasonably simple overall or portmanteau measure which (i) is firmly based in theory, (ii) does not require a great deal of detailed empirical data and (iii) is robust in the sense of rarely being far off the mark. Fortunately, at least if the Hicks-Kaldor argument is accepted and the comparison is in terms of economic efficiency, a framework for such a measure seems to be available.

---

[1] The number of subsets of rules from a set of n different rules is given by $2^n - 1$.

The rest of this paper presents this framework. It is based on fundamental principles regarding the relationship between property rights and economic efficiency. It is therefore solidly based on theory. It does not require extensive empirical investigation and seems to be highly robust. It is relatively easy to apply. It, therefore, seems to go a long way toward satisfying the requirements for a practical empirical measure for comparing fisheries management systems.

11.2 The Problem of Fisheries is Caused by Low Quality Property Rights

The fisheries problem manifests itself as depressed fish stocks, excessive fishing capital and fishing effort and low profitability of the fishing operations. In short it appears as economic waste, which, due to the great potential value of many fish resources, is often very great (World Bank, 2008).

The essence of the fisheries problem may be illustrated as in the Fig. 11.1 (Anderson, 1986; Arnason, 2007a). In this figure the curves labelled 'Sustainable revenues' and 'Costs' measure true social benefits and costs on a sustainable basis. As illustrated in the figure equilibrium or sustainable net benefits are maximized at effort level optimum sustainable yield (OSY). In accordance with many commercial fisheries, these net benefits are drawn so as to constitute a high proportion of the sustainable revenues. At that effort level sustainable biomass is also quite high, so biological conservation concerns are to a substantial extent met. Without proper management, however, so-called open access or common property fisheries tend to converge to effort level CSY (competitive sustainable yield) at which the net economic benefits are zero and the sustainable biomass is much smaller, perhaps even close to the point of biological collapse.

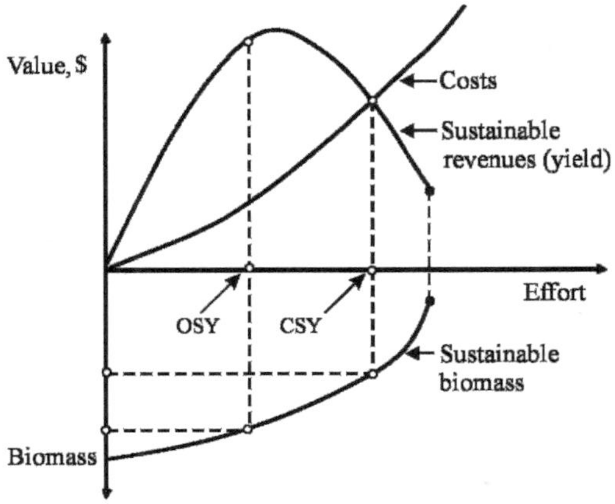

Fig. 11.1 The Equilibrium Fisheries Model

Until recently most of the world's marine fisheries and many fresh water fisheries have exhibited a strong tendency to converge to a point like CSY in Fig. 11.1 (World Bank and FAO, 2008). The reason for that has been well established in the literature (Gordon, 1954; Scott, 1955; Turvey, 1964; Hardin, 1968). The reason is that in these fisheries a number (usually quite a high number) of people have the right to extract harvest from common fish stocks. This has been referred to as the common property or common pool arrangement (Hardin, 1968; Ostrom, 1990). Naturally, as long as participation in the fishery generated income above the alternative, i.e. positive profits in Fig. 11.1, people exploited this right. As a result, whenever net benefits in the fishery were positive (effort below CSY in Fig. 11.1) aggregate fishing effort increased and vice versa. Thus, only at a point like CSY in Fig. 11.1 where there are no net economic benefits would the fishery come to a rest. This is the so-called tragedy of the common property or common pool social arrangement (Hardin, 1968).

As pointed out by Scott in 1955, this would not happen if there was only one fisherman, a sole owner as Scott put it, in the fishery. In fact, as Scott showed, a sole owner would operate the fishery at the efficient effort level OSY in Fig. 11.1. Thus, a necessary condition for CSY to happen is this common right of many fishers to exploit the resource.

So, the social arrangement of open access or common right to a fishery turns out to be necessary for the emergence of the fisheries problem illustrated in Fig. 11.1. On the other hand exclusive individual rights to the fishery – Scott's sole ownership – are sufficient for efficient utilization. Full exclusive individual rights of course amount to perfect property rights (see Section 11.2). Access rights shared by many, while certainly rights, may, on the other hand, be seen as extremely weak property rights. In this sense the fisheries problem is caused by low quality property rights. Weak property rights in the form of common access lead to total inefficiency. Strong property rights in the form of sole ownership lead to full efficiency.

11.3 The Basic Theory of Property Rights and Property Rights Quality

A property right specifies the rights someone (the owner) has with regard to something (the property). A property right generally consists of a bundle of rights or characteristics (Alchien, 1965; Demsetz, 1967). Property rights are in other words, a multi-dimensional phenomenon. The number of distinguishable characteristics that make up any given property right may be high. However, according to Scott (1989, 2000) the most crucial property rights characteristics are:

- Security
- Exclusivity
- Permanence
- Transferability

More specifically, the content of these characteristics are as follows:

Security. A property right may be challenged by other individuals, institutes or the government. Security here refers to the ability of the owner to withstand these challenges and retain his property right. It is perhaps best thought of as the probability that the owner will be able to hold on to his property right.

Exclusivity. This characteristic refers to the ability of the owner to utilize and manage his property without outside interference and to exclude others from doing the same. An individual's personal belongings, such as his clothes, usually have a very high degree of exclusivity. By contrast, the right to the enjoyment of a public park has almost zero exclusivity. The right of a fisherman to go out fishing has exclusivity the falls with the number of other fishermen holding the same right. It should be noted that *enforceability*, i.e., the ability to enforce the exclusive right, is an important aspect of exclusivity.

Permanence. Permanence refers to the duration of the property right. This can range from zero, in which case the property right is worth nothing, to infinite duration. Leases are examples of property rights of a finite duration. Note that there is an important difference between an indefinite duration, which does not stipulate the duration of the property right, and property right in perpetuity, which explicitly stipulates that the property right lasts forever. The duration of a property right may seem related to security – if a property right is lost, then, in a sense, it has been terminated. Conceptually, however, the two characteristics are quite distinct. Thus, for instance, a rental agreement may provide a perfectly secure property right for a limited duration.

Transferability. This refers to the ability to transfer a property right to someone else. For any scarce (valuable) resource, this characteristic is economically important because it facilitates the optimal allocation of the resource between competing uses. An important feature of transferability is *divisibility*, i.e., the ability to subdivide the property right into smaller parts for the purpose of transfer. Perfect transferability implies both no restrictions on transfers and perfect divisibility.

11.3.1 Graphical Representation of Property Rights

As suggested by Scott (1989), it is helpful to visualize the characteristics of property rights as measured along the axes in four-dimensional space. This is illustrated in Fig. 11.2. Obviously, if more than four characteristics are needed to describe a property right, the number of axes in the diagram would be correspondingly increased.

A given property right may feature the different property rights characteristics to a greater or lesser degree. It is convenient and totally unrestrictive to measure the degree to which a given characteristic is featured on a scale of 0 to 1. A measure of zero means that the property right in question features none of the characteristic. A measure of unity means that the property right features the characteristic fully. Given this we can draw a picture of perfect property rights, i.e., a property right that

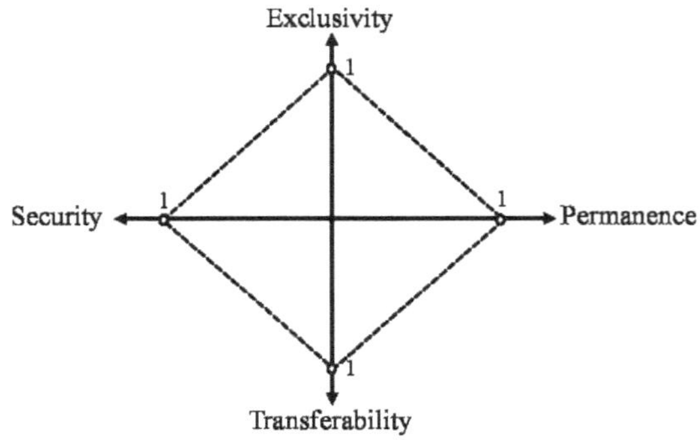

Fig. 11.2 Property Rights Characteristics

features all the property rights characteristics fully, as a rectangle in the space of the four property rights characteristics. This is illustrated in Fig. 11.2.

We refer to the map of the property rights characteristics as in Fig. 11.2, as the *characteristic footprint* of a property right. Obviously, the characteristic footprint of a perfect property right represents the outer bound for that of any property right. It follows that the characteristic footprint of any actual property right must be completely contained within that of the perfect property right as illustrated in Fig. 11.3.

11.3.2 Quality of Property Rights: The Q-measure

The fact that any real property right must be contained within the characteristic footprint of a perfect property right suggests the ratio of the area enclosed by the footprint of a real property right to that of the perfect one as a simple measure of the quality of any real property right. This measure has the convenient property of always being between zero and one. In addition, it satisfies the requirement that the closer the characteristic footprint of a real property right is to that of a perfect property right, the higher is the measure. Furthermore, it is easy to calculate and generalizes in a straightforward manner to any number of property rights characteristics.

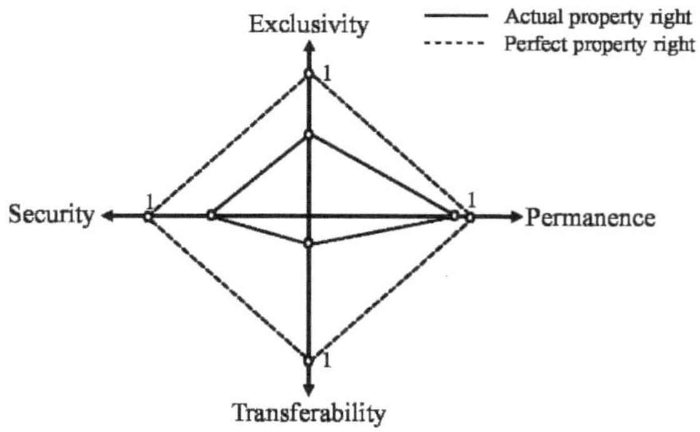

Fig. 11.3 Characteristic Footprints of an Actual and a Perfect Property Right

Thus this measure has many useful properties. However, it also has the significant limitation of treating all property rights characteristics equally.

To remedy this, the so-called Q-measure of property rights quality has been developed (Arnason, 2000). A general formula for the Q-measure is:

$$Q \equiv \left(\prod_{i=1}^{N} x_i^{a_i}\right) \cdot \left(w_1 + \sum_{j=N+1}^{M} w_{2,j} \cdot x_j^{a_j}\right). \tag{11.1}$$

This Q-measure applies to M property rights characteristics. The first N, x_i, $i =$ 1,2. . .,N, are essential, i.e. characteristics that render the Q-measure zero and, consequently, the property right worthless if they are zero. Hence the multiplication represented by the symbol $\prod_{i=1}^{N}$. The remaining M-N characteristics denoted by $x_j, j =$ N+1,N+2,. . .,M, are non-essential. Even if they are all zero, the Q-measure would not necessarily be zero. The exponents, a_I, $i =$ 1,2. . .,M are all positive. They essentially define the importance of the respective characteristic to the property right. The first N of them (those for the essential characteristics) measure the percentage change in Q when the respective characteristic increases by 1%. The weights, w_1 and $w_{2,j}$, are also positive and sum to unity. They essentially define the relative importance of the non-essential characteristics relative to those that are essential.

It is easy to check that since all characteristics are measured between 0 and 1, the Q-measure takes values in the interval [0,1]. A Q-value of zero means that the property right has no quality; it is worthless. A value of unity means that the property right is perfect.

In the simple case of the above four property rights characteristics, the Q-measure is defined by the expression

$$Q \equiv S^{\alpha} \cdot E^{\beta} \cdot P^{\gamma} \cdot (w_1 + w_2 \cdot T^{\delta}),\ \alpha,\ \beta,\ \gamma,\ \delta,\ w_1,\ w2 > 0 \text{ and } w_1 + w_2 = 1, \tag{11.2}$$

where S denotes security, E exclusivity, P permanence and T transferability. Note that in this version of the formula the first three property rights characteristics are considered essential and the fourth, transferability, non-essential.

Obviously, to apply the Q-measure defined by (11.1) and (11.2) the values for the relevant property rights characteristics have to be determined as well as the values of the exponents and the weights. This is the empirical work that is needed to apply the Q-measure.

11.4 Property Rights and Economic Efficiency

From the writings of Adam Smith (1776) to the modern theory of economic growth Barro and Sala-i-Martin (1995) it has been well established that the two basic sources of economic efficiency, i.e. maximum production of economic benefits per person, are:

(1) Specialization
Specialization, in this context, should be interpreted broadly. It comprises the division of labour between persons with each specializing in what he does best, specialization between firms with the most efficient firms carrying out the production of each commodity, and specialization between countries with each producing the goods in which it has comparative advantage.
(2) Accumulation of capital
This should also be interpreted broadly. Capital here refers to all assets that contribute to production and well-being. This, obviously, includes physical as well as human and biological capital.

A fundamental prerequisite for both specialization and accumulation is the existence of property rights. Specialization obviously requires trade. If there is no trade, people will be forced to be self-sufficient, i.e. to produce all their needs themselves. Specialization will not be possible. Trade, in turn, requires property rights. This, of course, is obvious. After all, trade is nothing but a transfer of property rights. So, without property rights there can be no trade. Hence, we must conclude that without property rights, there can be very little or no economic specialization.

Accumulation of capital obviously requires property rights. No one is going to save valuables in the form of physical capital, natural resources or even human capital unless he enjoys adequate property rights over his accumulation. There are two reasons for this. First, accumulation of capital necessarily means sacrifice of current consumption. Therefore, to accumulate one must be reasonably sure of not only retaining possession of the accumulated assets but also gaining from their existence.[2] Without property rights, this of course is not possible. Second, even if some people decided to accumulate nevertheless, this accumulation would be seized by others and, in order to avoid a similar fate, quickly consumed. So without property rights there will be (i) no accumulation and (ii) what capital there might exist will be quickly seized and squandered.

These results apply to natural resource utilization just as any other economic activity. With strong property rights, natural resources are likely to be used so as to produce the maximum flow of net benefits over time, preserved as appropriate and, in the case of renewable resources such as fisheries, even enhanced (Costello, Gaines, & Lynham, 2008). Without property rights such as under the common property or common pool arrangement, the opposite is likely to occur. Little if any net economic benefits will flow from the resources, they will be excessively exploited and most likely subject to a continual process of deterioration.

These arguments provide us with two points on the relationship between the quality of property rights and economic efficiency: With nonexistent property rights, economic efficiency is bound to be very low. With perfect property rights, full economic efficiency is attained. What about the points in between?

The relationship between property rights or varying quality and the resulting economic efficiency in natural resource use was examined in Arnason (2007). The basic

---

[2]This, of course, assumes something less than perfectly altruistic individuals.

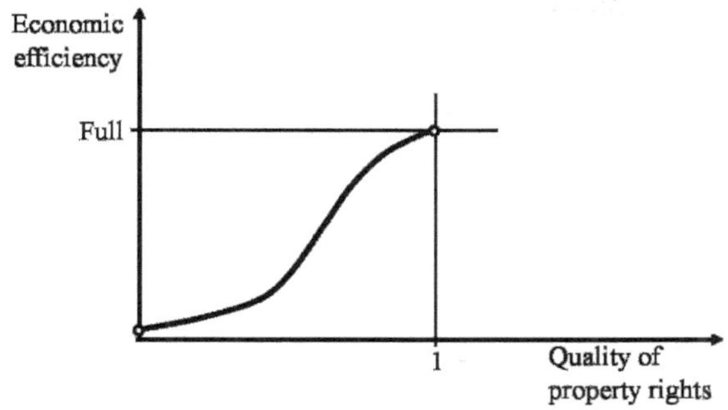

Fig. 11.4 Possible Relationship Between Property Rights Quality and Economic Efficiency

result is that economic efficiency is monotonically rising in the quality of the property rights involved. This is obviously an important policy result. According to the result any improvement in property rights quality will lead to an increase in economic efficiency and vice versa. Therefore, if the aim is to increase economic efficiency in natural resource use, the manager should install as strong private property rights as possible. Apart from this monotonicity, the exact form of the relationship could not be determined in general. However, numerical experiments suggested that the relationship might be like that illustrated in Fig. 11.4. Low quality property rights do not increase efficiency very much, but the efficiency gains start to take off as the property rights value approaches 0.5 and higher.

From the perspective of the current chapter the basic result of a monotonically increasing relationship between the property rights quality and economic efficiency is of great importance for it provides us with a shortcut to compare the economic efficiency of fisheries management systems. All that is needed is to assess the quality of the property rights involved and calculate the corresponding property rights value e.g. with the help of the Q-measure. The higher quality these property rights (e.g. their Q-value), the more efficient is the extraction activity likely to be and vice versa.

Readers reasonably knowledgeable about fisheries management theory may wonder how output taxes, a well-known method to generate efficiency in common property/common pool fisheries (Arnason, 2007), fits into this relationship between property rights and economic efficiency. The answer is that it fits very well. An entity entitled to charge at will for extraction from a resource is equivalent to a sole owner. In fact, a sole owner of a fishery may well choose to operate exactly by renting out extraction rights just as the sole owner of a strawberry field sometimes does.

11.5 Application to Fisheries Management Systems: Examples

Let us now consider how the above theory can be used to compare the efficiency of fisheries management systems. For this purpose let us consider four fisheries management systems as follows:

Table 11.1 Property Rights Characteristics: Quality

	Licences	IQs	ITQs	TURFs
Security	1.00	1.00	1.00	1.00
Exclusivity	0.05	0.67	0.67	0.95
Permanence	1.00	1.00	1.00	1.00
Transferability	0.00	0.00	1.00	1.00

(1) A licensing system
(2) An individual quota (IQ) system
(3) An individual transferable quota (ITQ) system
(4) A territorial user right (TURF) system

As explained above, a great number of variants of these basic systems are available. We will consider specific variants with property rights quality along the four basic dimensions as summarized in Table 11.1. Briefly, it will be assumed that all four systems are 100% secure and permanent. This means that whatever property rights are embodied in the fisheries management systems are fully secure and permanent assets. The systems however are taken to differ fundamentally in terms of exclusivity and transferability. The exclusivity in this case refers not only to the asset itself, i.e. the licences, IQs, ITQs and TURFs, but to each fisher's exclusive rights to the underlying resource in terms of his extraction volume, extraction methods and control over the fish stocks and their aquatic habitat. In this sense the exclusivity of licences is very small – perhaps 1/20 as in Table 11.1, while the exclusivity of the IQs and ITQs is considerable – possibly 2/3 as in Table 11.1, and very high or possibly 19/20 for the TURFs. As regards the TURFs, it should be mentioned that the implicit assumption is that the fish stock in question as well as the relevant aquatic habitat is largely confined to the TURFs. Obviously, if that were not the case, the exclusivity of the TURF would be correspondingly reduced. The fisheries management systems are assumed to differ also in terms of transferability. The licences and IQs are taken to be non-transferable. By contrast, the ITQs and TURFs are taken to be fully transferable.

Regarding Table 11.1, it is important to recognize that to assign numbers to the various characteristics of property rights is in general a nontrivial empirical task. To do this well in practice requires solid knowledge and understanding of the fisheries management systems in question, as well as clear understanding to the property rights characteristics themselves. A certain compensation is that it normally turns out the resulting Q-measure is not particularly sensitive to a degree of inaccuracy in this respect and, moreover, by employing the Q-measure, it is relatively easy to conduct a sensitivity analysis on the results or even fully fledged stochastic simulations to obtain confidence intervals.

The characteristic footprint of licences, ITQs and TURFs according to Table 11.1 is drawn in Fig. 11.5 – IQs are left out merely to avoid cluttering the diagram further. As is clear from the diagram, the TURFs (remember the assumption that the

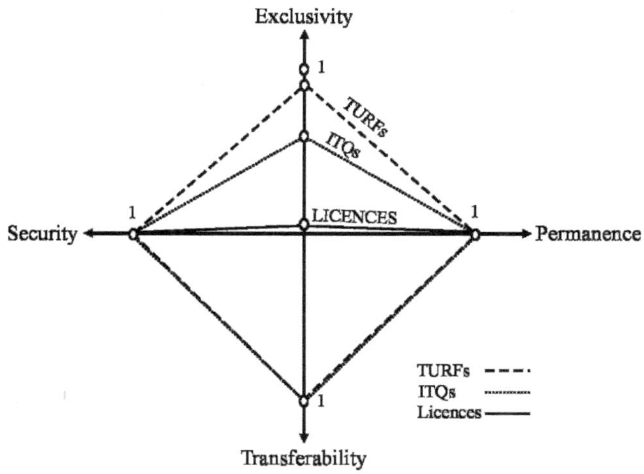

Fig. 11.5 Characteristic Footprints

fish stocks and ecosystem are largely confined to the TURFs) are extremely strong property rights. They have a characteristic footprint close to a perfect property right. The licences, by contrast, are obviously very weak property rights. The ITQs have quite high property rights quality but less than the TURFs. The reason is less exclusivity.

With the help of expression (2) we may obtain numerical values for the property rights quality of the four fisheries management systems. However, to do so we need to specify the parameters of the formula in expression (2). The parameter values adopted are listed in Table 11.2 below.

According to the specifications in Table 11.2, the elasticity of the Q-measure for the overall quality of the property right with respect to both security and permanence is 0.25. This means that as these variables increase by 1% the Q-measure increases by 0.25%. For exclusivity the corresponding elasticity is 0.5 or double that of secu-

Table 11.2 Parameters Adopted for Expression (11.2)

Parameters	Value	Characterization
α	0.25	Elasticity of Q w.r.t. security
β	0.50	Elasticity of Q w.r.t. exclusivity
γ	0.25	Elasticity of Q w.r.t. permanence
δ	1.00	
w_1	0.60	The relative importance of security, exclusivity and permanence
w_2	0.40	The relative importance of transferability

Table 11.3 Property Rights Q-Values

	Q-values
(1) Licences	0.13
(2) IQs	0.49
(3) ITQs	0.82
(4) TURFs	0.97

rity and permanence. The elasticity of the Q-measure with respect to transferability is variable ranging from minimum of zero (at zero transferability) to a maximum of 0.4 (at full transferability). These numbers are not based on empirical measurements but have been selected on the basis of the investigation of the relationship between property rights quality and economic efficiency contained in Arnason (2007).

On the basis of expression (2) and the numerical specifications in Tables 11.1 and 11.2, it is now straightforward to calculate the Q property rights values for the four fisheries management systems. The results are listed in Table 11.3.

The Q-values listed in Table 11.3, give quantitative measures for the property rights quality that was graphically presented in Fig. 11.5. As indicated in Fig. 11.5, the property rights value of the licensing system is very low. The practical inference is that this kind of a fisheries management system cannot be expected to result in substantial efficiency gains in fisheries. This outcome is in fact in good conformance with has been observed in fisheries licensing systems worldwide (OECD, 1997; National Research Council, 1999).

The Q-value for the IQ system is much higher signalling that ITQs constitute significant property right from the perspective of economic efficiency. The Q-value for the ITQ system is higher still signalling a fairly substantial property right. The difference between the two is accounted for by the transferability of the latter. On this basis it may be expected that both IQs and, in particular, ITQs lead to substantial efficiency gains in fisheries. This, again, is in accordance with real life observations (OECD, 1997; National Research Council, 1999; Hatcher, Pascoe, Banks, & Arnason, 2001).

Finally, the Q-value for the TURF system is very high. In fact, according to this number the TURF system, as specified, is close to being a perfect property right. Consequently, the economic efficiency associated with TURFs with the exclusivity as assumed above may be expected to be very high. Once again this fits the available observations (OECD, 1997; National Research Council, 1999). Indeed, it is usually found that TURFs with a high level of exclusivity exhibits a strong tendency to turn fishing into farming or, in this case, aqua/mariculture (Harte, 2000).

11.6 Conclusions

Due to the high number and complexity of fisheries management systems it is simply not feasible to compare and assess them on the basis of empirical investigation only. Fortunately the theory of property rights and economic efficiency appears to provide a foundation for comparing the efficiency of fisheries management systems in a relatively simple manner.

It can be shown that economic efficiency in resource utilization increases monotonically in all the characteristics of property rights (Arnason, 2007). Accordingly, any property rights measure which also increases monotonically in property rights characteristics will also be a measure of the efficiency of the associated natural resource utilization. The *Q*-measure explained in this paper provides one such measure. Although, this has not been fully tested empirically, it appears that the *Q*-measure may provide a relatively simple, quick and robust way to assess the relative efficiency of different fisheries management systems that may be proposed. Not only can the *Q*-measure be used to compare different options for fisheries management systems. It can also be used to assess the efficiency impacts of alterations to existing systems – after all, such alternations define new systems.

It is important to recognize the limitations of the *Q*-measure, however. Three such limitations are obvious. First, the *Q*-measure does not eliminate the need for empirical investigation It only reduces it. It is still necessary to understand the fisheries management system at hand and how its properties can be translated into property rights characteristics along the relevant property rights dimensions. This is in general a non-trivial task. Second, to apply the *Q*-measure requires the determination of the parameters of the *Q*-equation. Of course, fixed parameter values, such as the ones used in this study, can be employed. However, since the exact relationship between property rights characteristics and economic efficiency may differ from one fisheries situation to the next, better results would be obtained by adjusting these parameters to the situation at hand. This suggests the need to further empirical investigations. Third, the *Q*-measure is designed to gauge the economic efficiency of the fisheries management system. However, as discussed in the paper, other social impacts of the fisheries management system such as social stability and social equity are also relevant. These things are not directly accounted for by the *Q*-measure. However, as discussed at length in the introduction, maximum efficiency, i.e. the maximization of net economic benefits, also maximizes in a fundamental sense society's ability to meet other social considerations including stability, equity and so on.

References

Alchien, A. (1965). Some economics of property rights. *Il Politico, 30*, 816–829.

Anderson, L. G. (1986). *The economics of fisheries management*. Baltimore: Johns Hopkins University Press.

Arnason, R. (2000). Property rights as a means of economic organization. In R. Shotton (Ed.), *Use of property rights in fisheries management*. FAO Fisheries Technical Paper 401/1. Food and Agriculture Organization of the United Nations. Rome: FAO.

Arnason, R. (2007a). Fisheries management: Basic principles. In P. Safran (Ed.), *Fisheries and Aquaculture*, in *Encyclopedia of Life Support Systems (EOLSS)*, Developed under the auspices of the UNESCO, EOLSS Publishers, Oxford, UK.

Arnason, R. (2007b). Property rights quality and economic efficiency of fisheries management regimes: Some basic results. In T. Bjorndal, D. Gordon, R. Arnason, & R. Sumaila (Eds.), *Advances in fisheries economics*. Oxford: Blackwell.

Barro, R., & Sala-i-Martin, X. (1995). *Economic growth*. New York: McGraw-Hill.
Costello, C., Gaines, S. D., & Lynham, J. (2008). Can catch shares prevent fisheries collapse. *Science, 321*, 1678–1681.
Demsetz, H. (1967). Toward a theory of property rights. *American Economic Review, 57*, 347–359.
Gordon, H. S. (1954). Economic theory of a common property resource: The fishery. *Journal of Political Economy, 62*(2), 124–142.
Hardin, G. (1968). The tragedy of the commons. *Science, 162*, 1243–1247
Harte, M. (2000). Fisher participation in rights-based fisheries management: The New Zealand experience. In R. Shotton (Ed.), *Use of property rights in fisheries management*. FAO Fisheries Technical Paper 401/1. Food and Agriculture Organization of the United Nations. Rome: FAO.
Hatcher, A., Pascoe, S., Banks, R., & Arnason, R. (2001). *Future options for UK fish quota management*. CEMARE Report 58. University of Portsmouth. Portsmouth, UK.
Hicks, J. R. (1939). Foundations of welfare economics. *Economic Journal, 49*, 696–712.
Kaldor, N. (1939). Welfare propostitions of economics and interpersonal comparisons of utility. *Economic Journal, 49*, 549–552.
National Research Council. (1999). *Sharing the fish: Toward a national policy on individual fishing quotas*. Committee to Review Individual Fishing Quotas. Ocean Studies Board, Commission on Geosciences, Environment and Resources, National Research Council. National Academy Press. Washington, DC.
OECD. (1997). *Toward sustainable fisheries: Economic aspects of the management of living marine resources*. Paris, France: Organization for Economic Cooperation and Development.
Ostrom, E. (1990). *Governing the commons: The evolution of institutions for collective actions*. New York: Cambridge University Press.
Scott, A. D. (1955). The fishery: The objectives of sole ownership. *Journal of Political Economy, 63*(2), 116–124.
Scott, A. D. (1989). Conceptual origins of rights based fishing. In P. A. Neher, R. Arnason, & N. Mollett, (Eds.) *Rights based fishing*. Dordrecht: Kluwer Academic Publishers.
Scott, A. D. (2000). Introducing property in fisheries management. In R. Shotton (Ed.), Use of property rights in fisheries management. FAO Fisheries Technical Paper 404/2. Food and Agriculture Organization of the United Nations. Rome: FAO.
Smith, A. (1776). *An inquiry into the nature and causes of the wealth of nations*. Edition by E. Cannan 1977. Chicago: University of Chicago Press.
Turvey, R. (1964). Optimization and suboptimization in fishery regulations. *American Economic Review, 54*(1), 64–76.
World Bank & FAO. (2008). *The sunken billions: The economic justification for fisheries reform*. Washington DC: World Bank.

Chapter 12
Conclusion: The Innovation Evaluation Framework

Douglas Clyde Wilson, Kjellrun Hiis Hauge, and Martin Aranda

Abstract This chapter constitutes the report of the Innovation Evaluation Framework. It includes a reflection on the indicators of innovations and management performance used in the CEVIS project in terms of their conceptual definitions, operational definitions, and proxies indicators. These reflections are based on how these operational definitions actually were measured when making the comparisons to test hypotheses. The substantive content of the IEF is then summarized though a discussion of the CEVIS project's main conclusions. Finally, the chapter includes a discussion on the contribution of the project and its learning achievements.

Keywords Biological robustness · Cross-disciplinary · Economic efficiency · Effort management · Innovation evaluation framework · Fisheries management costs · Multi-disciplinary · Participatory governance · Rights-based management · Rule-based management · Social robustness · Trans-disciplinary

12.1 Introduction

This concluding chapter comprises the Innovation Evaluation Framework (IEF). As described in the introduction, the IEF was conceived as two things at the beginning of the CEVIS project. First it was to address a number of practical questions about the management innovations and how they related to the desired outcomes. So in one sense the entire book is the IEF and here we summarize the results of the various chapters for policy makers and other interested parties. CEVIS' main aim was to make specific contributions to the policy debate based on the available data. The usefulness of these contributions will be determined by the course of the fisheries policy discussion. The second aspect of the IEF was to be a reflection on the

D.C. Wilson (✉)
Innovative Fisheries Management – An Aalborg University Research Centre,
North Sea Centre, Willemoesvej 2, Hirtshals 9850 Denmark
e-mail: dw@ifm.dk

K.H. Hauge, D.C. Wilson (eds.), *Comparative Evaluations of Innovative Fisheries Management*, DOI 10.1007/978-90-481-2663-7_12,

indicators that we chose to use when we made these evaluations. Abstracting the IEF in this way provides insights into how different disciplines approach the problem of using science to clarify policy debates. It also offers lessons about various kinds of indicators and their usefulness and limitations in addressing policy choices. To create a logical flow for the chapter we present these two things in the reverse of the order in which they were created. We begin in Section 12.2 with the reflection on the indicators and finish in Section 12.3 with a summary of the project's main substantive conclusions.

The CEVIS project was conceived nearly five years ago. It set out to explore how science could help to evaluate some of the suggestions for policy changes being made in the then current discussions. Five years is a long time in fisheries policy. The innovations we elected to examine have followed policy trajectories that will determine their eventual fates much more than any impact this book might have. Rights-based management is being widely adopted, but also resisted in a few places like Scotland. Because of this contention, we might make a direct contribution, perhaps in the form of better clarifying some of the questions. Effort management is already widely used and stakeholder participation in fisheries management has become widely accepted in Europe. Indeed, it has become a norm, albeit one whose form is contested. Rule-based systems remain an ideal much as they were five years ago, but this ideal is gradually being approached and has become deeply integrated in how fisheries management is conceived, especially in relation to its scientific support base. If CEVIS were being built today it would be addressing more recently heard questions such as how to implement results-based management, reverse the burden of proof, and integrate fisheries into marine spatial planning.

12.2 The Abstracted Innovation Evaluation Framework

As described in the introduction, the strategy CEVIS adopted was to set up a learning process where we would not only evaluate potential fisheries management innovations; we would also explore different ways that different disciplines could work together to achieve this. We began with a cross-disciplinary[1] effort to develop hypotheses. After considerable discussion and experimentation – most intensively into the idea of using a fuzzy sets approach to case study research (Ragin, 2000) – we finally rejected the idea of trying to create a trans-disciplinary research

[1] A discussion of cross-, multi- and trans-disciplinary scientific cooperation is found in the introductory chapter. In our usage of the terms, multi-disciplinary means that scientists work on a common problem while remaining within their own discipline's methods and ways of framing questions. Cross-disciplinary means that scientists from more than one discipline work together on the problem, but still remain within one of the discipline's methods and frames. Trans-disciplinary means that common concepts and methods are developed that reflect the theories and methods of more than one discipline.

framework and then ask everyone to stick within this framework. It became clear that such a strategy would be very difficult for the individual disciplines to implement in practice and would yield results that were either too abstract or trivial to have any practical usefulness to policy makers. Instead, we decided on a multi-disciplinary strategy where separate disciplines sought to test specific questions about the innovations, while remaining open to the possibility of trans-disciplinary investigations where two or more disciplines could join in a single investigation where this seemed a promising approach. This meant that disciplinary teams used their state of the art methods and best available data to test hypotheses that linked whatever aspects of the innovations and outcomes they were able to get some answers for. This strategy meant that several possible relationships between innovations and outcomes were not examined at all, either because data was not available or because suitable methods have yet to be developed.

It is these focussed, disciplinary and trans-disciplinary research activities that produced the IEF as it actually emerged. Each discipline was asked first to choose and test whatever relevant hypotheses they could that linked the innovations to the outcomes. No matter what the discipline, testing hypotheses requires thinking through the relationship between the concepts you want to examine and the methods that you will use to do so. Testing hypotheses always involves some kind of comparison. This can be a simple comparison before and after, or something more complex. For example, if your hypothesis is that introducing effort-based management increases biological robustness you need to find a way to compare an effort-based management situation with a non-effort-based situation and examine what happened with biological robustness.

Linking the concepts to the concrete test means thinking through three things:

a. The conceptual definitions – what do you mean by effort-based management and what do you mean by biological robustness?
b. The operational definitions – how are the definitions of the two concepts actually represented in the particular comparison that you are making to test the hypothesis?
c. The proxy indicators – how are these operational definitions actually measured when making the comparisons to test the hypothesis?

This is the approach we used to abstract a summary of the IEF as a discussion of indicators. When the research was completed we asked each of the disciplines to report not just their results but also this process of developing proxy indicators to test hypotheses. The aim of the hypothesis testing was to get indications through the comparisons that are sufficiently strong and relevant to be used in a policy debate. An overview of the results is offered in Table 12.1 and the details of these results are discussed in the next section. What this abstraction of the IEF really reveals is the different ways that these indicators aid, or fail to aid, our understanding of these policy questions.

Table 12.1 Concepts and Indicators for the Innovation Evaluation Framework

Concepts	Operational definitions	Proxies	Discipline
Biological robustness	Maintenance of an adequate SSB	Median SSB; Probability that population falls below minimum SSB; Percentage of years where management targets were met; Years required to get a population above B_{lim}	Biology
Social robustness	Stakeholder acceptance	Respondents' reports of perceived levels of compliance with management measures; General acceptance expressed by respondents	Social science
	Institutional learning	Documentation and respondents' reports of specific instances of learning and problem solving	
Economic efficiency	Pareto efficiency = maximum resource rent after costs, which includes all external costs and earnings	Landed value minus costs minus the social costs of the change in the fish resource abundance; Net present value disregarding the social costs of the stock abundance	Economics
Cost effectiveness	Amount and type of effort needed to implement management innovation in terms of administration, research and enforcement costs	Records and perceptions of changes in expenditures	Economics
Participatory management	Greater cooperation in monitoring fishing activities by improved reporting of catches	100% observer coverage of discards implying improved stock assessment; Reduced level of underreporting implying improved stock assessment	Economics
	NWWRAC	Respondents' understanding and documentary information from before and after implementation	
	Decreased variability and bias in discard estimates	Increased sampling directed to possible sources of bias; Increased sampling directed to higher precision; Gradually decreasing levels of catch underreporting	Biology

Table 12.1 (continued)

Concepts	Operational definitions	Proxies	Discipline
	Dutch Biesheuvel Groups	Respondents' understanding of system before and after implementation in terms of the quality and breadth of participation	Social science
	NWW RAC	Respondents' understanding of system before and after implementation in terms of the quality and breadth of participation	
Rights based management	UK Producer Organization management of fixed quota allocations	Respondents' impression of system changes before and after implementation both overall and in terms of its (a) economic impact (b) practicality (c) qualities of initial allocation (d) impact on exit and entrance to fishery	Economics and social science
	Territorial use rights in fisheries (TURFs) for anchovy	Respondents' impression of system changes before and after implementation both overall and in terms of its (a) economic impact (b) practicality (c) qualities of initial allocation (d) impact on exit and entrance to fishery	
	Faroe Islands fishing days system	Respondents' impression of system changes before and after implementation both overall and in terms of its (a) economic impact (b) practicality (c) qualities of initial allocation (d) impact on exit and entrance to fishery	
	Northern hake ITQ	Respondents' impression of system changes before and after implementation both overall and in terms of its (a) economic impact (b) practicality (c) qualities of initial allocation (d) impact on exit and entrance to fishery	
	North Sea pelagic fishery	Respondents' understanding of system in respect to costs	Economics

Table 12.1 (continued)

Concepts	Operational definitions	Proxies	Discipline
	Dutch ITQ system	Respondents' understanding and documentary information from before and after implementation	
	Danish ITQ system	Respondents' understanding and documentary information from before and after implementation	
Rule-based systems	Cod recovery plans in the Baltic	Administrators' perceptions of system impacts	Economics
	Allocation of annual quota	TAC based on EU long-term management plans for plaice; Sensitivity to misreporting based on yearly 10% F reduction; An indirect proxy under which fishers continue until they have finished the TAE and do not report the extra catch	Biology
	Limitations on landings	Landings in terms of live weight of each species implying that the landing with the lowest net profit per unit TAC will be dropped first when restrictions are imposed	Economics
Effort management	The number and size of fishing vessels allowed (fishing capacity controls), the amount of time the vessels are allowed to operate (vessel usage and activity controls), or the product of capacity and activity (fishing effort controls)	Vessel such as the number of vessels is kept constant; Days-at-sea such as the days with the lowest net profit are dropped first when effort restrictions are imposed	Economics
	Restrictions on overall capacity or restrictions of access for a certain capacity to given resources in given area value (MPAs)	Catch Per Unit Effort (CPUE) per ICES square per species such that CPUE will be zero in the closed area	
	Polish area closures and days at sea system	Respondents' understanding and documentary information from before and after implementation	

Table 12.1 (continued)

Concepts	Operational definitions	Proxies	Discipline
	Swedish MPA system	Respondents' understanding and documentary information from before and after implementation	
	Limitation of effort by the fishing fleet	Days at sea (DAS; horsepower days at sea), seasonal days at sea; Area closed for fishing; Sensitivity to misreporting based on a reduction in assessed effort by 10%; An indirect proxy under which fishers stop fishing before exhausting their catch limits and then report their total catch	Biology
	Faroese fishing-days system	Respondents' understanding of system before and after implementation	Social science
	Area and seasonal closures in Baltic		
	EU days-at-sea programme		

12.2.1 Biological Robustness

The biologists defined the idea of biological robustness both as property of a management system and as an objective of a management innovation. As a property of a management system they approached biological robustness as 'the ability of a management strategy to account for uncertainty and error in the biological knowledge'. The strength of this approach is that it led to some very creative and testable hypotheses while capturing very important elements of biological robustness from a management perspective. However, it missed the many other aspects of biological robustness related to the overall ecological resilience of the system.

As an objective within their hypothesis tests they chose as an operational definition 'the maintenance of an adequate spawning stock biomass (SSB)'. The maintenance of an adequate SSB is central to giving fisheries management advice within a precautionary approach. The main weakness is that they continued along the line of a single species understanding of biological robustness, something which must also be understood as an ecosystem property. The proxy indicators they chose for this concept when testing their hypotheses were: (a) the probability that population falls below some minimum SSB; (b) median SSB; (c) the percentage of years where management targets for SSB were met; and (d) the years required to get a population above B_{lim}.

The main strengths of the first proxy indicator for this operational definition, the probability that a population falls below minimum SSB, are that it is generally accepted, and well understood in management discussions. As a probability it is especially valuable in evaluating if the management approach being used is precautionary.

The second proxy indicator they used for this operational definition, median SSB, is one of the items most often used to communicate to fishermen and managers. A mean provides information on benefits for the fishermen in the long run. The third proxy indicator they used for this operational definition, the percentage of years where management targets were met, is even easier to discuss with stakeholders than the median SSB.

The fourth proxy for this operational definition, the years required to get a population above B_{lim}, has a linkage to resilience of the system and is easy to use for managers making long-term and medium term plans.

An important limitation with all the four proxies, which stems from the operational definition, is that they contribute nothing to our understanding of what is going on with non-target species, secondary effects through food web, and other ecological effects. This implies both that other impacts of fishing are not taken into account and that only limited ecological effects are taken into account in evaluating the indicators. However, the four indicators together cover some very relevant aspects for fisheries management and they all suit the ICES advisory framework very well. Also, since the proxies are used in simulation studies, various impacts can be modelled where their specific effect on an indicator can be evaluated.

12.2.2 Social Robustness

The social scientists evaluating the impact of the innovations on the social robustness of management chose two operational definitions of social robustness.

The first is 'stakeholder acceptance', which is the legitimacy of the management system for the stakeholders as expressed through both their statements and behaviour. The proxy indicators for evaluating hypotheses using this definition were interview respondents' reports of perceived levels of compliance with management measures and the degree of general acceptance of the management system as expressed by respondents.

The second operational definition is 'institutional learning'. This is how well the management system adapts in response to external changes. Such changes can be in either natural or social systems. The proxy indicator for evaluating hypotheses using this definition was specific instances of learning and problem solving as expressed either by interview respondents or by documentary reports. A challenge with applying this concept is grasping when institutional learning has not taken place. Although this is necessary for considering a system's capacity for institutional learning, it is impossible to imagine all possible courses that a system could have taken through institutional learning.

A general challenge in empirical studies is to discover and disentangle the various sources that may have affected the proxy indicator of concern. Both proxies are interview-based indicators and so they are simply people's opinions and, furthermore, both of them seek to capture inherently subjective phenomena. Both management institutions and institutional learning are made up of people interacting on the basis of a set of shared understandings that are to a degree opaque even to the interacting participants. This double subjectivity is a research limitation that is intrinsic to the phenomena under study. Done correctly, interviews are an excellent tool for exploring these meanings. The trick is to try to engage the respondent in a conversation and to ask as few specific questions as possible. By keeping to a broad discussion using very open questions about the phenomenon under study, especially early in the interview, one is able to uncover how the respondent is defining the object of study. If the early questions are too specific the respondent is only to the investigators definitions of what is going on frame the responses too rigidly.

12.2.3 Economic Efficiency

The concept of economic efficiency was used only by the economists who were charged to evaluate the management innovations in this particular respect. They had an operational definition ready in the form of 'Pareto efficiency'. Pareto efficiency is an equilibrium measure that requires that if someone is better off, no one must be worse off. The underlying assumption of this approach is that the benefit from an action for those being better off compensates the losses of those who lose out from the measure. Of course, this does not happen in practice, but it is still a reasonable way to evaluate the economic efficiency of the management action as it can be argued to reflect an overall gain for the society.

The economists made use of two proxy indicators to represent Pareto efficiency. The first is the landed value of the catch net, both the economic costs to the vessel and the social cost of the change in resource abundance. The main strength of this proxy is that it is a composite measure that in one single number summarizes the key economic characteristics of the management system and makes it comparable to other systems. All advantages are counted against all disadvantages. The main weakness of this proxy indicator is that it is a complex theoretical, socio-economic, equilibrium measure in which advantages are counted against disadvantages making the indicator difficult to apply in practice.

The second proxy that the economists used for Pareto efficiency was net present value (NPV) a measure that does not consider the social costs of the stock abundance. This is also a well defined and broadly accepted composite measure that in one single number summarizes the key economic characteristics of the management system and makes it comparable to other systems. Its weakness is that it is dependent on the time horizon for which it is measured. The NPV does not require, in the same way that the overall idea of Pareto efficiency does, that the system is in equilibrium.

12.2.4 Costs of Management

The economists investigating the costs of management used empirical information. Their operational definition 'the amount and type of effort needed to implement management innovation in terms of administration, research and enforcement costs'. The major drawback of this definition is that it encompasses only overall costs rather than cost effectiveness in respect to benefits. Clearly we cannot get at cost effectiveness without understanding costs, but without a measureable definition of 'effectiveness' based on specific objectives cost, effectiveness cannot be understood even in terms of respondent perceptions. It was also evident from the investigation that the data does not exist in a form that allows the impacts of the various changes and innovations to be disaggregated. Indeed, in many cases the changes in operating costs to the European management system were so strongly affected by the implementation of the Data Directives that any influence from other innovations were effectively swamped. With many different changes going on simultaneously it was not possible to really evaluate the disaggregated costs of any of them. The proxy indicator of this effort was information on aggregate cost figures gathered from respondents.

12.2.5 Participatory Governance

The concept of participatory management was investigated by all three disciplines.

The biological team approached participatory governance through the lens of understanding the overall biological robustness of a management strategy as its ability to account for uncertainty and error in the biological knowledge. The hypotheses that this approach made possible did not just involve trying to reduce uncertainty by calling for more information, but also evaluating how the system can respond to the uncertainty that is always there as an ongoing management problem. They estimated the effects of variation in discard sampling and accuracy of landing data, which can be very useful for improving the accuracy of stock assessments. The operational definition of participation that this strategy led to was increased information about stock conditions. Thinking of participation as a way to increase the accuracy and reliability of fisheries information suggests many possibilities for application. They chose to look at two possible aspects. The first is the reduction of uncertainty in observations through the use of fishermen to collect data. The second is the reduction of non-reported catches as a result of increase in compliance. Of course this increase in compliance is only theoretical as there is no data to quantify the increment.

The hypothesis tests made use of three proxy indicators of this definition within the simulation model used for the tests. The first was increased sampling directed to possible sources of bias. This indicator is measureable and makes it possible to mimic discard sampling. This allowed them to get an idea of the extent of the deviations of under- and overestimations of discards. The indicator is limited in that it is only measureable through discard observations though observers or by making use of the new technologies for observing on-deck behaviour remotely.

Their second proxy indicator was similar. This was increased sampling directed to higher precision. This proxy made it possible to mimic an enlargement of the sampling process and therefore to evaluate whether a self-sampling programme would increase the accuracy of stock assessments or not.

The third proxy indicator of participation used by the biologist was gradually decreasing the levels of catch underreporting. The second and third proxy indicators have the same basic limitation as the first proxy when used in an empirical test.

The economists looking at economic efficiency took a similar approach to participatory governance. They gave participation the operational definition of 'greater cooperation in monitoring fishing activities by improved reporting of catches'. Understanding participation this way is useful in that it highlights one of the core goals of participatory approaches. That usefulness is limited, however, by the fact that even the fishers do not fully grasp the impact of fishing on the stock because of the damage to fish that escape the gear without being brought on deck. Furthermore, this approach understands the 'greater' in greater cooperation to be a function of the number of participants, but the effort required to monitor free riders also increases with the number of participants so the marginal contribution of each participant falls and at some point increasing numbers may actually decrease cooperation.

The economists also placed the proxy indicators into simulation models to evaluate the importance of participation. The first of these was a simulated 100% observer coverage of discards, which resulted in an improved stock assessment within the model. This allowed an evaluation of the impact on fish stock abundances of the improved stock assessments that the cooperation implied. The second proxy indicator was a reduced level of underreporting implying improved stock assessment, which was also linked to improved stock assessments in the models. For neither of the proxy indicators were they able to simulate the associated monitoring and control problems.

The economists charged with evaluating the impacts of the innovations on the costs of management took a somewhat similar approach to that of the social scientists. Their operational definition of participation was 'the implementation of the North West Waters Regional Advisory Council (NWWRAC)'. They represented this in their hypothesis testing through their respondents' understanding as well as documentary information about cost-related changes that happened through the creation of the NWWRAC.

The social scientists' operational definitions of participatory management were 'the implementation of the Dutch Biesheuvel groups' and again 'the implementation of the NWWRAC'. What they actually examined in their hypothesis testing was also the understandings communicated by respondents in interviews. However, in both cases they narrowed their investigations to the respondents' understanding of the system before and after implementation in terms of the quality and breadth of stakeholder participation.

12.2.6 Rights-Based Management

The rights-based management innovation was evaluated by the social scientists and by the economists addressing impacts on the costs of management. In all instances the operational definition of rights-based management was a specific implementation of a rights-based system. The proxy indicator for each one was the respondents' impressions of the situation before and after implementation. Within this overall shared approach the social scientists had some specific areas of emphasis in their interviews. They focussed on perceptions of the economic impact of the rights-based system. They also investigated the practicality of its implementation especially in terms of the perceived fairness and smoothness of the initial allocation of rights. Finally they focussed on questions of the impacts of the rights-based system on the ability of new fishers to enter the industry and how well the system facilitated exit from the fishery, especially in the context of the reduction of fishing capacity.

12.2.7 Rule-Based Management

Rule-based management was examined by the two economists teams and the biologists. The economists evaluating the costs of management studied the consequences of the cod recovery plans in the Baltic, and the plans themselves became the operational definition of the concept and they were evaluated through the perceptions of relevant administrators.

The biologists and the economists investigating economic efficiency assumed some kind of rule-based management because their simulation models require fishing impact to be modelled. Many of their hypotheses were related to a comparison between a TAC regime with various forms of effort regimes (see below). TAC management was represented in a number of different ways by the different disciplines when evaluating the innovations. The biologists used as a straightforward operational definition: the allocation of annual quota in terms of kilograms and single species. In one comparison they used as a proxy the TAC-based EU long-term management plan for plaice. As discussed above they represented TAC management more indirectly as 'sensitivity to misreporting based on yearly 10% F reduction' when they were making a comparison with effort management. The strengths and weaknesses of this approach are discussed below in the section on effort management.

The other discipline to make explicit comparisons with TAC management was economics. They used limitation on landings as their operational definition, which is both well defined and widely accepted within economics. The clear disadvantage of focussing on landings is that it excludes discards of fish. The actual proxy they used was 'landings in terms of live weight of each species implying that the landing with the lowest net profit per unit TAC will be dropped first when restrictions are imposed'. The advantage of this approach is that landings are easy to observe and fishermen's behaviour is included according to economic theory. However, it does not take into account that catch composition of fleet segments differ from the composition of TACs fixed for each species. It would be more accurate to link profit to sustainability when direct information about total catches becomes available.

12.2.8 Effort Management

Effort management appeared in the hypotheses of three disciplines, economics, biology and sociology. The economists made use of two operational definitions of the concept. The first was 'controls on the number and size of fishing vessels allowed, the amount of time the vessels are allowed to operate or the product of capacity and activity'. For the economics team this was a straightforward way to proceed because the most important production factors in fishing are vessels and sea days. These are both clearly measurable. They pointed out, however, that effort is a composite measure that includes a number of production factors, not all of which are measurable.

The economists' first operational definition has two proxy indicators. The number of vessels is a relatively straightforward one. This indicator is easy to observe in a simple sense. However, vessels are platforms for fishing effort and so actual fishing effort will vary in complex ways among them. Moreover, there are many frequently changing external factors that will have an impact on the potential effort of a particular vessel. The other proxy that they used was 'a production function where trips with lowest net revenues per day at sea are dropped first' as a proxy of effort when they were comparing it to total allowable catch/quota (TAC) management. In this comparison they were using a proxy for TAC management based on dropping the lowest net profit per unit TAC. This proxy is well defined and observable and reflects fishers' behaviour following standard economic theory.

The economists' second operational definition was 'restrictions on overall capacity or restrictions of access for a certain capacity to given resources in given area value' which was used in hypotheses in respect to marine protected areas (MPAs). The team characterized this definition as useful but data demanding. Stock information is usually gathered for the overall fishing grounds and it is often not possible to find data on differences in the stock inside and outside the MPA. Moreover the definition cannot take into account the migration and diffusion of stocks in and out of the MPAs. They only used one proxy for this definition, which is the catch per unit effort (CPUE) per ICES square per species, setting the CPUE to zero in the closed area. The indicator is estimated using landings and effort. It has the advantage of being able to assess the impact on the fish stocks in some detail, but discards are ignored when using the CPUE measure that only considers recorded landings.

The biologists chose the simpler 'limitation of effort by the fleet' as their operational definition of effort management. The first proxy for this concept used in their models was horsepower days at sea. This is a clearly defined indicator that is already used in management discussions. Its main weakness is that it does not include effects of improved fishing efficiency. The second proxy they used for effort limitation was hectares of areas closed for fishing. Their evaluation of the proxy was similar to that of the economists. It is a clear enough indicator; however, the mobility of the fish limits the indicator's usefulness for understanding fishery impacts. Furthermore, it cannot account for reallocations of fishing effort outside the boundaries of the closed area.

In comparisons involving TAC and effort-based decision rules the biologists used some interesting proxies for both systems in their simulation analyses. In one case they used 'sensitivity to misreporting based on a reduction in assessed effort by 10%' as a proxy for effort management. The comparisons were made with a similar proxy for TAC-based decision rules. This allowed them to evaluate the ways in which different management rules operate under uncertainties created by missing data. The obvious limitation is that the methods they used cannot be tested against real data, as misreporting is not recorded.

In another case they based the comparisons on proxies that were based on expectations of fisher behaviour. The simulations were based on setting limits with a TAC-based rule, a total allowable effort (TAE)-based rule or a combination. In these combination comparisons, effort management was represented by the indirect proxy: when the effort advice is lower than the TAC advice, fishers stop fishing before exhausting their catch limits and then report their total catch. This indirect proxy for TAE was used in a comparison with TAC in which TAC was represented by the indirect proxy: fishers continue until they have finished the TAE and do not report the extra catch. This approach again allowed some interesting investigation of system level behaviour and the assumptions about fishers' motivations make sense from a standard economic perspective but are not tested against actual data.

The social science operational definitions and proxies are, of course, quite different. As the research is interview-based the operational definitions are all simply the ways the interview respondents understood the system being discussed, so the purpose of the interview is to draw out and understand these meanings. In terms of the hypotheses being tested, the proxy indicators in all hypotheses are descriptions of the situation before and after the implementation of the effort-based system.

12.2.9 Reflections of the IEF

The major contrast seen in the abstracted IEF above is between the social scientists approach to hypothesis testing that relied mainly on before and after evaluations of the implementation of measures and the simulation modelling carried out by the biologists and the economists. The underlying point of quantification is the transparency of reasoning and comparison that careful measurement makes possible. The limit on this power is finding and using comparable concepts and measurements, for example measuring participatory governance as 'gradually decreasing levels of catch misreporting'. This proxy indicator hardly captured the concept of participation, in anything approaching the richness yielded, for example, by respondents' insights about what happened before and after the innovation was carried out. It did allow the team, however, to develop a very specific and useful recommendation about how one aspect of participation should be structured.

Quantification is one axis around which multi-disciplinary scientific cooperation should be designed because it is a useful tool. It is also dangerous because so many relevant facts, values and interests either are not or cannot be measured. A more insidious danger is that modelling policy questions can actually distort policy debates. Because of both the convenience and the great rhetorical power that goes with quantification (Funtowicz & Ravetz, 1990; Porter, 1995), calculable values are given greater weight in the debates then they inherently merit. Economic efficiency is perhaps the most salient example. Other mechanisms that release entrepreneurial creativity are also important for economic growth, but efficiency receives far more policy attention than things that are more difficult to quantify such as risk spreading or network development. The value of economic efficiency is also often tied to the notion of an abstract 'society' for which is argued to be a general benefit, this obscures the critical fact that the gains from policies to increase efficiency accrue to specific groups. These distributional impacts are harder to measure, making them easier to ignore.[2] The inherent reductionism of quantification strongly qualifies its ability to aid policy debates. Like all powerful tools it must be used very carefully.

One arena where CEVIS, and several other EU projects,[3] has explored this careful use is through developing simulation modelling starting with the premises of good governance practice. On the one hand, simulation modelling is particularly vulnerable to the problems of quantification. Simulation modelling frames policy-relevant realities as an input to science and therefore, models address limited aspects of important concepts. In the CEVIS process we found that it was very difficult even for scientists to reflect on exactly how the models mirrored reality when the concepts and their proxy indicators directly represented policy options. The development of the simulations sometimes reflects the very same beliefs about causality that gave initial rise to the policy recommendations that the simulations are evaluating. For example, at one point in the IEF when the hypotheses being explored were related to the input of information into management, the forms of management were measured by assumptions about sensitivity to misreporting. The point here is not that simulations can become circular and hence meaningless, although this is, of course, possible. In this case, in fact, the approach provided some illuminating insights about the robustness of the overall management system. Assumptions always have to be made, both in simulation and in policy. Simulation modelling offers the possibility of clarifying assumptions because of its very dependence on using assumptions to represent reality. It is becoming a more and more common tool in marine management where it acts as a boundary object (Star & Griesemer, 1989) between marine scientists and stakeholders. It is an activity that different parties can do together

[2]An argument defending the importance of economic efficiency can be found on the early pages of Chapter Eleven.

[3]Please see http://flr-project.org/doku.php and http://www.efimas.org/

through such things as participatory modelling (Wilson & Pascoe, 2006) where different scenarios are proposed and then simulated. This can increase dialogue and help participants to clarify their objectives to one another. However, done in such a multi-disciplinary and even multi-stakeholder context the simulations unavoidably take on different meanings for different participants, even while aiding communications. Seen from this perspective the inclusion of beliefs about policy implications in model construction suggests that to be really effective participatory modelling must include not only the testing of alternative scenarios but the deconstruction of the simulation itself.

Bio-economic simulation modelling is the one place that CEVIS was able to achieve a truly trans-disciplinary approach that yielded useful results. In this case the trans-disciplinary approach achieved its promise of yielding more useful information for decision-makers than either biology or economics could yield on their own. Because of the low abundance of many stocks in EU waters, management innovations are often aimed at solving problems with threatened stocks. This means that biological objectives are central in analyses of economic implications for management innovations. In this book the economic analyses are based on the same simulation tools as the biological analyses and thus demonstrate how these disciplines can be linked in a trans-disciplinary manner.

In contrast to the model-based approaches, the social sciences avoid the reduction of important concepts to their measureable aspects, but do so at a cost of comparisons where what is being compared is never precisely defined. While all measurement in science is to some degree subjective, subjectivity is the very subject matter of social science. Shared meaning is the constituent material of institutions. Interviews that seek to explore the meanings respondents attribute to institutions as widely and sensitively as possible are the *sine qua none* of researching any institutional phenomena. There are other ways of performing institutional research, of course, some very dependent on statistical techniques, but all of them depend on translating subjective phenomena into comparable units for hypotheses testing. A reader finally evaluates these hypotheses tests purely on the basis of coherence (Habermas, 1990). First the internal coherence of the argument and second the coherence of the argument with the reader's own experience. Carefully done, systematic social science often achieves both these forms of coherence and in doing so can make an important contribution to policy debates.

Almost all social science indicators in the IEF relied on perceptions of before and after conditions related to the introduction of innovations. Such comparisons are of great interest to policy makers and indeed to other stakeholders who want to get an idea of the implications of the innovations. This kind of hypothesis testing cannot make predictions because there are too many other factors and avenues of interpretation. In the CEVIS experience, especially in the first cross-disciplinary phase where social science methods were used by mixed teams to evaluate innovations outside of Europe, these before and after comparisons proved very valuable in framing the questions that the entire project would investigate in the second phase. Furthermore, the key concept of social robustness could not be measured in any other way.

12.3 The Innovations and Their Impacts

This section summarizes some selected substantive lessons from the CEVIS project. The work done by CEVIS, as reflected in the chapters of this book, is extensive and complex. This summary is not meant as an alternative to reading the more detailed discussions of issues of interest. However, it does provide a shortcut to the main findings for busy people interested in general fisheries management policy. The section is organized by the four main innovations and offers a few general findings about each.

12.3.1 Participatory Governance

1 *Participation can increase the quality of many aspects of fisheries management.* Participation in fisheries management may take many forms, and in this book they include consultation in regulation questions, local management, fishermen's contribution to the knowledge base, and consultation on the overall objectives and means. Participatory governance can increase the quality of many aspects of fisheries management, including increased support of the system and better conflict management. The case studies on New Zealand, Canada, Alaska and the Faroe Islands revealed pride among management stakeholders. It is worth noting, however, that in some cases after a set of institutional changes the group of stakeholders that remains involved is smaller than before and the voices of those who are excluded are no longer heard. Even in broadly participatory programmes the highest positive regard from stakeholders for the system will be from the representatives of stakeholder groups rather than grassroots members. Of course, they are the voices of their group and are the ones that managers have to deal with most directly.

The Baltic case shows what may happen when innovations or new forms of regulations are implemented in a top-down fashion. The management system lacks acceptance and trust and fishermen comply much less frequently with the rules. A fundamental distrust may make any change difficult to accept and thereby hamper institutional learning. Participatory governance may thus help manage conflicts, which are increasing and spreading with the advent of spatial management being carried out in the context of broader marine spatial planning. Participation and trust can also create institutional contexts in which it is easier for people to behave responsibly and thereby have a positive effect on biological robustness. The Community Management Boards in Canada demonstrated increased responsibility for the resource and improved the commitments to scientific advice. A similar sense of responsibility was observed in the Alaska case, where scientific advice enjoyed trust and respect in the participatory TAC-setting process in the Fisheries Management Council.

2 *Excluding the broader civil society may reduce gains from participation.* While almost all CEVIS cases included some form of participatory governance, both European and non-European cases include examples where representation of orga-

nizations from civil society is limited. The civil society may be less relevant in the direct management of the fishery. For example direct participation by civil society in the Biesheuvel Groups in the Netherlands or the Community Management Boards in Canada, where day-to-day conservation is acted out, would have less impact on reaching fisheries management goals than it would in the European Regional Advisory Councils where broader conservation goals are set. A relevant issue for future Europe is to discuss the role of environmental NGOs and civil society in general and at what scale their influence is most relevant. The Alaska case exemplifies that environmental organizations have used campaigns and court cases as tools to influence public opinion. However, they also expressed a wish to have a voting member on the North Pacific Fisheries Management Council, implying that they did consider a role in negotiations as fruitful for achieving their objectives.

3 *Participation is important in science and data collection as well as management.* Participatory governance can also imply changes in the role of science from simply 'telling the answers' to cooperation with stakeholders on the knowledge production and evaluation. Icelandic fishermen decide half of the sampling locations for the scientific ground fish survey. Alaska stakeholders evaluate factors to ensure optimum yield. Canadian stakeholders work closely with the same set of scientists over long periods of time in facilitating stock assessments. The fishing industry in the Faroe Islands has a central role in evaluating the scientific advice for effort regulations, and the fishing industry in New Zealand has the responsibility to provide the necessary scientific basis for quota decisions. Participants in all of these exercises report that they increase the trust in scientists and confidence in their results, while scientists report that they are able to maintain scientific quality.

In CEVIS, the EU cases on the interface between science and stakeholders focus on the quality of catch data, i.e. discard data and illegal landings. Cooperation in these cases implies improvement of data in the scientific stock assessments. In terms of biological robustness, the studies on discard data suggest that it may be more important to identify and address possible sources of bias than to increase the sample sizes, but that biological robustness may not be affected when only immature fish is discarded. Cooperation to improve the catch data can also improve the economic performance of the fishing fleet. These results were conditioned on a TAC regime as the simulations indicated a slightly negative effect on economic results in an effort scenario. Getting proper data on management costs has been a challenge, but in the Spanish Basque case, the administration costs increased when the RAC was created. It is too early to conclude whether the increased costs will be permanent, or whether these are implementation costs.

12.3.2 Rights-Based Approaches

4 *Transferable rights increase economic efficiency.* Increases in the qualities of fishing rights such as transferability, security and durability clearly increase economic efficiency. This is shown theoretically (Chapter 11) and empirically in the cases

where individual transferable quotas (ITQs) have been implemented. These characteristics have developed a sense of ownership and have generated an involvement in management and enhancing of competitiveness. Further, it appears that the rights-holders are more concerned about protecting the resources and environment. An obvious benefit with rights-based systems is that it makes planning easier for rights owners. In Iceland and Alaska, this planning has resulted in efficiency gains, especially with regard to processing. This is particularly so in the latter case as the management system moved away from a race-for-fish. In Nova Scotia those Community Management Boards that do not allow transfers of IQ among members have had many more problems dealing with exits from the fishery than those that do.

The case studies show several examples where rights are given in exchange for increased responsibilities of the rights owners. The Alaskan cooperatives, the Biesheuvel group and the Canadian Community Management Boards were given the responsibility to do local level management, while the New Zealand industry had to provide and pay scientific advice. The extra burden has been possible to bear economically. In several of the cases the profitability of the fisheries due to stronger rights has enabled the industry to shoulder additional management services, and hence reduces costs to the public.

5 *Rights-based management programmes can and should have a flexible design.* There are many aspects to take into consideration when designing a rights-based management programme, including the nature of the property right, management units, determination of total allowable catch, monitoring and enforcement, need for other regulations, rent extraction and cost recovery and initial allocation. The Iceland and New Zealand cases illustrate that ITQ systems can develop over time so that sufficient flexibility should be built into the ITQ systems to be able to amend and adjust rules. In New Zealand the initial allocation was in fixed tonnage, which had to be changed to an allocation in percentages of the quota.

The case studies show that rights-based management systems change over time and that flexibility of the system combined with institutional learning improve this process. The systems of the North Sea, the Faroe Islands and the Western Shelf demonstrated capacities for institutional learning and for keeping a fairly high stakeholders' acceptance among the commercial actors. However, the institutional learning within the rights-based management RBM systems was mostly geared towards making rights more tradable and/or secure or exclusive. Future learning may thereby be reduced since rights-holders will want to maintain the value of their investment in the rights. The ITQ system in Nova Scotia has reduced potentials for adaptive management by locking ecological realities that evolve either naturally or as a result of greater scientific understanding into hard institutional boxes. A fish stock is an ecological reality that is hard to define and that interacts with other ecological realities. Property rights are powerful social constructs with strong implications for policy. Their treatment is much more likely to be determined by courts according to the principles and precedents of property and finance, than by marine managers seeking to take an ecosystem approach.

The initial allocation of quotas has proven to be especially difficult regarding legal aspects, where national rules of equal treatment, the right to a free choice of occupation and the protection against deprivation of property have challenged ITQ systems. Actors that have not received rights may perceive the system to be unfair. The equity problem was partly solved in the Alaska case by buy-out programmes. In both the Canada and Alaska case though, most of the controversy in relation to the rights-based system stemmed from the initial allocation.

6 *Transferability of rights has social costs that it is possible, but difficult, to mitigate.* When rights-based management is introduced it may be an important policy goal to avoid the concentrations of quota either geographically, or in numbers of owners, or both. As in the New Zealand and the Iceland case, the ITQ system in Nova Scotia has intensified the organizational and geographical concentration of the industry. It has also shifted more of the burden of reducing excess capacity to crew members than is perhaps fair. Attempts to reduce these negative impacts through the design of the system and closely related policies have had mixed results and remain controversial. Limits on transfers across groups have reduced concentration in the North Sea and Canadian cases. In the UK some mechanisms have been deployed to favour retiring skippers by maintaining their rights even when they leave the trade. These mechanisms are, however, criticized for creating a class of 'slipper skippers'. Furthermore, when nations aim at protecting fishing communities and own national interests, care must be taken to avoid infringement on European Community law and the EC Treaty. State aid of various forms and ways to shield quotas from being bought by other nationals may not comply with existing laws and agreements. Limits on transferability create a definite cost in economic efficiency. This is directly reflected in the prices of individual quotas, which are lower where transferability is limited than where it is not. Determining what the cost in efficiency actually is for some degree of limits on transferability remains a critical research question.

7 *Transferable rights do not reduce capacity but rather make rapid capacity reduction smoother and more humane.* The New Zealand case shows that ITQ systems do not necessarily reduce capacity; capacity was reduced in both the Canadian ITQ system and the Alaska cooperative case, but the reductions cannot be directly traced to the ITQ system. In Alaska there was a buy-out and scrapping programme while the main engine for the further reduction was collaboration within the cooperatives. The tendency of some stakeholders and even the general public, which we found particularly in the Icelandic and Canadian cases, to use ITQs to explain all the changes in population and employment patterns over the past two decades is a gross oversimplification. The basic lesson seems to be that it is the enforcement of restricted quotas or other fishing opportunities that is the real driver of a reduction in fishing capacity. While not minimizing the problems of equity and pain involved in initial rights allocations, transferable property rights do make the radical capacity reductions that are sometimes required less chaotic and more humane by providing alternatives to bankruptcy as the mechanism for exits from the fishery that are being made unavoidable by the enforcement of restrictions on fishing.

12.3.3 Effort Control

8 *Carefully designed MPAs increase biological robustness but with economic costs.* Simulations suggest that MPAs generally have a negative effect on the profitability of most fleets over a period of 10 years. MPAs create increased costs because of fewer options in fishing locations while at the same time reducing short-term catches. There was one exception to this: some small fleet segments do show increased profit, likely based on advantageous location vis-à-vis the MPA. Further simulations indicate that spatial and/or temporal closures as a supplement to either TAC systems or effort control improve biological robustness. However, the robustness is very closely linked to how the effort is re-allocated between fleet-segments, areas, and seasons, and is also sensitive to the assumptions in relation to fleet specific catchability. Evaluations of the effect of closures thus require high resolution information on the actual effort allocation by vessel and about fleet behaviour.

9 *Effort control increases biological robustness when the link between effort and mortality is controlled.* Simulation studies indicated that effort-based management is more biologically robust than TAC regulations, but that these results are conditioned on allowing sufficient year-to-year variation in effort. Explanatory factors are that advice for TAC-based management is more sensitive to knowledge uncertainties and that effort control results in less discards.

In the case of direct effort management, biological robustness is found to be conditioned on monitoring and controlling the link between fishing effort and fishing mortality. Such a control is challenged by the dynamics of species and fleets, but also environmental factors, all of which influence the relationship between effort and fishing mortality. An effort regime can account for such influences, e.g. by including additional measures on allocation of effort in certain seasons and/or areas. The Faroese case is a counter example where a failure to monitor and control increases in capacity has hampered biological robustness.

12.3.4 Decision Rule Systems

10 *Adaptive rule-based systems can increase biological robustness.* Implementing an adaptive approach in harvest control rules has the potential to improve the biological robustness in TAC regimes. This was shown by a simulation study where the TAC was adjusted within the fishing season by including the most recent information. In addition, long-term catches increased. Given the worldwide struggle to implement the ecosystem approach, the management of Alaska groundfish offers a rather pragmatic contribution: an upper limit to all catches in a given ecosystem. The more complex Traffic Light approach in Canada was tried and put aside because it was too complex to give clear guidance, however it is being experimented with again in shrimp management. The Alaska case suggests that TAC regulations can provide a precautionary harvest of groundfish, but that the success of a TAC regime

also depends on management measures to make a harvest control effective. The TAC setting process is supported by most stakeholders, the exception being the environmental NGOs who call for greater consideration to reducing the ecosystem impacts of fishing. The same is true in the Regional Advisory Councils, where EU stakeholders are getting a role suggesting and evaluating decision rules, but where environmental NGOs also feel that their participation could be strengthened.

12.4 Final Remarks

CEVIS was an experiment in how experts on fisheries from many disciplines could use science to inform fisheries policy debates. The challenge was to find ways for the different disciplines to work together to answer policy-related questions. During the first phase of the project we formed cross-disciplinary teams to travel to areas of the world where the innovations of interest had been implemented. These groups' reports inspired much of the more focussed multi-disciplinary and trans-disciplinary work in the second phase.

The cross-disciplinary approaches worked very well. These consisted mainly of social scientists and economists working with biologists carrying out social science type research that involved interviews with stakeholders, biologists and other fisheries management professionals. These activities were mostly in the first phase of the project, but contributions to the work done on social robustness in the second phase were also made by biologists. The cross-disciplinary contributions of the project advisors were also a very valuable contribution. The disciplinary mix led to richer questions and discussion topics than would otherwise have been the case. Seven of the ten summary results from CEVIS outlined above were contributed to either entirely or in part by cross-disciplinary work. The cross-disciplinary fruits were not only the result of the formation of diverse teams, the insights gained about rights-based management were the result of empirical social science work grounded and guided in part by theoretical economics.

Significant CEVIS products were the results of trans-disciplinary work involving biologists and economists. Importantly, the trans-disciplinary work that was successful was based on many years work in several different projects developing bio-economic modelling for fisheries. In contrast, the main attempt to develop a trans-disciplinary approach within CEVIS, which was based on trying to create a common approach to defining the key concepts using abstract categories that were new and unfamiliar, was not successful. The bio-economic modelling made a major contribution. Not only was the bulk of the work on effort management carried out this way, the simulation modelling that framed participation in terms of information provision produced valuable insights on ways to organize the provision of information. The main lesson seems to be that trans-disciplinary cooperation is possible and valuable, but is achievable when there is a similarity of analytic method and a long-term commitment to turning that similarity into practical products rather than common but abstract categories. Trans-disciplinary cooperation where methods are not similar is more demanding and would require totally new methodological approaches.

The CEVIS project was an interesting experiment in having many different kinds of scientists attack a set of important, but scientifically speaking rather diffuse, problems. The project yielded, in our opinion, some useful results. It also contributed to creating a cross-disciplinary team that has gained important experience and learned valuable lessons about using science to contribute to managing policy problems.

References

Funtowicz, S. O., & Ravetz, J. R. (1990). *Uncertainty and quality in science for policy*. Dordrecht, The Netherlands: Kluwer Academic Publishers.

Habermas, J. (1990). *Moral consciousness and communicative action*. Cambridge: MIT Press.

Porter, T. M. (1995). *Trust in numbers: The pursuit of objectivity in science and public life*. Princeton: Princeton University Press.

Ragin, C. C. (2000). *Fuzzy-set social science*. Chicago: The University of Chicago Press.

Star, S. L., & Griesemer, J. R. (1989). Institutional ecology, 'translations' and boundary objects: Amateurs and professionals in Berkeley's Museum of vertebrate zoology, 1907–39. *Social Studies of Science, 19*(3), 387–420.

Wilson, D. C., & Pascoe, S. (2006). Delivering complex scientific advice to multiple stakeholders. In L. Motos & D. C. Wilson (Eds.), *The knowledge base for fisheries management: Vol. 36. Developments in Aquaculture and Fisheries Science Series* (pp. 329–353). Amsterdam: Elsevier.

Index

K.H. Hauge, D.C. Wilson (eds.), *Comparative Evaluations of Innovative Fisheries Management,* DOI 10.1007/978-90-481-2663-7_BM2,

Kolibri 360 GmbH
Ferdinand-Jühlke-Straße 7
99095 Erfurt, Deutschland
produktsicherheit@kolibri360.de